CONTROL AND LEARNING IN ROBOTIC SYSTEMS

CONTROL AND LEARNING IN ROBOTIC SYSTEMS

JOHN X. LIU
EDITOR

Nova Science Publishers, Inc.
New York

For permission to use material from this book please contact us:
Telephone 631-231-7269; Fax 631-231-8175
Web Site: http://www.novapublishers.com

NOTICE TO THE READER

The Publisher has taken reasonable care in the preparation of this book, but makes no expressed or implied warranty of any kind and assumes no responsibility for any errors or omissions. No liability is assumed for incidental or consequential damages in connection with or arising out of information contained in this book. The Publisher shall not be liable for any special, consequential, or exemplary damages resulting, in whole or in part, from the readers' use of, or reliance upon, this material.

This publication is designed to provide accurate and authoritative information with regard to the subject matter covered herein. It is sold with the clear understanding that the Publisher is not engaged in rendering legal or any other professional services. If legal or any other expert assistance is required, the services of a competent person should be sought. FROM A DECLARATION OF PARTICIPANTS JOINTLY ADOPTED BY A COMMITTEE OF THE AMERICAN BAR ASSOCIATION AND A COMMITTEE OF PUBLISHERS.

LIBRARY OF CONGRESS CATALOGING-IN-PUBLICATION DATA

Control and learning in robotic systems / John X. Liu (editor).
    p. cm.
Includes bibliographical references and index.
ISBN 1-59454-356-9
1. Robotics. 2. Robots--Control systems. I. Liu, John X.
TJ211.C637   2005
629.8'92--dc22   2005010776

Published by Nova Science Publishers, Inc. ✦ *New York*

CONTENTS

PREFACE

Robotics began as a science fiction creation which has become quite real, first in assembly line operations such as automobile manufacturing, airplane construction etc. They have now reached such areas as the internet, ever-multiplying-medical uses and sophisticated military applications. Control of today's robots is often remote which requires even more advanced computer vision capabilities as well as sensors and interface techniques. Learning has become crucial for modern robotic systems as well. This new book deals with control and learning in robotic systems.

Remote control of robots over the Internet is attracting more and more attention from both academia and industry. Physical interaction with remote environments over IP networks poses many technical challenges which are still outstanding. It is widely recognized that the most challenging difficulties are associated with Internet transmission time delays, delay jitter, bandwidth variation and bandwidth limitation. To deal with these problems, the majority of current work is concentrated on developing advanced remote control algorithms and/or interface techniques whereas data communication between the human operator and the remote robot through the Internet is often treated as given conditions. Chapter 1 describes a novel data communication architecture, for which the core is the trinomial transport protocol, to facilitate the remote control of Internet robots. With a proper choice of protocol parameters, the transmission rate of the proposed scheme is very smooth when available bandwidth is stable and the rate adapts to the network rapidly when bandwidth varies. The scheme provides relatively minimized delay and jitter. The mathematical analysis shows that the trinomial protocol is both TCP-friendly and intra-protocol fairness convergent. The simulations confirm the mathematical analysis and the experimental implementation validates the feasibility of the introduced scheme.

Decision making and/or control is the process by which free variables (options, actions, decisions, control signals, . . .) of a system are evaluated and selected in order to realize some goals. Reactive decision making (and/or control) that is decision making in situation where things are not going as expected (reorganization of emergency services during a catastrophe event in order to ensure efficiency ; adjustment of production rate when demand of a product deviates significantly from its estimated value; reallocation of resources/time to activities in order to maintain the required quality of service when customers arrival rate at a service deviates from its nominal value ; . . .) because of external uncontrollable effects such as state of the nature change and/or other decision makers' actions is common in practice. Problems of decision making in presence of uncertainty due to the randomness of the state of the world are well formulated and solved using the so called state of the world model for decision and those involving many decision makers with conflict interests are formulated and solved using non cooperative game theory. Classical assumptions concerning these decision making models (known probability distribution of the state in the case of state of the world model for decision making and perfect knowledge of the game that is knowledge of other players strategy and goals in the case of game theory for instance) restrict their application in practical situation. Indeed, in practical situation the environment will be constantly changing and so decision strategy or control law must be constantly adapted that is what we call *reactive decision making or control*. To efficiently react, at least a partial knowledge of environment change or other players behavior will be of great importance. This can be done by setting up an *intelligence system* (system that collects and transforms data to knowledge and possibly knowledge to recommended decision or control action) consisting in sensors (in broad sense including spying and monitoring for instance) to collect information about the environment and processing capabilities to process this information (data analysis, features extraction, information and sources fusion, . . .) in order to retrieve knowledge about environment and send it to the decision maker or controller. This fresh knowledge will modify decision maker (or controller) belief about expected state of the world and/or other actors intention and strategy that will allow to react efficiently by changing decision or control that was initially planned. In Chapter 2 we will concentrate on modeling how to use the knowledge from intelligence system by developing an integrated framework based on influence diagrams (in the case of decision making) and mathematical programming (in the case of control) as mathematical tools to support decision making and/or control in reactive context.

The rings event is one of men's apparatus gymnastics. The ring exercises have the characteristics that the apparatus, namely rings, can move in all directions freely, i.e. free-floating characteristics at gripping point. This is a different point from other gymnastic events, therefore its motion property is totally unique. So far the rings have been studied mainly in the field of biomechanics. These studies mainly discuss the results obtained by motion measurements of actual gymnasts' performance. On the other hand, we describe researches on "the rings gymnastic robot" using robot technology in Chapter 3. In an approach using the rings gymnastic robot, it is possible to analyze performances according to body type and characteristics for each gymnast. It is also capable of developing feats that nobody has ever done. These are not realized in the conventional approach based on the above-mentioned motion measurements. Purpose of studies on the rings gymnastic robot is to understand ring exercises through the robot and to apply the acquired skill to gymnastic coaching. In other words, it is to acquire the performance skill through the robot and to transfer the skill from the robot to gymnasts (to improve their motions) respectively. In these studies, fuzzy control is adopted. Because how to generate forces required in performances is useful for the gymnastic coaching as indicated in the area of sports biomechanics. In addition, if the control method is represented as "if-then" rules, it is easy for gymnasts to understand the knowledge to realize the performance as compared to reading the time-series data. In this chapter, we discuss not only derivation of a three-link model of the robot but also a way of acquiring the performance skill, which is represented by fuzzy control rules. These three-link model-based analyses are necessary for our approach as a preliminary step toward analyses with simulations based on a three-dimensional model and experiments. In order to realize an objective performance, it is divided into basic exercises. A fuzzy controller is given for each basic exercise and one of them is selected according to the situation. Parameters of the fuzzy control rules are optimized by genetic algorithms (GAs). Simulation results show that the acquired skill is effective for realizing the objective performance.

Chapter 4 concerns a class of uncertain systems described by classical mathematical models or by relational knowledge representations with unknown parameters. In recent years the concept of so called uncertain variables has been introduced and developed as a tool for analysis and decision making problems in this class of uncertain knowledge-based systems. The formal descriptions and methods based on the uncertain variables may be used for the systems characterized by an expert presenting his/her knowledge on the unknown parameters and for the intelligent uncertain systems with learning algorithms.

The comprehensive description of the uncertain variables and their applications in different kinds of uncertain and intelligent systems with details and examples has been presented in the book: Z. Bubnicki, *Analysis and Decision Making in Uncertain Systems* (Springer Verlag, 2004). The purpose of this chapter is to describe shortly basic concepts and definitions concerning uncertain variables, and a brief review of new problems and results in this area. The definitions of uncertain logics and variables, and their applications to basic analysis and decision problems are presented in Secs 2, 3, 4. The Secs 5–10 are devoted to new problems and results, in particular, to stability and stabilization of uncertain systems and learning systems in which the learning process consists in *step by step* knowledge validation and updating. The chapter is completed with a list of other problems and practical applications (Sec. 11).

In Chapter 5, a new planning and control scheme for intelligent robotic systems is presented. Robotic systems obtain environmental information from perceptive sensors and respond to the perceptions to execute tasks through decision and control process. In perceptive frame, the evolution of the robotic system is driven by perceptive references, which is directly related to sensory measurement of the system output, instead of the reference of time. The hybrid hierarchical perceptive framework of robotic systems has continuous and discrete layers, and it also has continuous and discrete perceptive references. The discrete layers enable the robot systems to operate at higher levels for improving the robotic intelligence, namely, planning and modifying original tasks and actions based on the perceptions through switching the low level perceptive controllers. Using hybrid automata and hybrid formal language theory, a hybrid perceptive control theory for modeling and analyzing the motion planning and control of mobile robots has been developed. Hybrid and perceptive automata are able to exhibit the robot behaviors both continuous and discrete in perceptive frame. The discrete expressions of hybrid language can be accepted by discrete part of the automata, while the continuous part of the language performs the control of the continuous physical system. Stability can be guaranteed for the switched systems based on the dynamical properties. In contrast to a continuous perceptive reference, the hybrid perceptive framework is able to deal with unexpected events which can block the evolution of the continuous or discrete reference, and the reference keeps evolving during the unexpected events. At the discrete levels, the hierarchical architecture of the hybrid automata improves the formal language so that the automata are able to treat the unexpected events as the disturbances of a tracking control system, it can return back to execute the original task after processing the unexpected events; at the continuous level, the system can make a discrete switch to prevent the references from stagnation. The experimental results, given by a tele-operation system consisting of a phantom joystick and a mobile manipulator, show the effectiveness of the proposed perceptive control theory. The controller is stable and able to process unexpected events and return back to the desired task. In the autonomous control mode, the hybrid perceptive references can not be stopped by blocking the continuous reference, including obstacles, in the tele-operation mode, the control performance is not affected by the block of discrete reference due to communication time delay.

Chapter 6 concerns with modeling and solution of a Master Production Scheduling problem characterized by multi-timeframe, multi-product and multi-resource. The considered problem is a kind of scheduling problems with setup carryover and very expensive setup cost, and thus, it is expected that each individual product is set up once along the scheduling periods. Therefore, the main issue for the MPS problem is how to schedule production in a continuous mode of periods. Giving the demand quantity, production capacity and operational principles, the considered problem was formulated into a mixed-quadratic-binary program (MQBP). This MQBP model can be transformed into a mixed-binary program and solved by the existent software packages. However, this optimization approach is time-consuming, especially in large scaled problems. To facilitate subsequent solution analysis, this MQBP model was decomposed into two inter-dependent sub-models of a quadratic binary constraint (QBC) and a linear program (LP). In order to satisfy both speed and accuracy requirements, a two-phased heuristic approach was proposed. In Phase 1, the search space of the QBC sub-model was reduced into a preliminary scheduling pattern and thereby a reference model from the assignment problem is formed. In Phase 2, according to the permutation property in the reference model, a stochastic global optimization procedure that incorporates a genetic algorithm with neighborhood search techniques was applied to obtain a desirable solution. Numerical evidence has shown that the proposed pattern-based heuristic approach is effectively applicable for an uninterrupted production schedule.

Reinforcement learning is a fundamental process by which organisms learn to achieve their goals from their interactions with the environment. The crucial issue in reinforcement learning applications is how to set meta-parameters, such as the learning rate, "temperature" for exploration, in order to match the demands of the task and reduce the learning time. In Chapter 7, we propose a new method, based on evolutionary computation. The basic idea is to encode the metaparameters of the reinforcement learning algorithm as the agent's genes, and to take the metaparameters of best-performing agents in the next generation. First, we investigated the influence of metaparameters on the agent learned policy by considering a battery capturing task. Then, by utilizing the capturing behavior, we considered a more complex task where the Cyber Rodent robot has to survive and increase its energy level. Our main focus was to see the effect of metaparameters and initial weight connections on learning time. The results show that appropriate settings of metaparameters found by evolution have a great effect on the learning time and are strongly dependent on each other. Furthermore, we verified in the real hardware of Cyber Rodent robot that metaparameters evolved in simulation are helpful for learning in real hardware.

In Chapter 8, a new Voronoi-based technique to model the environment is presented. The objective is to build a unique environment model, which can be used to solve different problems in robotics such as localization, path-planning and navigation. The proposed model combines geometric and topological approaches to take advantage of each method. This topo-geometric model is obtained from the Voronoi Diagram, which is built using the measurements of a laser telemeter. Apart from the topological and geometric information extracted from the Local Voronoi Diagram, a symbolic information, which represents distinctive places typical of indoor environments (HALL, CORRIDOR, etc.), is also provided to the model. The technique used to learn and recognize these places is HMM (Hidden Markov Models). Different algorithms based on topo-geometric maps are also proposed to solve the problems of localization, navigation and path-planning. Experimental results show the effectiveness of these algorithms. However, the models based on Voronoi Diagrams present a problem when the number of objects detected by the sensor is minor to two. In this case, no Voronoi Diagram is generated. This implies that the robot has not got any information to navigate through the environment. We solve this problem by generating a Virtual Voronoi Diagram which guarantees the continuity of model.

Chapter 9 presents a novel multi-stage learning controller for designing a neural and fuzzy reasoning based switching controller, in order to control underactuated manipulators. In the first phase of learning, parameters of both antecedent and consequent parts of a fuzzy indexer are optimized by using evolutionary computation. Design parameters of the fuzzy indexer are encoded into chromosomes, i.e., the shapes of the Gaussian membership functions and corresponding switching laws of the consequent part are evolved to minimize the angular position errors. Such parameters are trained for different initial configurations of the manipulator and the common rule base is extracted. In the next stage of training, neural network based compensator is designed for reducing the remaining nonlinear dynamics. Then, these trained fuzzy rules with compensator parameters can be brought into the operation of underactuated manipulators. Simulation results show that the new methodology is effective in designing controllers for underactuated robot manipulators.

The aim of Chapter 10 is to compare adaptive abilities in hybrid genetic algorithm. For the hybrid genetic algorithm, a rough search technique and iterative hill climbing technique are applied to genetic algorithm. The rough search technique is used to initialize the population of the genetic algorithm, and the iterative hill climbing technique is to find a better solution within the convergence area of the genetic algorithm loop.

For constructing the adaptive abilities, we use a fuzzy logic controller and compare it with two conventional heuristics. The proposed fuzzy logic controller and two conventional heuristics can adaptively regulate the rates of the crossover and mutation operators in the genetic algorithm during genetic search process. They are tested and analyzed under the hybrid genetic algorithm proposed in this paper. Finally, a best algorithm is recommended.

Evaluation of the quality of attributes (features) is an important issue in the machine learning, data mining and in a research of intelligent systems. There are several important tasks in the process of data analysis e.g., feature subset selection, constructive induction, decision and regression tree building, and visualization, which contain an attribute evaluation procedure as their (crucial) ingredient. Relief algorithms are general and successful attribute evaluators.They are able to detect conditional dependencies between attributes and provide a unified view on the attribute evaluation in regression and classification. In addition, their quality estimates have a natural interpretation. While they have commonly been viewed as feature subset selection methods that are applied in prepossessing step before a model is learned, they have actually been used successfully in a variety of settings, e.g., to select splits or to guide constructive induction in the building phase of decision or regression tree learning, as the attribute weighting method and also in the inductive logic programming. Chapter 11 describes how, why and when Relief algorithms work, their properties and parameters. It provides answers to questions: what kind of dependencies they detect, how do they scale up to large number of examples and features, how to sample data for them, how robust are they regarding the noise, how irrelevant and redundant attributes influence their output, and how different metrics influences them. It also lists some of their successful applications.

In: Control and Learning in Robotic Systems
Editor: John X. Liu, pp. 1-20

ISBN 1-59454-356-9

Chapter 1

REAL-TIME DATA STREAMING FOR INTERNET ROBOTS

Peter Xiaoping Liu[*]

Department of Systems and Computer Engineering, Carleton University,
Ottawa, ON, K1S 5B6, CANADA

Max Meng[**]

Department of Electronic Engineering,
Chinese University of Hong Kong, Hong Kong

Abstract

Remote control of robots over the Internet is attracting more and more attention from both academia and industry. Physical interaction with remote environments over IP networks poses many technical challenges which are still outstanding. It is widely recognized that the most challenging difficulties are associated with Internet transmission time delays, delay jitter, bandwidth variation and bandwidth limitation. To deal with these problems, the majority of current work is concentrated on developing advanced remote control algorithms and/or interface techniques whereas data communication between the human operator and the remote robot through the Internet is often treated as given conditions. This paper describes a novel data communication architecture, for which the core is the trinomial transport protocol, to facilitate the remote control of Internet robots. With a proper choice of protocol parameters, the transmission rate of the proposed scheme is very smooth when available bandwidth is stable and the rate adapts to the network rapidly when bandwidth varies. The scheme provides relatively minimized delay and jitter. The mathematical analysis shows that the trinomial protocol is both TCP-friendly and intra-protocol fairness convergent. The simulations confirm the mathematical analysis and the experimental implementation validates the feasibility of the introduced scheme.

Key Words— Internet, Distributed Computing, Remote Control, Data Communications.

---

[*] E-mail address: xpliu@sce.carleton.ca
[**] E-mail address: max@ee.cuhk.edu.hk

I   Introduction

In the last decade the Internet has blossomed into an engine of communications with staggering implications for both institutions and individuals. Currently, however, the Internet is mainly used as a means of sending e-mail and getting information from remote databases [1]. The underlying technologies continue to advance rapidly such that both wired and wireless networks are becoming increasingly accessible, affordable, and powerful. The current trend is for people not to be tethered to a desktop any longer; instead they can use the Internet wherever and the industry is responding with small mobile Internet devices – palm-tops, personal digital assistants (PDA), multi-function cellular phones etc[2]. As a result, network-based applications are quickly expanding into new areas and network-based remote operation systems such Internet robots are attracting more and more attention from both academia and industry.

The first Internet telerobot appeared in 1994 [3]. By 2001, more than forty such systems had been put online around the world, allowing users to visit museums, tend gardens, navigate undersea, float in blimps or handle protein crystals via the Internet [1, 5]. These systems, however, are largely experimental and not ready to provide real-world services. Physical interaction with remote environments over the Internet poses many technical difficulties which are still outstanding. For example, new technologies are needed for coordinating simultaneous users, for handling unexpected errors, for ensuring trustworthiness (reliability, security and privacy) etc. Particularly, how to cope with problems associated with time-varying transmission delay and not-guaranteed bandwidth (i.e., bandwidth limitation and variation) is an unavoidable and fundamental issue facing Internet robotic systems.

In order to overcome the above problems, most of the current work is concentrated on developing advanced *remote control algorithms* [3, 4] and *interface techniques* [5, 6]. The *data communication* between the human operator and the remote robot through the Internet, however, is usually treated as given conditions and rarely touched. For example, most systems employ directly one of the two currently available transport protocols: TCP (Transmission Control Protocol) [7] and UDP (User Datagram Protocol) [8].

We know that TCP was originally designed for the reliable transmission of static data, such as files, emails etc., over low-bandwidth, high-error-rate networks [7]. When a packet gets lost or corrupted, it's retransmitted. The cost of guaranteeing reliable delivery is great delay and delay jitter. In addition, TCP transmission rate changes drastically with time, which also makes TCP inappropriate for the tasks of remote control.

Compared to TCP, UDP supplies minimized transmission delay and jitter by omitting connection setup process, acknowledgement, and retransmission [8]. From the comparison shown in Fig. 1, it is clear that UDP outperforms TCP in terms of both average delay and delay jitter.

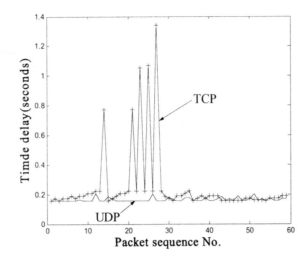

Fig.1. A comparison of delay and delay jitter between UDP and TCP, where one TCP and one UDP flow share the same network bottleneck

However, UDP is merely a raw protocol [8, 9]. The UDP sender just pushes datagrams into the net at a specified (usually fixed and constant) rate and the UDP receiver accepts incoming datagrams off the network. Neither the sender nor the receiver is aware of network states such as bandwidth availability and network congestion. Consequently, UDP is unable to adjust its transmission rate adaptively according to network conditions, which usually leads to two undesirable polar situations:

- The UDP rate is too high when the network is congested. At this scenario, the network is easy to be driven to congestion collapse by such UDP flows [9-13]. It is also unfair for TCP flows since UDP senders persist in pushing data into the traffic at a high rate whereas TCP senders back off [9];
- The UDP rate is too low when there is extra bandwidth, which will be extremely undesirable since the extra bandwidth is wasted.

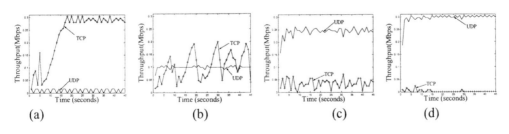

Fig.2. UDP interacts with TCP

The simulation results given by Fig.2 show how UDP flows affect TCP ones when one TCP flow and one UDP flow share a network bottleneck. In Fig.2(a) where the UDP sending rate is very low, TCP takes bandwidth very fast and consumes most bandwidth in the end. It is very undesirable situation for UDP in this case. For the scenario shown in Fig.2(b) where the UDP sending rate is proper, TCP and UDP share the bottleneck rather fairly and their average throughputs are close. For the case shown in Fig.2(c) where the UDP sending rate is too high, TCP starves by shrinking its sending widow size. The UDP flow is *not* TCP-friendly and thus it is unfair for the TCP flow. For the scenario shown in Fig.2(d) where the UDP rate is extremely high, TCP drops its rate to nearly nothing and the network is collapsed for TCP traffic.

From above illustrations, it is clear that neither TCP nor UDP is appropriate for real-time applications. In fact, there is overwhelming work in developing new mechanisms/protocols for streaming multimedia over the Internet, such as the TCP-friendly Rate Control Protocol (TFRC) [10], the Rate-based Adaptation Protocol (RAP) [11], the Loss-delay Adjustment Protocol (LDA) [12], the Square-Increase/Multiplicative-Decrease(SIMD) Protocol[13], the real-time transport protocol (RTP) etc. Although both multimedia and Internet robots are real-time applications, they are different in several ways. First, most current multimedia systems account for variations in delay and bandwidth by buffering data [14, 15]. The ordinary buffer mechanism, however, is generally unacceptable for the remote control of Internet robots since timeliness is critically important in this paradigm: feedback and command signals should be delivered immediately once they are available. Secondly, the requirements on the curve of transmission rate are different under the condition of bandwidth variation. For multimedia, moderate variation in transmission rate is generally acceptable and will be absorbed by the buffer. To avoid flooding or emptying the buffer, the data rate should not change significantly over the *entire* duration of the application [10, 11]. For Internet robots, the requirement for rate smoothness is much stringent. Also, when available bandwidth increases, it is usually more desirable for the transmission rate to adapt to the increase and to get to the new steady state rapidly without large overshoot.

The above discrepancies between Internet multimedia and Internet robots call for different communication mechanisms for Internet robots, which is already recognized by many researchers [1, 5, 23-25]; however, so far, little work has been reported in the literature. This is the primary motivation of this work. In this paper, we introduce a novel end-to-end data communication mechanism for the remote control of Internet robots. The core of the scheme is the trinomial protocol that is a source-based transport protocol with a TCP-compatible congestion and rate control mechanism. By choosing the values of the three protocol parameters properly, the desirable transmission curve can be obtained and the protocol is both TCP-friendly and intra-protocol fairness convergent. In addition, since it has no retransmission mechanism; delay and delay jitter are minimized. The paper is organized as follows. Section II introduces the theoretical model of the data transmission mechanism. Particularly, the constraints on the trinomial protocol to be TCP-friendly are derived. In Section III, the implementation issues of the scheme, such as congestion signaling, roundtrip time estimation and rate adjustment are described. The simulation study and experimental implementation are presented in Section IV and Section V respectively. The trinomial protocol is compared with both TCP and UDP, which are employed by most current Internet robotic systems, in terms of data rate smoothness, TCP-friendliness, Intra-protocol fairness convergence, time delay, jitter, and loss rate. In the last, conclusions are drawn in Section VI.

II The Proposed Data Communication Mechanism

A   The Trinomial Protocol

First, let us tentatively consider a class of linear transport protocols:

$$Increase:\ x(t+1) = b_I(t)x(t) + a_I(t) \tag{1}$$

$$Decrease:\ x(t+1) = b_D(t)x(t) + a_D(t) \tag{2}$$

where, *Increase* refers to the increasing policy of transmission rate when there is no network congestion; *Decrease* refers to the decreasing policy when network congestion occurs; $x(t)$ is the transmission rate at time instant t; $a_I(t)$, $b_I(t)$, $a_D(t)$, and $b_D(t)$ are coefficients that are independent of $x(t)$. For simplicity, we drop the independent variable t for all the coefficients hereafter. It should be mentioned that nonlinear algorithms different from (1) and (2) are also possible; however, nonlinear schemes tend to be sensitive to system parameters and difficult to implement [16,17]. Consequently, we only consider linear protocol candidates.

In order to fulfill the requirements of optimal convergence to both efficiency and fairness, for the spectrum of protocols given by (1) and (2), the following constraints should be satisfied [18]:

For *increase* policy, $a_I \geq 0$ and $b_I = 1$,

$$\text{i.e.,}\ x(t+1) = x(t) + a_I,\ \text{where}\ a_I \geq 0 \tag{3}$$

For *decrease* policy, $a_D = 0$ and $0 < b_D < 1$,

$$\text{i.e.,}\ x(t+1) = b_D x(t),\ \text{where}\ 0 < b_D < 1 \tag{4}$$

Inspired by the SIMD protocol presented by Jin [13], we propose a novel rate-based transport protocol as follows:

$$Increase:\ S_t = S_0 + (\frac{t}{\alpha})^\gamma \quad \alpha \geq 1, and\ \gamma \geq 0 \tag{5}$$

$$Decrease:\ S_{t+} = (1-\beta)S_t, \quad 0 < \beta < 1 \tag{6}$$

where *Increase* refers to the increase in sending rate as a result of the receipt of acknowledgements (there is no congestion); *Decrease* means the decrease in sending rate on the detection of packet loss (congestion occurs); t is the time instant whose unit is the number of roundtrip times (RTT's); S_t is the sending rate (packets/RTT) at time instant t; S_0 is the initial sending rate (packets/RTT) after the last decrease; and α, γ, and β are nonnegative constants. It is clear that the decrease policy given by (6) satisfies the constraint given by (4). By differentiating (5) with respect to t, we can get $S_t = S_{t-1} + \dfrac{r}{\alpha}(\dfrac{t}{\alpha})^{\gamma-1}$. Thus, a_t is equal

to $\dfrac{r}{\alpha}(\dfrac{t}{\alpha})^{\gamma-1}$ which is always nonnegative. Consequently, the increase policy, (5), satisfies the constraint given by (3) as well.

We call this protocol *the trinomial protocol* because there are three parameters (α, β, γ) which are adjustable. The values of α, β and γ will be determined by the requirements of specific applications and the constraint on the protocol to be TCP-friendly which we will discuss later on. The three parameters play different roles, which is illustrated by using Fig. 3

α: it appears in the denominator of an exponential term. The larger value of α, the smoother transmission rate can be expected locally in the steady state (A → B); however, too large α leads to a slow response to network bandwidth increase;

γ: it determines how fast the protocol probes and takes extra bandwidth. Along with α, it actually determines the speed of the transient response of the protocol. Greater γ means a quicker transient response as well as larger overshoot and oscillation in transmission rate;

β: it determines the cutback step magnitude of transmission rate when congestion occurs (the height from C to D). Consequently, it actually controls the global smoothness of the transmission rate.

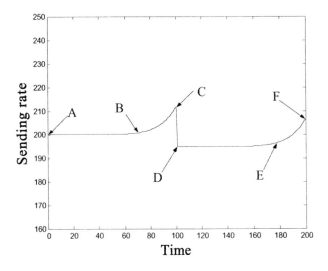

Fig.3: Illustration of the roles of α, β, and γ

The key idea underlying the trinomial protocol is that, by choosing the values of α, β and γ properly, it is possible that the transient response of the protocol to network bandwidth variations is quick whereas the transmission rate is smooth in the steady state. For instance, as to be demonstrated by the simulations shown later on, by choosing β=0.05 and γ=3 (the value of α is determined by the constraint (16) that will be discussed next), the transmission rate of the trinomial protocol is much smoother than that of TCP in the steady state whereas the transient response to bandwidth increase/decrease is fast.

B TCP-Friendliness

In the current Internet congestion control paradigm, routers play a relatively passive role. The function of routers in congestion control is just to signal the occurrence of congestion indirectly and implicitly by dropping data packets. It is thus the end-system (the transport protocol) that performs the crucial role of responding appropriately to these signals. Consequently, in order to maintain the stability of the Internet and to ensure fairness between flows, it is crucially important that a new protocol has a proper congestion control mechanism and coexists in harmony with the TCP protocol since dominant Internet applications are TCP-based. The Internet Engineering Task Force (IETF) argues that *any new protocols must be TCP-friendly* ([9, 19]). In this subsection, we are exploring this aspect of the trinomial protocol.

In [9], Floyd et al. defines that a non-TCP flow is TCP-friendly if its arrival rate does not exceed the arrival of a conformant TCP flow under the same circumstance, i.e., $\lambda_{max} = \sqrt{3}/(R\sqrt{2p})$ where λ_{max} is the maximum sending rate (packets/second) of the new protocol, R is the roundtrip time (RTT) and p is the probability of packet loss. Next, we derive the constraint on the parameters (α, β andγ) for the trinomial protocol to be TCP-friendly.

First, Let us consider the deterministic Internet model described in [18] where loss events occur periodically with a fixed probability p. The increase and decrease in sending rate are thus deterministic as shown in Fig 4. Each epoch has the same duration (The number of *RTTs* is the same: N_d). t_b is the time right after the last packet drop and t_e is the time just before another packet drop. H is the total number of packets sent in each epoch (between two successive drops). S_m is the maximum sending rate (at t_e) and S_0 *is* the minimum sending rate (at t_{b+}).

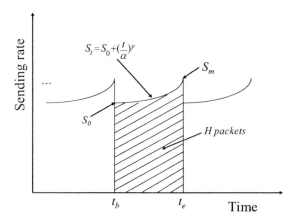

Fig.4. Deterministic model

According to the decrease rule (6), we have:

$$S_0 = S_{t_b} = (1-\beta)S_m \tag{7}$$

The duration (N_d) of each epoch in terms of RTT is:

$$N_d = t_e - t_b = \alpha(\beta S_m)^{1/\gamma} \tag{8}$$

and the number of packets sent (H) in each epoch is given by:

$$H = \int_0^{N_d} S_t dt = (1-\beta)S_m N_d + \frac{N_d^{r+1}}{(\gamma+1)\alpha^\gamma} \tag{9}$$

Substituting the N_d in (9) with (8), we have:

$$H = \alpha(1-\beta)\beta^{(1/\gamma)}S_m^{((\gamma+1)/\gamma)} + \frac{\alpha}{\gamma+1}(\beta S_m)^{((\gamma+1)/\gamma)} \tag{10}$$

The average sending rate λ (packets per second) is the total number of packets sent in each epoch (H) divided by the duration of the epoch in seconds ($T_d R$), where R is the RTT value. The packet loss probability, $p=1/H$. Thus, we can get:

$$\lambda = \frac{H}{N_d R} = \frac{1}{pRN_d} \tag{11}$$

By the definition of TCP-friendliness and (11), we have:

$$1.5N_d^2 \geq H \tag{12}$$

Combining (8), (10) and (12), we have:

$$\alpha \geq \frac{(r+1-r\beta)S_0^{((\gamma-1)/\gamma)}}{1.5(r+1)(\beta)^{(1/r)}(1-\beta)^{((r-1)/r)}} \tag{13}$$

Because the stochastic model gives a more accurate description of the real Internet, next, we consider the stochastic Internet model where congestion (loss) events occur randomly with a fixed probability p. Let H_i donate the number of packets sent in the ith epoch up to but not including the first lost packet. The probability that H_i equals to k (k is the number of packets that are sent successfully before the first loss event takes place in the ith epoch) is:

$$P[H_i = k] = (1-p)^k p \approx pe^{-pk}, \quad \text{for } p << 1, k = 0, 1, 2, \cdots \tag{14}$$

Let N_{di} be the duration of the ith epoch in roundtrip time (RTTs), $N_{di} = H_i / \overline{S}_i$, where $\overline{S}_i$ is the average sending rate in the ith epoch. Thus, the total increase in sending rate (ΔS_i) in the ith epoch from the beginning to the time instant when the first loss event occurs is:

$$\Delta S_i = (\frac{N_{di}}{\alpha})^\gamma = (\frac{H_i}{\alpha \overline{S}_i})^\gamma$$

It is very involving to derive $\overline{S}_i$ because H_i and $\overline{S}_i$ are inter-dependent. We use the time-averaged sending rate $\overline{S}$ to approximate $\overline{S}_i$ and $\Delta S_i = (\frac{H_i}{\alpha \overline{S}})^\gamma$ [13]. The expectation of ΔS_i is:

$$
\begin{aligned}
E[\Delta S_i] &= \sum_{x=0}^{\infty} \Delta S_i P[H_i = k] \\
&\approx \sum_{k=0}^{\infty} (\frac{H_i}{\alpha \overline{S}})^\gamma P[H_i = k] \\
&\approx \sum_{k=0}^{\infty} (\frac{k}{\alpha \overline{S}})^\gamma pe^{-pk} \\
&\approx \int_0^\infty (\frac{x}{\alpha \overline{S}})^r pe^{-px} dx \\
&= \frac{\Gamma(r+1)}{(\alpha p \overline{S})^r}
\end{aligned}
\tag{15}
$$

It is noticed that, under the deterministic model, $H_i = 1/p$, $N_{di} = H_i / \overline{S}$. Thus, for the deterministic model, we have $E[\Delta S_i] = (\frac{1}{\alpha p \overline{S}})^r$. To be TCP-friendly, the realization should work for both the deterministic and stochastic model. Thus, combining with (13), we get the constraint on the parameters (α, β and γ) as:

$$\alpha \ge \frac{(r+1-r\beta)\Gamma(r+1)^{(1/r)} S_0^{((\gamma-1)/\gamma)}}{1.5(r+1)(\beta)^{(1/r)}(1-\beta)^{((r-1)/r)}} \tag{16}$$

Consequently, if (16) is satisfied, the trinomial protocol is theoretically TCP-friendly.

III Implementation Issues

The trinomial protocol is a rate-based transport protocol. It works as follows: the trinomial source sends datagram packets with the header of sequence numbers and time stamps; the corresponding sink acknowledges each packet by providing end-to-end feedback; each acknowledgement (ACK) packet contains the sequence number of the corresponding delivered data packet; by using these ACKs, the trinomial sender detects loss events, samples the current round-trip time (*RTT*) and estimates *RTT* for the next packet; the trinomial sender adjusts its sending rate according to (17) and (19) which will be discussed next; before the rate is adjusted, the value of α is updated according to (16) as well.

A   Congestion Determination

To implement a transport protocol with congestion control mechanisms, a very important issue is how to determine whether the network is congested [26]. There are two principal approaches: the loss-based scheme [20] and the Explicit-Congestion-Notification-based mechanism (ECN-based) [28,29]. Although the ECN-based mechanism sounds more accurate than the loss-based scheme, most routers in today's Internet still adopt the DropTail queue management scheme and do not support the ECN-based mechanism. On the contrary, the loss-based mechanism does not rely on any explicit congestion signal from the network. Consequently, in today's Internet, packet loss seems to be the most feasible *implicit* congestion signal. Signaling congestion via packet drops has been proven to be a simple and robust mechanism by the success of TCP implementations. Hence, in the implementation of the trinomial protocol, we adopt the loss-based mechanism to determine whether the network is congested.

B   Rate Adjustment

Due to the real-time nature of Internet robots, the trinomial protocol is a rate-based protocol rather than a window-based protocol (such as TCP). For rate-based protocols, the data transmission rate is controlled by adjusting the inter-packet-gap (IPG) in the source [11]. When there is extra bandwidth (no loss event occurs), to increase sending rate, the value of IPG is updated iteratively according to the following:

$$IPG_{i+1} = \frac{IPG_i \times W}{IPG_i + W} \tag{17}$$

where W is the increase step height. If the transmission rate is updated every T seconds and the number of packets sent during each step is increased by k every step, the value of W should be T/k ([11] and [20]). In [21], it is suggested that rate-based protocols should adjust transmission rate not more than once per RTT. Changing the rate too often leads to unnecessary oscillation whereas infrequent change results in an unresponsive behavior [11].

Based on this consideration, the value of IPG is updated very RTT seconds. Combining with (5), we get:

$$W = R(\frac{\alpha}{i})^r \tag{18}$$

where R is the RTT value of the next packet. According to (6), to reduce transmission rate when congestion occurs, the value of *IPG* should be updated based on (19).

$$IPG_{i+1} = IPG_i / \beta \tag{19}$$

C   RTT Estimation

From the above subsection, we know that the transmission rate of the trinomial protocol is controlled by adjusting the IPG value. According (18), it relies on the estimate of the RTT value for the next packet to determine the IPG value. Too large RTT estimate causes a conservative protocol whereas too small RTT estimate results in an aggressive protocol. Consequently, RTT estimation is another key issue about the implementation of the trinomial protocol. The model we use for estimation is an adaptive autoregressive (AR) model as shown in (20).

$$\hat{R}_{n+1} = a_{n1}R_n + a_{n2}R_{n-1} + \cdots + a_{nk}R_{n-k+1} + a_{n(k+1)}R_n^L + a_{n(k+2)}R_n^H \tag{20}$$

where, $\hat{R}_{n+1}$ is the RTT estimate for the next packet; $R_n, R_{n-1}, \cdots, R_{n-k+1}$ are the actual RTT measurements for the k latest packets, R_n^L and R_n^H are the upper and lower RTT boundaries for the next packet respectively; $a_{n1}, a_{n2}, \cdots, a_{nk}$ are the coefficients of the model which are updated at each step according to

$$a_{ni} = C - (R_{n-i} + AC)(R_n - AC) / (B - A^2C), \tag{21}$$

where, $A = \sum_i R_{n-i}$, $B = \sum_i R_{n-i}^2$ and $C = 1/(k+2)$

    Readers who are interested in the above RTT estimation algorithm and the derivation of (21) may refer to our previous work [30] for details.

IV Simulation Studies

In this section, we present the simulation results. We use the widely adopted network simulation tool, *ns-2*, which was developed by DARPA through the VINT project at LBL, Xerox PARC, UC Berkeley, and USC/ISI [22], for simulation studies. In the simulation of the trinomial protocol, $\gamma=3$ and $\beta=0.05$. The value of α depends on the concurrent initial sending rate of each increase epoch and is updated according to (16).

A   TCP-Friendliness

In this simulation, n trinomial and n TCP flows share a bottleneck network link and the value of n is varied from 1 to 100.

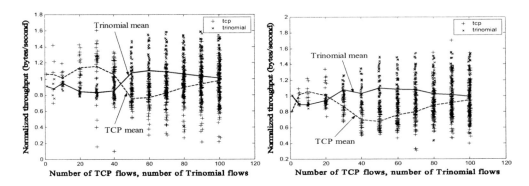

Fig.5. The trinomial protocol competes with TCP

From the results shown in Fig.5, we can see that although the variance of normalized throughputs for individual flows is quite large, the mean normalized throughputs of both TCP and the trinomial protocol mostly fall into the range of [0.7, 1.2] when the number of flows, n, varies from 1 to 100. As n increases, the mean normalized throughputs of both TCP and the trinomial protocol converge to the same value. Compared to other multimedia-oriented protocols, such as TFRC [10], RAP [11], LDA [12] and SIMD [13], the trinomial protocol is rather fair to TCP in a wide range of traffic. As illustrated in Fig2, UDP does not provide this harmony with TCP.

The reasons that the variance is quite large for individual trinomial flows may lie in the following two main facts:

- Errors in RTT estimation are unavoidable. As mentioned in the preceding section, the value of the RTT estimate for the next data packet determines the transmission rate of the trinomial protocol directly. A larger RTT causes a lower transmission rate whereas a smaller RTT results in a higher transmission rate;
- The trinomial protocol and TCP use different implementation mechanisms. TCP is window-based whereas the trinomial protocol is rate-based;

B   Transmission Rate

From the results shown in Figs 6 and 7, both TCP and the trinomial protocol respond to bandwidth variations properly. To be specific, both protocols are capable of taking extra bandwidth when the network is not full whereas backing off when available network bandwidth drops. In other words, both TCP and the trinomial protocol converge to new steady states. However, it is evident that the oscillations in the sending rate of the trinomial protocol at the steady state are much smaller than those of TCP.

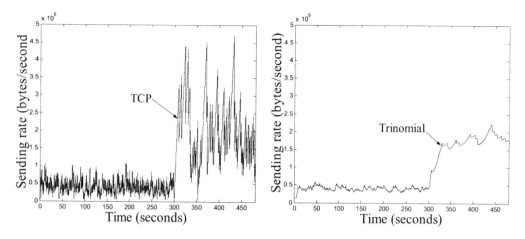

Fig.6. Response to bandwidth increase

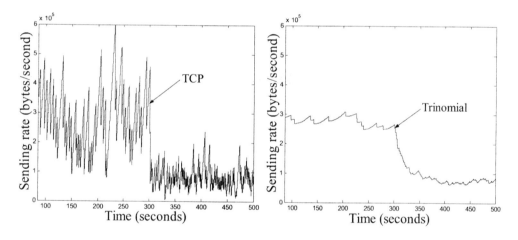

Fig.7. Response to bandwidth decrease

C   Time Delay and Data Packet Loss

In the simulation of time delay and packet loss, we use a network scenario where three flows (1 TCP, 1 UDP and 1 trinomial) to be examined share a bottleneck with a number of background TCP flows. From the simulation results given by Fig. 8, we can see that the trinomial protocol presents similar performance to UDP. Their delays are almost constant and their average delays are much smaller than that of TCP. From Fig.9, the trinomial protocol also provides similar performance to UDP on packet loss rate, which is better than TCP as well. This result was not expected before simulation. The reason maybe that they have different implementation mechanisms: TCP is window-based whereas both the trinomial protocol and UDP are rate-based.

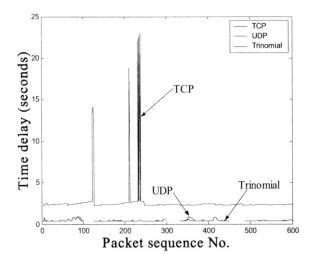

Fig.8. Time delay comparison

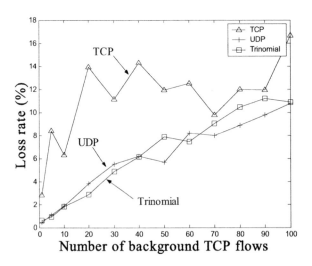

Fig.9. Packet loss rate comparison

D   Protocol Convergence

From the results shown in Fig.10(a), the two trinomial flows sharing a network link converge to the same sending rate although their initial sending rates are different, which is as the two TCP flows do in Fig.10(b). In other words, the trinomial protocol is intra-protocol fairness convergent as TCP. As explained in Section 1, UDP flows do *not* converge since UDP does *not* respond to changes in network states.

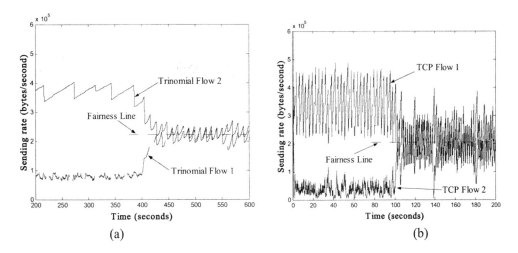

Fig.10. Protocol convergence

V   Experimental Implementation

The introduced data communication mechanism is implemented on an Internet mobile robot developed. The primary objective of the experiments is to verify the feasibility of the scheme. The hardware configuration of the Internet mobile robot mainly consists of a commercial *ActiveMedia* Pioneer 2 PeopleBot (P2PB) mobile robot. On the head of the robot, a *Sony* D30/31 pan-tilt-zoom (PTZ) video camera is used to provide live visual feedback on robot's immediate environment. There are two sonar arrays for range measurement. The P2PB mobile robot is connected to the Internet through a pair of wireless LAN adaptors (BreezeCom's BreezeNet PRO 11 indoor wireless Ethernet adaptors).

A   Date Categorization

Various types of information need to be exchanged between the robot and the human operator. Generally, there are four classes of data [24]:

- Administrative data (such as access control, user validation, configuration data etc.): small packet size, once-for-all transmission, no real-time requirement, but requiring reliable delivery;
- Control commands (such as desired translation velocity and rotation angle etc.): small packet size, non-periodic transmission, real-time delivery is required, and the most lately information is preferred should packets were lost;
- Image data (the most important and costly information feedback): large packet size, periodic transmission, real-time delivery is required. requiring significant bandwidth, and the most lately information is preferred should packets were lost; and
- Other information on the scene and the robot: (such as position and speed of the robot, range data etc.): small packet size, periodic transmission, real-time delivery is required, and the most lately information is preferred should packets were lost.

From the above categorization, it is clear that data communications in the remote control of Internet robots require *real-time* delivery except for the once-for-all administrative data. Consequently, in the implementation, we employ the trinomial protocol introduced for the transmission of all real-time information. The TCP protocol is adopted for the transmission of administrative data for which a TCP connection is opened when a teleoperation session is started and closed once the teleoperation is geared up.

B   Software Architecture

The data communication scheme is realized by a client-server software architecture for robot control and feedback information display. There are two servers, the video server and the control server, and the two corresponding clients, the video applet and the control applet. The web browser is a Java-enabled web browser and the web server is a Linux Apache web server. A brief block diagram of the functional software structure is shown in Fig. 11.

On the client side -- the web browser, the video applet is responsible for live image decoding and display etc., and the control applet is for intercepting and interpreting human control commands (mouse-click events in the control panel) and display of other information feedback, such as robot position, speed and ranging etc.

On the server side -- the P2PB mobile robot, the video server is in charge of video grabbing, compression, encoding and transmission, while camera pan-tile-zoom control, motor control and sensor information sampling are taken care of by the control server.

C   Experiments

The setup of the environment of the mobile robot is shown in Fig. 12. In the experiments, by using a Java-enabled web browser, such as Microsoft Internet Explorer or Netscape Communicator, on a regular PC, the users successfully navigated the P2PB mobile robot from point O_0 to point O_F via the Internet. Fig. 13 shows a sequence of the snapshots of the P2PB mobile robot when it was remotely guided via the Internet. We have conducted the experiments in different day times -- morning, noon, afternoon and night -- trying to capture the "rush hour" of the Internet traffic. Every time, the user succeeded to navigate the mobile robot through the experimental environment.

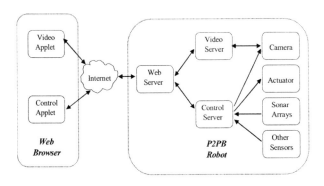

Fig.11. System software architecture

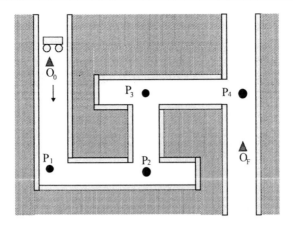

Fig.12. The objective of the experiment is to guide the mobile robot from point O_0 to O_F remotely

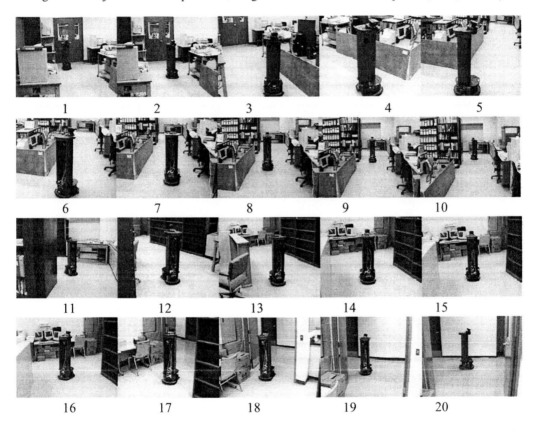

Fig.13. A sequence of robot images showing how the P2PB mobile robot is teleoperated over the Internet by the user to move through a clustered environment

VI Discussions, Conclusions and Future Work

In this paper, a novel data communication mechanism for the remote control of Internet robots is introduced. The core part of this mechanism, the trinomial protocol, is analyzed theoretically and its performances, such as transmission curve, TCP-friendliness, intra-protocol convergence, packet loss rate and time delay are evaluated by using simulation studies. The results confirm the theoretical model very well. Furthermore, the experimental implementation on an Internet mobile robot shows the feasibility of the data communication mechanism.

The experiments are mostly carried out on the Internet in Canada (which is much fast) due to the logistics reason; however, it does show the feasibility of the proposed data communication mechanism. For real-world applications, the scheme may need to be extensively tested on the Internet with greater diversity, irregularity and heterogeneity in the future. Extensive real-world experiments would help us to identify some of actual issues that may exist in the Internet and cannot be easily captured.

Wireless links are necessary in many applications of Internet robots, e.g. Internet mobile robots. Like most other transport protocols, the trinomial protocol relies on packet loss as an indictor to determine whether the network is congested. If packets get lost, the network is considered congested. For wired network connections, this inference mechanism holds very well. However, for wireless connections, this inference does not always work very well because packet loss in wireless links does not necessarily mean congestion. This problem is actually faced by all congestion control protocols, such as TCP. Unfortunately, there is no effective and simple ways to solve this problem at the time being. Current approaches to distinguishing congestion loss from wireless loss require cooperation from intermediate nodes on the path, which is however infeasible in most cases [32]. The desirable approach should be an end-to-end mechanism that does not require intermediate hosts to take any specific actions. From the literature, this problem has not been well solved [33, 34]. Consequently, it would be one extended topic (challenging, but with great potentials for a wide spectrum of applications) of this data transmission scheme if wireless connections are involved in the application.

References

[1]   K. Goldberg and R. Siegwart (editors), Beyond Webcams: An Introduction to Online Robots, The MIT Press, Cambridge, Massachusetts, USA, 2002

[2]   B. Giordano and R. Buckley, "Trends in colour imaging on the Internet," in *Proc. 9th International Congress of the AIC*, SPIE Vol. 4421-4422, June 2001, Rochester, USA

[3]   K. Goldberg *et al.*, "Desktop teleoperation via the World Wide Web," *Proc. IEEE Int. Conf. Robotics and Automation*, Nagoya, Japan, 1995, pp654-659

[4]   K. Taylor and B. Dalton, "Issues in Internet telerobotics," in *Proc. Int. Conf. Field and Service Robotics*, Canberra, Australia, 1997, pp151-157

[5]   C. Sayers, *Remote Control Robotics*, Springer Verlag, New York, 1998

[6]   T. Fong and C. Thorpe, "Vehicle teleoperation interfaces," *Autonomous Robots*, vol.11, no.1, pp. 9-18, 2001

[7]   J. Postel, RFC 793: Transport Control Protocol, DARPA Internet Program Protocol Specification, September 1981

[8]   J. Postel, RFC 768: User Datagram Protocol, August 28, 1980

[9]   S. Floyd and K. Fall, "Promoting the use of end-to-end congestion control in the Internet," *IEEE/ACM Trans. Networking*, , vol.7, no4., pp458-472, 1999

[10] J. Padhye, J. Kurose, D. Towsley and R. Koodli, "A model based TCP-friendly rate control protocol," in *Proc NOOSDAV*, 1999, pp137-151

[11] R. Rejaie, M. Handley, and D. Estrin, "RAP: An end-to-end rate-based congestion control mechanism for realtime streams in the Internet," in *Proc. IEEE Infocom*, March 1999, pp1337-1345

[12] D. Sisalem and H. Schulzrinne, "The loss-delay adjustment algorithm: A TCP-friendly adaptation scheme," in *Proc. International Workshop on Network and Operating System Support for Digital Audio and Video (NOSSDAV)*, (Cambridge, England), Jul. 1998, pp. 215—226

[13] S. Jin, L. Guo, I. Matta and A. Bestavros, "TCP-friendly SIMD congestion control and its convergence behaviour," in *Proc. ICNP'2001: 9th IEEE International Conf. Network Protocols*, Riverside, CA, November 2001

[14] H. Schulzrinne *et al.*, RFC1889: RTP: A Transport Protocol for Real-time Applications, *Internet Engineering Task Force*, 1996

[15] W. Tan and A. Zakhor, "Real-time Internet video using error-resilient scalable compression and TCP-friendly transport protocol," *IEEE Trans. Multimedia*, vol.1, no.2, pp.172-186, 1999

[16] K. Ramakrishnan and R. Jain, "A Binary Feedback Scheme for Congestion Avoidance in Computer Networks with Connectionless Network Layer," in *Proc. SIGCOMM'88*, August 1988, pp. 303-313

[17] R. Jain, K. Ramakrishnan, D. Chiu, "Congestion Avoidance in Computer Networks with a Connectionless Network Layer," in *DEC-TR-506*, reprinted in C. Partridge, Ed., *Innovations in Internetworking*, published by Artech House, October 1988

[18] D.-M. Chiu and R. Jain, "Analysis of the increase and decrease algorithms for congestion avoidance in computer networks," in *Computer Networks and ISDN Systems*, No.17, 1989, pp1-14

[19] S. Floyd, *RFC 2914:* Congestion Control Principles, Best Current Practice, September 2000

[20] V. Jacobson, "Congestion avoidance and control," *ACM SIGCOMM*, August 1988, pp 314-329

[21] J. Mahdavi and S. Floyd, "TCP-friendly unicast rate-based flow control," Technical note sent to the end2end-interest mailing list, January 1997. http://www.psc.edu/networking/papers/tcp-friendly.html

[22] The Network Simulator - ns-2, http://www.isi.edu/nsnam/ns/

[23] P. Fiorini and R. Oboe, "Internet-based telerobotics: Problems and approaches," http://citeseer.nj.nec.com/148450.html, 1997

[24] Lasso and T. Urbancek, "Communication architectures for web-based telerobotic systems," in *IEEE Mediterranean Conference on Control and Automation, Dubrovnik, Croatia*, June 27-29, 2001

[25] P. X. Liu, M. Meng, J. Gu and Simon X. Yang "A Study of Internet Delay for Teleoperation Using Biologically Inspired Approaches," *International Journal of Robotics and Automation*, vol. 17, no.4, pp186-195, 2002

[26] R.. Jain, "A delay-based approach for congestion avoidance in interconnected heterogeneous computer networks," *ACM Computer Communication Review*, vol. 19, no. 5, pp.56--71, 1989

[27] R. Rejaie, An End-to-End Architecture for Quality Adaptive Streaming Applications in the Internet, Ph.D. Dissertation, University of Southern California, 1999

[28] K. Ramakrishnan and S. Floyd, RFC 2481: A proposal to add explicit congestion notification (ECN) to IP, January 1999

[29] J. Heinanen, F. Baker, W. Weiss, and J. Wroclawski, Assured forwarding PHB group, *IETF Internet-Draft*, January 1999

[30] P. X. Liu, M. Meng and J. Gu, "Internet Roundtrip Delay Prediction Using the Maximum Entropy Principle," *Journal of Communications and Networks*, vol. 5, no.1, 2003, pp.65-72

[31] P. X. Liu, M. Meng, X. Ye and J. Gu, "A Rate-based Protocol for Internet Robots," in *Proc. 2002 IEEE World Congress on Intelligent Control and Automation* (WCICA 2002), Jun. 10-14, 2002, pp59-65

[32] H. Balakrishnaa, V. Padmanabhan, S. Seshan, and R. Katz, "A comparison of mechanisms for improving TCP performance over wireless links," in *ACM SIGCOMM'96*, Aug. 1996

[33] S. Biaz and N. Vaidya, "Discriminating congestion losses form wireless losses using inter-arrival timers at the receiver," in *IEEE Symposium ASSET'99*, RiChardson, TX, USA, March 1999

[34] S. Cen, P. Cosman, and G. Voelker, "Eng-to-end differentiation of congestion and wireless losses," in *Proc.ACM Multimedia Computing and Networking 2002*, San Jose, CA, Jan. 23-24, 2002. SPIE vol. 4673, pp. 1-15

[35] H. Schulzrinne et al., RFC1889: RTP, A Transport Protocol for Real-Time Applications, 1996

In: Control and Learning in Robotic Systems
Editor: John X. Liu, pp. 21-58

ISBN 1-59454-356-9
©2005 Nova Science Publishers, Inc.

Chapter 2

MODELING FOR REACTIVE CONTROL AND DECISION MAKING IN UNCERTAIN ENVIRONMENT

Ayeley P. Tchangani[1,2]*

[1]Dpt. GEII, IUT de Tarbes, Université Toulouse III - Paul Sabatier,
1 rue Lautréamont, BP 1624, 65016 Tarbes Cedex,
France. Fax: +33 (0)5 62 44 42 19
[2]Laboratoire Génie de Production, 47 Avenue d'Azereix,
BP 1629, 65016 Tarbes Cedex, France.

Abstract

Decision making and/or control is the process by which free variables (options, actions, decisions, control signals, ...) of a system are evaluated and selected in order to realize some goals. Reactive decision making (and/or control) that is decision making in situation where things are not going as expected (reorganization of emergency services during a catastrophe event in order to ensure efficiency ; adjustment of production rate when demand of a product deviates significantly from its estimated value; reallocation of resources/time to activities in order to maintain the required quality of service when customers arrival rate at a service deviates from its nominal value ; ...) because of external uncontrollable effects such as state of the nature change and/or other decision makers' actions is common in practice. Problems of decision making in presence of uncertainty due to the randomness of the state of the world are well formulated and solved using the so called state of the world model for decision and those involving many decision makers with conflict interests are formulated and solved using non cooperative game theory. Classical assumptions concerning these decision making models (known probability distribution of the state in the case of state of the world model for decision making and perfect knowledge of the game that is knowledge of other players strategy and goals in the case of game theory for instance) restrict

*E-mail address: tchangani@geii.iut-tarbes.fr, tchangani@caramail.com

their application in practical situation. Indeed, in practical situation the environment will be constantly changing and so decision strategy or control law must be constantly adapted that is what we call *reactive decision making or control*. To efficiently react, at least a partial knowledge of environment change or other players behavior will be of great importance. This can be done by setting up an *intelligence system* (system that collects and transforms data to knowledge and possibly knowledge to recommended decision or control action) consisting in sensors (in broad sense including spying and monitoring for instance) to collect information about the environment and processing capabilities to process this information (data analysis, features extraction, information and sources fusion, ...) in order to retrieve knowledge about environment and send it to the decision maker or controller. This fresh knowledge will modify decision maker (or controller) belief about expected state of the world and/or other actors intention and strategy that will allow to react efficiently by changing decision or control that was initially planned. In this paper we will concentrate on modeling how to use the knowledge from intelligence system by developing an integrated framework based on influence diagrams (in the case of decision making) and mathematical programming (in the case of control) as mathematical tools to support decision making and/or control in reactive context.

Keywords: Reactive Decision Making, Reactive Control, Intelligence System, Influence Diagrams, Mathematical Programming, Decision Support System.

1   Introduction

Decision making or control (selection of some free variables called *decision variables* or *control variables* of a system to realize some objectives) problems in literature are formulated using different mathematical tools according to their nature and they can be classified into two main categories, *static* or *dynamic (sequential),* by considering the effect of *time* on the outcome of decisions. In the literature, the term control is almost reserved for dynamic case with control variables belonging to a continuous set whereas decision making is generally used for discrete decision variables case.

In the framework of static decision making, it is supposed that decision made at a given time instant do not influence future decisions ; at each time instant or period decision maker have to choose *decision/control variables* or to take some *actions* (production level for a given period, stock level, investment budget level, stop/set up a machine, make further diagnosis, etc.) in order to realize some objectives (maximize profit/minimize cost, avoid failure, attain a goal, etc.) when respecting some constraints due to necessarily limitation of available resources (budget limitation, lack of qualified manpower, etc.) and possible technical specifications. These problems are well formulated as mathematical programming [5] ; when uncertainties (unavoidable in practice) are considered, it leads to terminology such as stochastic mathematical programming [20, 21], fuzzy mathematical programming [20], bayesian linear programming [18], etc. In the cases where uncertainties are due to the nature, these problems are formulated as the so called *state of the world model* for decision making [19, 21]. Problems of decision making involving many decision makers with conflict interests are formulated and solved using *noncooperative game* theory [21]. In traditional game theory it is assumed that each decision maker, player or agent has a well defined utility function that he/she attempts to maximize and moreover each agent is assumed to have a complete knowledge of other player strategies. It is well known that these assumptions are rarely realized in practical situation ; indeed players most of time do not have a clear formulation of their utility and do have just an approximate idea of other players strategies and desires. Decisions are then made by attempting to anticipate the intension of opponents with possible use of sensors (in large including spying, monitoring, etc.).

In sequential decision making case (air traffic control, control of autonomous surveillance aircraft, logistics planning and scheduling, equipment diagnosis and repair, etc.), decision at a certain instant does influence future decisions and the purpose is to choose a sequence of decisions or controls that realize a desired objective over an horizon. These problems are mainly formulated as optimal control [10] in the case where a well defined model of system dynamics do exist or Markov decision processes (MDP) [3, 4, 17] in the case where the dynamics of the system are defined stochastically that is the outcome (next state of the process) of a decision is defined by means of a probability distribution. In the case of MDP where the state of the system is not completely observed at each instant these models are known as partially observable Markov decision processes (POMDP) [9].

Decision making or control models presented so far are suited for the context where necessary information for decision making is available at each instant or at least easily estimated and the objective is to ensure a long term performance (maximizing profit for a given period, transferring a mobile from point A to point B in minimum time or with minimum energy consumption for instance). Hypothesizes of disposing of necessary information at the time decision is made do not hold in reactive or real time decision situation (*emergency* or *crisis*) where decision process must be completed before a given deadline (with relevant information arriving *sequentially*) otherwise the decision becomes irrelevant for the pursued goal. For instance in the case of noncooperative game where a player set up some sensors to assess actual behavior of its opponent (think of a government agency that is fighting a terrorist group), information from different sensors may arrive asynchronously to the decision maker and decision must be made in real time context. But in many domains and during a crisis or emergency necessary informations for decision making are prioritized, that is all attributes of the nature or opponents intension do not have the same importance and decision maker can content himself with information at hand as soon as it concerns important condition variables or attributes related to the pursued goal. So to construct a reactive or real time decision support system one must first identify important attributes that condition the decision and then construct relationship between these attributes and the decision/control variables. The non avoidable uncertainty that underlies this relationship suggests to use mathematical tools dealing with uncertainties such as *bayesian networks* [8, 15] that deals with probabilistic uncertainty. In this paper, bayesian nets and their extensions, *influence diagrams* [7, 8] in one hand and mathematical programming in other hand, will be used to construct a reactive decision and control support system for use in an uncertain environment (randomness of the state of the world and/or noncooperative opponents). Existing experimental data and expertise can be easily integrated to the system by taking advantage of bayesian nets learning capabilities. The main idea expressed in this paper is the existence of an intelligence system that permanently observes the environment and supply decision maker/controller with fresh information as resumed by Figure 1. In nominal conditions, a stationary policy or control law can be established using formal or non formal methods such us heuristics and the purpose of intelligence system will consist in estimating how far actual conditions differ from nominal ones and supply this information to decision maker or controller that will react by applying an update decision/control instead of planned one.

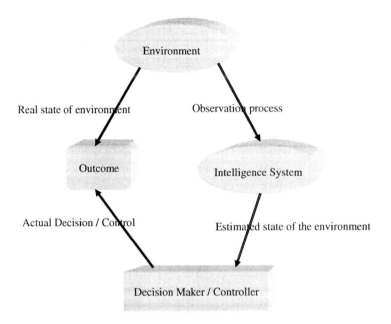

Figure 1: Integrated framework for intelligent decision making or control in uncertain environment.

In this paper we will concentrate more on how to use actual estimated deviation from nominal conditions to generate a reactive decision or control.

The remainder of this paper is organized as follow: in the next section, process of modeling decision making problem is considered including the presentation of the mathematical tool (influence diagrams) used for this purpose, the established model and criteria as well as some potential application ; third section will consider the problem of reactive control with an organization similar to previous section (mathematical tool, established model, potential applications) and finally section four presents conclusions of this work.

2   Modeling for Reactive Decision Making

The problem of decision making (choosing free variables or actions in a discrete set in order to optimize some objectives in presence of uncertain environment) that we consider here is determined by following attributes:

- $\mathcal{A}$, finite and discrete set of actions or decisions that can depend on time (here we suppose that it is stationary) ; actual action is noted $A(t)$;

- $\mathcal{S}$, finite and discrete set of the states of the system ; actual state is noted $S(t)$;

- $\mathcal{X}$, finite and discrete set of the state of the world ; actual state of the world is noted $X(t)$;

- $\mathcal{B}$, finite and discrete set of possible opponents actions ; actual action of the opponents is noted $B(t)$;

- $\mathcal{O}$, the set of the outcomes mainly in the case of partial observed environment ; actual outcome is noted $O(t)$

- $\mathcal{I} : \mathcal{O} \times \mathcal{A} \to \mathbb{R}$, a possible numerical objective function that must be optimized ;

- a model of system dynamics, how the state of the system changes over time given historical information. We will consider here that the dynamics of the system form a Markov process that is the next state depend only on actual information (no memory) ; in this case the model is completely determined by the transition matrix given by (1)

$$P_{ij}^{a,b,x} = \Pr\{S(t+1) = j/S(t) = i, A(t) = a, B(t) = b, X(t) = x\}. \tag{1}$$

If a profile of the state of the world $X^*(t)$ and a profile of opponents actions $B^*(t)$ are known then the transition matrix $P_{ij}^{a,b,x}$ becomes just a non stationary MDP transition matrix $P_{ij}^{a}(t)$ and an optimal policy π^* defined by (2)

$$\pi^* : \left[0, \; \infty \; \right[\times \mathcal{S} \to \mathcal{A}, \, (t,s) \mapsto \pi^*(t,s) \tag{2}$$

can be obtained by any formal or non formal method (we suppose it). Then if environmental conditions are not changed, one just applies this policy at each instant. But in practice $X(t)$ and $B(t)$ will necessarily change over time around their nominal profiles $X^*(t)$ and $B^*(t)$ and the decision maker for efficiency must react to this changes ; but in general decision maker will not observe actual $X(t)$ and $B(t)$ at the time he/she must react but only estimates $(\overline{X}(t)$ and $\overline{B}(t))$ of these variables if an intelligence system is constructed. In the following paragraphs we will show that this reactive decision making process is reducible to constructing an influence diagram of the system. Let us first introduce briefly influence diagram concepts.

2.1 Influence Diagrams

An influence diagram is a simple visual representation of a decision problem. Influence diagrams offer an intuitive way to identify and display the essential elements, including decisions, uncertainties, and objectives, and how they influence each other. It shows the dependencies among the variables more clearly than the decision tree. An influence diagram or decision graph [7, 8] consists of a direct acyclic graph (DAG) known as its *structure* that depicts relationships among variables in a decision problem and conditional probabilities distribution of each node given evidence on its parents (nodes that have a direct arc into the considered node) known as its *parameters*. An influence diagram has 3 types of nodes as shown by Figure 2 with following meaning:

- *chance nodes* (oval) represent uncertain variables (environment) that influence the decision problem ;

- *decision nodes* (rectangle) represent choices open to decision maker ;

- *value nodes* (diamond) represent attributes (most of time numeric) the decision maker cares about.

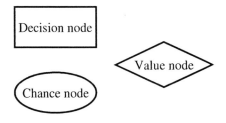

Figure 2: Different nodes of an influence diagram.

Influence diagrams are an extension of bayesian networks (BN) (see [2], [8], etc.) that are DAG consisting with chance nodes only. Bayesian Networks (BN) derive from convergence of statistical methods that permit one to go from information (data) to knowledge (probability laws, relationship between variables, ...) and Artificial Intelligence (AI) that permit computers to deal with knowledge (not only information). The terminology BN comes from work by Thomas Bayes [1] in eighteenth century. Its actually development is due to [15]. The main purpose of BN is to integrate uncertainty in expert system. Indeed, an expert, most of time, has only an approximative knowledge of the system that he or she formulates in terms like: *A* has an influence on *B* ; if *B* is observed, there exists a great chance that *C* occurs ; and so on. In the other hand, there are data (measurements for example) that contain some information which must be transformed into relationships between variables. In influence diagrams, an arc or edge relating two chance nodes is called a *relevance arc* because it indicates that the value of one variable (source node) is relevant to the probability distribution of the other node (destination node), arcs from decision nodes to chance nodes are known as *influence arcs* meaning that the decision influences the outcome of the chance node and arcs into decision nodes (from chance nodes) are called *information arcs* meaning that the outcome of the chance node will be known at the time decision is made. Decision nodes are ordered in time that is there is a direct link between all decision nodes. Finally, arcs from chance or decision nodes into value nodes represent functional links. Relevant arcs may mean many things depending on the problem at hand such as: implication, correlation, causality, ... Figure 3 shows an example of influence diagram that represents following decision problem. To monitor a machine, some *sensors* are put on it in order to give information about its *actual state*. According to this information one decide whether to stop the machine for *diagnosis* or not. Stopping machine for diagnosis as well as letting the machine operate in bad state has some *cost*. Notice that the value node representing cost node can be replaced by a chance node to represent a qualitative outcome.

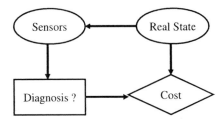

Figure 3: An example of an influence diagram.

This mathematical tool will be used as the backbone to construct a decision support system for reactive decision making in uncertain and/or conflicting environment.

2.2   Decision Making in Uncertain and Conflicting Environment

2.2.1   State of the World Model for Decision

In many decision-making problems, the reward or outcome of a given decision depends on the state of the world. Classically, such decision-making problems are formulated in terms of state-of-the-world decision-making model [21] with possible extension such us that one considered in [19]. The principle (roughly) is as follow: there are n possible actions or decisions $\{a_1, a_2, ..., a_n\}$ and m possible states of the world $\{x_1, x_2, ..., x_m\}$. If an action a_i is decided and the observed state of the world is x_j then decision maker receives a reward $r(a_i, x_j)$. This decision making problem can be simply represented by the influence diagram of Figure 4.

Different algorithms are used in the literature to solve this decision-making problems, that is deciding the best action (or optimal action) knowing that the state of the world will be observed a posteriori ; most popular algorithms are: Hurwicz, Savage and Laplace algorithms that are recalled below.

- Hurwicz algorithm. This algorithm selects an action that maximizes a convex combination of maximum and minimum reward with regards to state of the world, that is

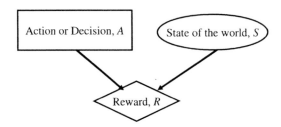

Figure 4: An influence diagram representation of state-of-the-world decision-making model.

the optimal action a^* is selected as (3)

$$a^* = \arg\max_{a_i} \left(\lambda\max_{x_j}(r(a_i,x_j)) + (1-\lambda)\min_{x_j}(r(a_i,x_j)) \right), \ 0 \le \lambda \le 1. \qquad (3)$$

For $\lambda = 0$, we have Wald's algorithm (or MaxiMin reward algorithm) that shows an extremely prudent behavior with regards to risk ; when $\lambda = 1$ we have the MaxiMax reward algorithm.

- Savage or MiniMax Regret algorithm. The regret matrix is defined as follow: call $i^*(j)$ the optimal action index if we know that the state of the world is x_j, then the regret of taking action a_i when the state of the world is x_j is given by $r\left(a_{i^*(j)},x_j\right) - r(a_i,x_j)$ and so the minimax regret algorithm selects as optimal action a^*, one that minimizes the regret, namely

$$a^* = \arg\min_{a_i} \left(\max_{x_j}(r\left(a_{i^*(j)},x_j\right) - r(a_i,x_j)) \right). \qquad (4)$$

- Laplace or Maximum Expected Value: here it is supposed that each state x has probability $p(x)$ of occurrence and the optimal action is selected as (5)

$$a^* = \arg\max_{a_i} \left(\sum_{j=1}^{m} p(x_j)r(a_i,x_j) \right). \qquad (5)$$

Previous algorithms are easy to implement and computationally efficient ; their main drawback is that they give an off-line (open loop) type solution. They do not consider a possible evidence on environmental events (absence of information link from chance node X to decision node A on Figure 4) in order to have a reactive behavior during decision process. If the state of the world were known in advance, the decision would be obvious. In most practical cases decision makers will try to estimate the state of the world by using some sensors (to be understood in large context including humans) which information is used to estimate the most probable state of the world before making decision ; this new model of decision making can be represented by the influence diagram of Figure 5 where $\overline{X}$ is the estimation of actual state X in response to information supplied by sensors.

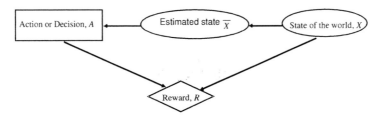

Figure 5: An intelligent influence diagram representation of state-of-the-world decision-making model.

2.2.2 Conflicting Decision Making: Game Theory

Game theory is a branch of mathematical analysis developed to study decision making in conflicting situations. Such a situation exists when two or more decision makers who have different objectives act on the same system or share the same resources. Game theory provides a mathematical process for selecting an optimum strategy (that is, an optimum decision or a sequence of decisions) in the face of an opponent who has a strategy of his own. In traditional game theory one usually makes following assumptions.

- Each decision maker (agent, player) has available to him well-specified choices (decisions) or sequences of choices.

- Every possible combination of choices available to the players leads to a well-defined state.

- A specified outcome(reward, payoff or consequences) for each player is associated with each state. In the case of zero-sum game, payoffs are numerical values and sum of payoffs to all players is zero in each state.

- Each decision maker has perfect knowledge of the game and of his opposition ; that is, he knows in full detail the rules of the game as well as payoffs and strategies of all other players.

- All decision makers are (*substantive*) rational ; that is, each player, given two alternatives, will select the one that yields him the greater payoff.

The last two assumptions, in particular, restrict the application of game theory in real-world conflicting situations. Nonetheless, game theory has provided a means for analyzing many problems of interest in economics, management science, and other fields. Substantive rationality suppose that the decision makers seek maximum utility and have all cognitive resources to this end and this is not the case in majority of cases. Noncooperative games are those games where players can not cooperate. One solution concept in the case of non constant sum game is the so-called *Nash equilibrium*, named after John Nash, that is a set of strategies, one for each player, such that no player has incentive to unilaterally change her action. Players are in equilibrium if a change in strategies by any one of them would lead that player to earn less than if she remained with her current strategy. For games in which players randomize (mixed strategies), the expected or average payoff must be at least as large as that obtainable by any other strategy. In a constant sum game (noncooperative game), the sum of all players' payoffs is the same for any outcome. Hence, a gain for one participant is always at the expense of another, such as in most sporting events. Given the conflicting interests, the equilibrium of such games is often in mixed strategies. Since payoffs can always be normalized, constant sum games may be represented as (and are equivalent to) zero sum game in which the sum of all players' payoffs is always zero. Maximin optimization in this case is a rational solution concept. For simplicity let us consider a two persons, players, decision makers or agents, A and B, noncooperative game that can model practical situation such as two armies in military operations, a government fighting against a terrorist group, ... Let suppose that A and B dispose of finite set strategies $\mathcal{A}$ and $\mathcal{B}$ respectively. To any couple (a, b) where $a \in \mathcal{A}$ and $b \in \mathcal{B}$ is associated a gain or reward $R(a, b)$ for player A that is loss for player B. This situation can simply be represented by influence diagram of Figure 6.

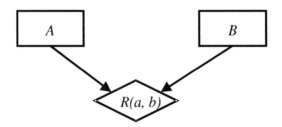

Figure 6: An influence diagram representation of a two players game.

A main drawback of this model is the absence of interaction between players because of perfect information assumption that is difficult to be realized in practice. For instance one can not reasonably talk about perfect information when two armies are fighting each other or when a state is fighting a terrorist group or in concurrence on a market. In fact, in the point of view of a given player, let say A, the behavior of other player(s) B can be considered as a state of the nature and so by using some "sensors" including spying for instance, player A can guest possibly behavior of the opponent B ; in terms of influence diagram, we get Figure 7 where $\overline{B}$ is the estimation of B given information delivered by sensors.

These two models for decision making in uncertain and conflicting environment will be integrated in a unique framework in the following paragraph.

2.3 Model for Reactive Decision Making in Uncertain and Conflicting Environment

The purpose of this section is to construct an integrated framework for making decision with state of the world decision model and noncooperative game model. We consider building a decision support system in the spirit of decision maker A, with *actions* or *decisions* set

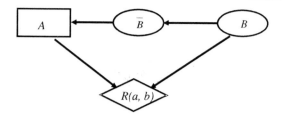

Figure 7: An intelligent influence diagram representation of a two players game (in the point of view of A).

a , that is acting in an uncertain environment determined by a random variable X which takes its values called states in a set x . Moreover there are other decision makers whose interests are in conflict with A's interests ; in the spirit of A their behavior is summarized in a random variable B taking its values in a set $\mathcal{B}$. This model of decision making is known in the literature as *cybernetic system* [11]. The outcome for the decision maker A (when he/she makes decision a, B makes decision b and X is in state x) is a random variable denoted $O(a; b, x)$, where $a \in \mathcal{A}$, $b \in \mathcal{B}$ and $x \in x$, that takes its values in a set o ; to each outcome $o \in o$ and each decision $a \in \mathcal{A}$ one may associate a numerical value $R(o,a)$ that permit to numerically order elements of o and the goal of decision maker A will be to optimize $R(a, o)$ or any function related to the outcome. In practical situation actual state of the world or the strategy (behavior) of opponents can not be known. But as it was done in former section, we consider that, decision maker A dispose of sensors (in a broad sense including humans) that will give informations to estimate the actual state of the world $\overline{X}$ and actual strategy of opponent $\overline{B}$. Each sensor will have a partial representation of X or B and send this information asynchronously to $A's$ decision center. We suppose that there are n_X sensors to sense state of world represented by

$$\overline{x} = \left\{ \overline{X}_1, \overline{X}_2, ..., \overline{X}_{n_X} \right\}$$

and n_B sensors for sensing opponents' strategies determined by

$$\overline{\mathcal{B}} = \left\{ \overline{B}_1, \overline{B}_2, ..., \overline{B}_{n_B} \right\}.$$

This problem can be compactly described by an influence diagram ; in terms of influence diagram we have:

- $n_X + 1$ chances nodes to represent the state of world X and its images $\overline{X}_1, \overline{X}_2, ..., \overline{X}_{n_X}$ given by the n_X sensors ;

- $n_B + 1$ chances nodes to represent opponent's strategy B and its images $\overline{B}_1, \overline{B}_2, ..., \overline{B}_{n_B}$ given by the n_B sensors ;

- two chances nodes $\overline{X}$ and $\overline{B}$ resulting of information fusion from each category of sensors ;

- a decision node to represent decision maker A ;

- a chance node O to represent the outcome $O(a; b, x)$;

- and finally a possible value node to represent $R(o, a)$.

We ignore time here because the constructed influence diagram represent the configuration of decision problem at each instant so time is implicit in this model. Figure 8 summarizes this model in terms of an influence diagram. Notice that one may avoid using nodes $\overline{X}$ and $\overline{B}$ by letting sensors directly influence the decision node A thus considering estimation as part of decision process.

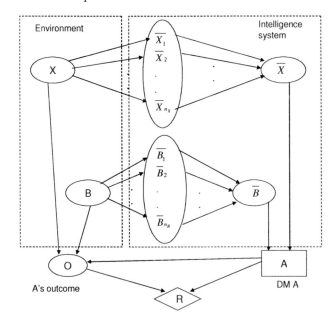

Figure 8: Integrated influence diagram for decision making in uncertain and conflicting environment in the spirit of DM A.

We will now define how this integrated model can be used for reactive decision making. Let suppose that a stationary policy $\pi_A^*(t, X^*(t), B^*(t))$ for A that optimize some long term performance in the presence of a known profile $X^*(t)$ of the state of the world and a known profile $B^*(t)$ of opponents actions is obtained by any formal or non formal method. The purpose of the reactive decision support system constructed so far is to efficiently react to any new evidence sensed by sensors by making the appropriate (in some sense) decision. The reactive decision rule or policy $\pi_A^r(t)$ of A at time t is then defined as follows.

Definition 1 *The reactive decision rule (or policy) of A at each instant t is defined by (6)*

$$\pi_A^r(t) = \begin{cases} \pi_A^*(t, X^*(t), B^*(t)) & \text{if nominal conditions are not changed} \\ \pi_A(t, \overline{X}(t), \overline{B}(t)) & \text{otherwise} \end{cases} \tag{6}$$

where $\pi_A(t, \overline{X}(t), \overline{B}(t))$ is defined to ensure some local performance.

2.3.1   Local Performance Criteria

According to the nature and application domain of the decision problem, different criteria may be used to define policy function $\pi_A\left(t,\overline{X}(t),\overline{B}(t)\right)$. Here is some criteria that can be used.

Failure avoidance/Goal attainment   For some applications in domain such as maintenance, medical treatment, air traffic control, and mainly for any system related to security of humans, the principal objective of a decision support system is to permit to act in the sense that avoid catastrophes. Thus, it is supposed that the outcome set o contain a particular subset $\mathcal{F} \subset o$ (possibly single element) to be avoided known as failure subset ; then the "appropriate" action a^* given actual sensors information is determined by (7)

$$\pi_A\left(t,\overline{X}(t),\overline{B}(t)\right) = a^* = \min_{a\in\mathcal{A}}\left\{\Pr\left\{O(t+1)\in\mathcal{F}\,/\overline{X}(t),\,\overline{B}(t)\right\}\right\} ; \tag{7}$$

the selected action is that one which minimizes the probability of immediate failure. In other applications the main purpose will be to act so that the outcome O attain a particular subset $g \subset o$ (possibly single element) known as the goal subset. In fact it is a dual criterion of failure avoidance criterion because avoiding particular subset is equivalent to trying to attain its complementary. In this case we have (8)

$$\pi_A\left(t,\overline{X}(t),\overline{B}(t)\right) = a^* = \max_{a\in\mathcal{A}}\left\{\Pr\left\{O(t+1)\in g\,/\overline{X}(t),\,\overline{B}(t)\right\}\right\}. \tag{8}$$

Maximum expected reward/Maximin reward   When outcome O is convertible to numerical value through $R(a,o)$ in the case of applications such as production management, marketing and any application where the outcome can be expressed in money like scale, criteria presented in section 3 can be used to select appropriate action. For instance when using maximum expected reward criterion, $\pi_A\left(t,\overline{X}(t),\overline{B}(t)\right)$ is determined by (9)

$$\pi_A\left(t,\overline{X}(t),\overline{B}(t)\right) = a^* = \max_{a\in\mathcal{A}}\left\{E_o\left\{R(a,\,o)/\overline{X}(t),\,\overline{B}(t)\right\}\right\} ; \tag{9}$$

where $E_o\left\{R(a,\,o)/\overline{X}(t),\,\overline{B}(t)\right\}$ denotes conditional mean or mathematical expected value of $R(a,o)$ with regards to outcome's probability distribution ; and in the case of maximin reward criterion, $\pi_A\left(t,\overline{X}(t),\overline{B}(t)\right)$ is determined by (10)

$$\pi_A\left(t,\overline{X}(t),\overline{B}(t)\right) = a^* = \max_{a\in\mathcal{A}}\left\{\min_{o\in o}\left\{R(a,\,o)/\overline{X}(t),\,\overline{B}(t)\right\}\right\}. \tag{10}$$

An example of potential application of the model constructed is considered in the following paragraph.

2.3.2   Application: Reactive Maintenance

We consider a small example in the field of maintenance. The maintenance of a machine is organized in two level: a planned cyclic policy level and a reactive level. The planned policy consists in stopping machines periodically for a systematic maintenance process. But to avoid a possible unpredictable failure of machine, an intelligence system is implemented to continuously monitor the machines ; this system consists in a number of sensors that sense some physical signals (vibration, electric signal, temperature, etc) and this information undergoes a data fusion process by a data fusion system that display the estimated status of the machine in terms of different level of warnings to the decision maker. The environment conditions (temperature, pressure, humidity, electromagnetism, ...) have an influence on machine status ; these conditions are not directly observed ; only an estimation is available. The planned maintenance policy is established in nominal environment conditions but these conditions can be better or worst ; on the basis of information displayed by data fusion system and estimated environment conditions, the reactive decision is to decide wether to ask an expert to diagnosis the machine or not. Stopping a machine in normal status for diagnosis or letting a machine fails before planned maintenance process has a cost. In terms of influence diagram we have following nodes:

- Actual machine status (wether it is undergoing failure or not), a chance node ; we suppose that this node has two states: Normal (the machine is functioning normally) and Undergoing_Failure (there are troubles on the machine that could lead to its failure).

- Sensors status in terms of warnings, a chance node ; the warnings correspond to machine status (green light for normal and red light for undergoing failure for instance) ; so this node has two states: Green and Red.

- Environment conditions, a chance node with states Good/Bad.

- Estimated environment conditions, a chance node with states Good/Bad.

- Next status, status of machine at next observation instant, a chance node ; it has two states: Failure and No_Failure specifying whether the machine fails or not.

- Decision, a decision node ; this node has two actions: Stop and Not_Stop.

- cost induced, value node.

At the time decision is made, decision maker is in possession of information about sensors warnings and estimated environment conditions so there are *information arcs* between these chance nodes and the decision node. Sensors warnings are influenced by machine status so there is *relevance arc* between the later node and former one ; there are also *relevance arcs* going from environment conditions node and machine node into future status node in one hand and from environment conditions node into estimated environment conditions node in other hand. Decision made at actual instant will influence next status of the machine so there is an *influence arc* from decision node to next status and finally there are *functional arcs* from decision node and future status node into cost node. The structure of the influence diagram representing this decision problem is given by Figure 9 drawn using Netica^{TM} a software package developed by Norsys Software Corp. (see [14]) to deal with bayesian nets and influence diagrams.

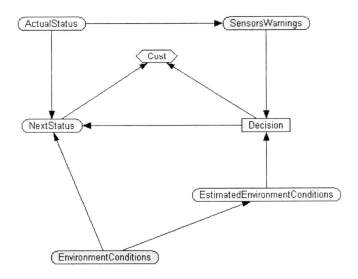

Figure 9: Influence diagram of recative maintenance problem.

The remaining task is to specify parameters that is conditional probability distribution of each node. Let suppose that sensors report right status of the machine in 95 % of cases , that is

$$\Pr\{SensorsWarnings = Green/ActualStatus = Normal\} = 0.95,$$
$$\Pr\{SensorsWarnings = Green/ActualStatus = Undergoing\_Failure\} = 0.05,$$
$$\Pr\{SensorsWarnings = Red/ActualStatus = Undergoing\_Failure\} = 0.95,$$
$$\Pr\{SensorsWarnings = Red/ActualStatus = Normal\} = 0.05,$$

and that conditional probability table for nodes estimated environment conditions and next status are given by Figure 10 (a) & (b) respectively.

Once the structure and parameters of the influence diagram representing the reactive decision model have been defined, any evidence concerning information nodes (sensors warnings node and estimated environment conditions node) is propagated to estimated probability of next status of the machine and to test the outcome of each decision and for this purpose packages such as Netica are very helpful. For instance Figure 11 shows a configuration where sensors report green status and estimated environment conditions are in nominal status ; we can see that without any decision (equal chance of stopping or not)

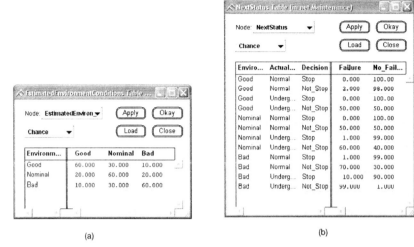

Figure 10: Conditional probabilities of EstimatedEnvironmentConditions node (a) and NextStatus node (b).

there are 22.3% of chances that machine fails, but if we decide to stop for inspection and reparation then this probability drop to 0.39% whereas if we decide not to stop then this probability rise to 44.2%.

Many configurations of this type can be tested. This application is just an academic example to show the feasibility of the idea expressed in this section ; this approach for reactive decision making will be more significant for a complex system with a great number of nodes and relationships.

3   Modeling for Reactive Control

Many processes in practice in domains such as engineering, economy, management, societal, .. can be described by dynamic system (11)

$$x(t+1) = f(x(t), u(t), w(t)), \ t \geq 0, \ x(0) = x_0, \tag{11}$$

where $x(t) \in x \subseteq \mathbb{R}^n$ is the state of the system at time t, $u(t) \in u \subseteq \mathbb{R}^m$ is the control signal and $w(t) \in w \subseteq \mathbb{R}^p$ is an external perturbation (including the state of the nature and possible opponents actions for instance) and f is the function representing process dynamics ; $\mathbb{R}$ and $\mathbb{R}^d$ denote the set of real numbers and the real space vector of dimension d respectively. Systems described by equation (11) have been widely studied in literature (see for instance [10], [11] and references therein) regarding the possibility of driving the state $x(t)$ from an initial condition $x(0)$ to a final set and ensuring simultaneously some performance indices in spite of uncontrollable perturbation $w(t)$. The study of linear case with Gaussian perturbation has reached a certain maturity nowadays [10]. But we argue that for majority of processes in domains such as operational research, management, economy,

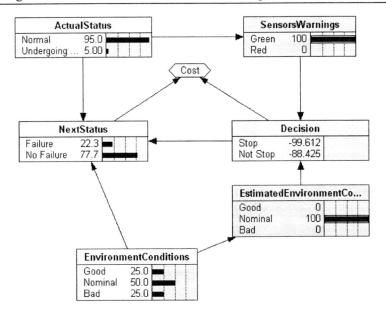

Figure 11: Example of simulation results of reactive maintenace problem.

manufacturing, environmental science, the hypothesis of Gaussian behavior of perturbation does not hold. For such processes a scenario consisting of a profile $w^*(t)$ of perturbation will be first suggested. From there, a controller $u^*(t)$ can be determined by non necessarily formal methods [11] (heuristic for instance) and from an estimated deviation $\overline{\delta w}(t)$ around the known profile $w^*(t)$ of perturbation at instant t, a correction $\delta u(t)$ of the control will be determined. In the following, we show that the determination of such correction can be reduced to a *mathematical programming* problem. But let us first introduce briefly this subject.

3.1   Mathematical Programming

Mathematical Programming is a branch of mathematics concerned mainly by methods and algorithms that permit a decision maker or controller to select variables (investment level, production level, parameters values of a system, etc.) in order to optimize a given performance index (maximize profit, minimize cost, maximize the availability of a system, etc.) and satisfying some constraints (budget limitation, lack of qualified manpower, etc.). Basically, a mathematical programming problem consists in the following :

$$\begin{cases} \min/\max\limits_{x} & f(x) \\ & G(x) \leq 0 \\ & x \in \mathcal{D} \subseteq \mathbb{R}^n \end{cases} \tag{12}$$

where $x \in \mathbb{R}^n$ is decision variables vector, $G(x) : \mathbb{R}^n \to \mathbb{R}^m$ is constraints vector, $f(x) : \mathbb{R}^n \to \mathbb{R}$ is the objective function to be optimized and $\mathcal{D}$ is the admissible set of decision variables.

System (12) can represent a wide range of practical problem in engineering, social sciences, economics, management sciences, etc. According to the nature of functions f, G, the set $\mathcal{D}$, and decision variables x, the problem is known as linear programming, nonlinear programming, integer programming, mix integer programming, convex optimization, quadratic optimization (see [5] and references therein for more details on theory, classification, algorithms, and applications). The practical importance of such decision aid systems encourage researches that have leaded to very efficient algorithms (see for instance [13]) and software [12]. In the following we will show that the determination of reactive controller at each instant defined previously can be reduced to a mathematical programming problem that can be solved using existing algorithms and software.

3.2 Model for Reactive Control

Let us consider the system described by equation (11). We suppose that a control law $u^*(t)$ that optimizes a certain performance index in the presence of a known perturbation profile $w^*(t)$ is given (obtained by any formal or non formal method) leading to the optimal state trajectory $x^*(t)$ that verifies (13)

$$x^*(t+1) = f(x^*(t), u^*(t), w^*(t)), \ x^*(0) = x_0. \tag{13}$$

Notice that for many applications (inventory level control, water pool level control, queues length control in queueing networks, etc.) perturbation signal (demand, arrival rate, etc.) are time dependent with in some cases a known general long term profile (we suppose this for the rest of this section). In practice the actual perturbation $w(t)$ will deviate from its long term profile $w^*(t)$. A deviation $\delta w(t)$ around the known profile $w^*(t)$, that is $w(t) = w^*(t) + \delta w(t)$ will result in a deviation $\delta x(t)$ around the desired state trajectory $x^*(t)$, that is $x(t) = x^*(t) + \delta x(t)$. The reactive control purpose is to generate $\delta u(t)$ around the long term control law $u^*(t)$ that will compensate the effect of $\delta w(t)$. But most of time, at the time instant t actual $\delta w(t)$ will not be known and we suppose that only an estimate $\overline{\delta w}(t)$ supplied by an intelligence system (sensors and data processing) will be available so that $\delta u(t)$ will compensate only its effect leading to an estimate $\overline{\delta x}(t+1)$ of $\delta x(t+1)$. Estimated deviation dynamics are given by

$$\overline{\delta x}(t+1) = g(t, \delta x(t), \delta u(t), \overline{\delta w}(t)).$$

The purpose of reactive control signal $\delta u(t)$ is to reduce $\overline{\delta x}(t+1)$ to 0 subjected to feasibility conditions that is solving for $\delta u(t)$ following problem (14)

$$g(t, \delta x(t), \delta u(t), \overline{\delta w}(t)) = 0, \ \text{s.t.} \ u^*(t) + \delta u(t) \in \mathcal{U} \ . \tag{14}$$

But this problem will not always have a feasible solution and we propose to implement the controller $u(t)$ defined by

$$u(t) = \begin{cases} u^*(t) + v & \text{if } v \text{ exists} \\ u^*(t) & \text{otherwise} \end{cases} \tag{15}$$

where v is the solution of the following mathematical programming problem (16)

$$\min_{v} {}_{\jmath} \left(g(t, \delta x(t), v, \overline{\delta w}(t)) \right), \ u^{*}(t) + v \in u \tag{16}$$

and ${}_{\jmath}$ is a certain criterion measuring some penalty of failing to apply the real reactive control and possible penalty regarding the importance of reactive control.

If f is a smooth function then $g(t, \delta x(t), \delta u(t), \overline{\delta w}(t))$ can be taken equal to

$$g(t, \delta x(t), \delta u(t), \overline{\delta w}(t)) = \overline{\delta x}(t+1) = A(t)\delta x(t) + B_u(t)\delta u(t) + B_w(t)\overline{\delta w}(t)$$

with

$$A(t) = \left(\frac{\partial f}{\partial x} \right)_{(x^*(t), u^*(t), w^*(t))},$$

$$B_u(t) = \left(\frac{\partial f}{\partial u} \right)_{(x^*(t), u^*(t), w^*(t))},$$

$$B_w(t) = \left(\frac{\partial f}{\partial w} \right)_{(x^*(t), u^*(t), w^*(t))}.$$

In this case when the criterion ${}_{\jmath} \left(\delta x(t), \delta u(t), \overline{\delta w}(t) \right)$ is quadratic type defined by

$$_{\jmath} \left(\delta x(t), \delta u(t), \overline{\delta w}(t) \right) = \delta u(t)^T R \delta u(t) + \overline{\delta x}(t+1)^T Q \overline{\delta x}(t+1)$$

$$R = R^T \geq 0, \ Q = Q^T \geq 0$$

(M^T denotes the transpose of the matrix M and $M \geq 0$ means that M is symmetric semi definite positive, all its eigenvalues are non negative) and the set u is a polyhedron, the implemented controller is given by equation (15) with v being the solution of following *quadratic programming* problem (17)

$$\min_{v} \left\{ \begin{array}{c} v^T \left(R + B_u^T(t) Q B_u(t) \right) v + \\ 2v^T B_u^T(t) Q \left(A(t)\delta x(t) + B_w(t)\overline{\delta w}(t) \right) \end{array} \right\} \text{ s.t. } u^*(t) + v \in u \tag{17}$$

and if $u = \mathbb{R}^m$ then

$$v = - \left(R + B_u^T(t) Q B_u(t) \right)^{-1} B_u^T(t) Q \left(A(t)\delta x(t) + B_u(t)\overline{\delta w}(t) \right)$$

if and only if $\left(R + B_u^T(t) Q B_u(t) \right)$ is always invertible.

To see the applicability of this approach let us consider some potential applications.

3.3   Examples of Potential Applications

3.3.1   Reactive Resources and Time Allocation in a Push-Type Queueing Network

Push type queueing networks [6] or networks driven by incoming activities as opposed to pull type networks which are driven by outgoing activities (demand) are suitable for modeling service systems such as hospitals, administrations, restaurants, registration activities in universities, etc as opposed to manufacturing systems. But it is shown that a pull type queueing network can be translated into a push type by adding virtual resources (see [6]). For these push type systems, the degrees of freedom (control variables) consist mainly in hiring and layoff of the resources (personnel for instance), use of part-time resources and overtime. Notice that we are mainly concerned here by *renewable resources* (a resource is said to be renewable if its total usage at every moment is constrained, example: personal, computer memory, buffer length, machines, etc.) and not *non-renewable resources* (a resource is said to be non-renewable if its total consumption up to any given moment is constrained, example: energy, money, etc.).

Let us consider a push type queuing network consisting with n_s stations with N_j renewable resources (for instance equivalent machines) at station j and n activities or classes with non stationary arrival rate. The dynamics of activity i queue length q_i is described by

$$q_i(t+1) = q_i(t) - \Psi_i(t) + \sum_j R_{ji}\Psi_j(t) + \lambda_i(t), \ q_i(t) \geq 0$$

or compactly by

$$q(t+1) = q(t) - (I_n - R^T)\Psi(t) + \lambda(t), \ t \geq 0, \ q(0) \tag{18}$$

where

- $q(t) \in \mathbb{R}_+^n$ is the state of system (18) where $q_i(t)$ represents the activity i queue length at time t. $\mathbb{R}_+^n$ denotes a n dimensional real vector space with non negative entries.

- $\Psi(t) \in \mathbb{R}_+^n$ is the output rate vector where $\Psi_i(t)$ is the quantity of activity i that is completed during time slot $(t, t+1)$; $\Psi_i(t)$ depends on manpower devoted to activity i that is $\Psi_i(t) = \mu_i(t)\zeta_i(t)$ where $\mu_i(t)$ and $\zeta_i(t)$ are service rate and time slot allocation (the percentage of time slot allocated) for activity i during $(t, t+1)$.

- $\lambda(t) \in \mathbb{R}_+^n$ is the arrival rate vector, $\lambda_i(t)$ is the quantity of activity i arriving from outside the system during time slot $(t, t+1)$.

- R is the routing matrix, R_{ij} is the percentage of activity i routed to activity j after completion during time slot $(t, t+1)$; we suppose that its spectral radius $\sigma(R)$ is less that 1 $(\sigma(R) < 1)$ and R^T denotes the transpose of R. I_n is n dimensional identity matrix.

- We denote by $c(j), \ j = \overline{1, n_s}$, all activities affected to station j and we suppose that the service rate $\mu_i(t)$ of activity $i \in c(j)$ is a function of quantity $n_i(t)$ of resources allocated to this activity, $\mu_i(t) = f_i(n_i(t))$ where f_i is positive non decreasing and $f_i(0) = 0$. A typical behavior of such function is given by Figure 12.

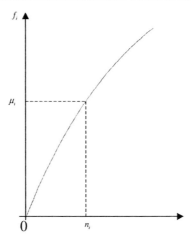

Figure 12: A typical behavior of servie rate as a function of allocated resources.

The output rate vector $\Psi(t)$ is then defined by

$$\Psi(t) = \begin{bmatrix} f_1(n_1(t))\zeta_1(t) \\ f_2(n_2(t))\zeta_2(t) \\ . \\ . \\ . \\ f_n(n_n(t))\zeta_n(t) \end{bmatrix}.$$

Notice that dynamics (18) could result from sampling a linear fluid push queueing network [6] with non constant arrival rate. Control problem can then be thought in two ways: resource allocation and time slot allocation.

Resource allocation problem   Here two subproblems can be considered depending on whether resource level of each station is constant or can be considered as a control variable.

Constant quantity of resources per station   It is supposed here that each station j has a constant resource level N_j and the problem is to determine the proportion $n_i(t) \geq 0$ of resources that must be allocated to activity $i \in c(j)$ for the whole time slot $(t, t+1)$ (thus $\zeta_i(t) = 1, i = \overline{1, n}$) so to optimize some performance index in general related to queue length $q(t)$ or completion time ; this allocation must verify

$$\sum_{i \in c(j)} n_i(t) \leq N_j.$$

Let suppose that a desired profile $q^*(t)$ is obtained by optimizing a certain objective index when a known profile $\lambda^*(t)$ of $\lambda(t)$ is given that leads to a feasible and optimal resources allocation $n^*(t)$ obtained for instance by solving

$$\mu^*(t) = -(I_n - R^T)^{-1}(q^*(t+1) - q^*(t) - \lambda^*(t)),$$
$$n_i^*(t) = f_i^{-1}(\mu_i^*(t)), \; i = \overline{1, n},$$
$$\sum_{i \in c(j)} n_i^*(t) \leq N_j.$$

Notice that $(I_n - R^T)^{-1}$ exists as we suppose that $\sigma(R) < 1$. A deviation $\delta\lambda(t)$ around the known profile $\lambda^*(t)$ will cause a deviation $\delta q(t)$ around the desired profile $q^*(t)$. To overcome this deviation we propose to modify the service rate $\mu^*(t)$ if possible by $\delta\mu(t)$. Dynamics (18) becomes then

$$q^*(t) + \delta q(t+1) = q^*(t) + \delta q(t) - (I - R^T)(\mu^*(t) + \delta\mu(t)) + \lambda^*(t) + \delta\lambda(t),$$
$$q^*(t) + \delta q(t) \geq 0, \; \forall \, t \geq 0$$

and dynamics of deviation are given by

$$\delta q(t+1) = \delta q(t) - (I - R^T)\delta\mu(t) + \delta\lambda(t).$$

But at the time the corrected term $\delta\mu(t)$ must be generated only an estimation $\overline{\delta\lambda}(t)$ of $\delta\lambda(t)$ obtained by an *intelligence system* is available and we propose to generate $\delta\mu(t)$ such that

$$\overline{\delta q}(t+1) = \delta q(t) - (I - R^T)\delta\mu(t) + \overline{\delta\lambda}(t) = 0,$$

that is

$$\delta\mu(t) = (I - R^T)^{-1}\left(\delta q(t) + \overline{\delta\lambda}(t)\right).$$

Then each station j solve following problem: find $n_i(t), \; i \in c(j)$ submitted to

$$f_i(n_i(t)) = \max(0, \mu_i^*(t) + \delta\mu_i(t)) \text{ and } \sum_{i \in c(j)} n_i(t) \leq N_j$$

that does not necessarily have a solution. If an upper bound $n_i^{\max}$ (with $\sum_{i \in c(j)} n_i^{\max} \leq N_j$) of resources that can be allocated to each activity i is known then $n_i(t)$ is determined by

$$n_i(t) = \min(n_i^{\max}, v_i) \text{ where } v_i = f_i^{-1}(\max(0, \mu_i^*(t) + \delta\mu_i(t))).$$

If there is no upper bound of resources allowable per activity, each station j may solve a mathematical programming problem of the form

$$\min_{n_i(t), \, i \in c(j)} \left\{ \sum_{i \in c(j)} \jmath_i(f_i(n_i(t)) - \max(0, \mu_i^*(t) + \delta\mu_i(t))) \right\}$$
$$\text{s.t. } n_i(t) \geq 0, \; \sum_{i \in c(j)} n_i(t) \leq N_j.$$

where $\jmath_i$ measures a cost that results from class i not being assigned the appropriate service rate. In the case where f_i are linear function given by $f_i(n_i(t)) = \alpha_i n_i(t)$ and $\jmath_i$ is defined by $\jmath_i(x) = |x|$, former mathematical programming problem is equivalent to following linear programming problem

$$\min_{n_i(t),\,\lambda_i} \Sigma \lambda_i$$

$$\text{s.t.} \begin{cases} \alpha_i n_i(t) - \lambda_i \leq \max\left(0,\, \mu_i^*(t) + \delta\mu_i(t)\right) \\ -\alpha_i n_i(t) - \lambda_i \leq -\max\left(0,\, \mu_i^*(t) + \delta\mu_i(t)\right) \quad,\ i \in c\,(j) \\ \Sigma_i n_i(t) \leq N_j,\ n_i(t) \geq 0,\ \lambda_i \geq 0 \end{cases}.$$

Variable quantity of resources per station   In the case where resource level N_j at each station is a variable, for instance decided by an upper level decision maker, the quantity

$$\delta N_j(t) = \sum_{i \in c\,(j)} \left(n_i(t) - n_i^*(t)\right)$$

of resources is allocated (hired, if $\delta N_j(t) > 0$) to station j or retrieved (laid off, if $\delta N_j(t) < 0$) from station j (with possible other performance index consideration) where $n_i(t)$ is given by

$$n_i(t) = f_i^{-1}\left(\max\left(0,\, \mu_i^*(t) + \delta\mu_i(t)\right)\right).$$

Time slot allocation   Here each station j uses all its resources for an activity $i \in c\,(j)$ during a percentage $\zeta_i(t)$ of the time slot leading to a constant service rate μ_i for each activity. The problem is then to find time allocation vector $\zeta(t)$ verifying $C\zeta(t) \leq 1$ where $C \in \mathbb{R}^{n_s \times n}$ is the constituency matrix defined by $C_{ji} = 1$ if activity i is performed by station j and 0 otherwise. Let suppose that a nominal or long term allocation given for instance by

$$\zeta^*(t) = -M^{-1}(I - R^T)^{-1}\left(q^*(t+1) - q^*(t) - \lambda^*(t)\right) \geq 0,\ C\zeta^*(t) \leq 1,\ t \geq 0.$$

exists where $M = diag(\mu_i)$; then the deviation $\delta\zeta(t)$ around the nominal allocation $\zeta^*(t)$ that ensures

$$\overline{\delta q}(t+1) = \delta q(t) - (I - R^T)M\delta\zeta(t) + \overline{\delta\lambda}(t) = 0$$

is determined by

$$\delta\zeta(t) = M^{-1}(I - R^T)^{-1}\left(\delta q(t) + \overline{\delta\lambda}(t)\right)$$

and the feasible reactive controller for each activity i is given by

$$\zeta_i(t) = \zeta_i^*(t) + u_i$$

where u_i is the solution of following mathematical programming problem

$$\min_{u_i,\,i \in c\,(j)} \left\{\sum_{i \in c\,(j)} \jmath_i(u_i - \delta\zeta_i(t))\right\}$$

$$\zeta_i^*(t) + u_i \geq 0,\ \sum_{i \in c\,(j)} \left(\zeta_i^*(t) + u_i\right) \leq 1.$$

where $\jmath_i$ measures a cost that results from class i not being assigned the appropriate time slot rate. Here gain if $\jmath_i$ is defined by $\jmath_i(x) = |x|$ then we have following linear programming problem

$$\min_{u_i,\,\lambda_i} \sum \lambda_i$$

$$u_i - \lambda_i \leq \delta\zeta_i(t), \; -u_i - \lambda_i \leq -\delta\zeta_i(t), \; i \in c\,(j)$$

$$\zeta_i^*(t) + u_i \geq 0, \; \sum_i (\zeta_i^*(t) + u_i) \leq 1, \; i \in c\,(j).$$

The time slot allocation is suited for a single station performing different services.

Remark 2 *The control problem presented so far can be organized hierarchically where the low level consisting with stations will supply actual deviation $\delta q(t)$ and estimated deviation $\overline{\delta\lambda}(t)$ of queues and arrival rate around their nominal value to a high level decision maker that will compute service rate variation $\delta\mu(t)$ or time slot allocation variation $\delta\zeta(t)$ and supply it back to low levels. Then each station is free to have its own objective subjected to the satisfaction of the high level request.*

Simulation results 1: Resource allocation System depicted on Figure 13 represents an organization that gives service to customers arriving at a non stationary rate $\lambda_1(t)$ in queue 1 at station 1 ; a proportion δ of these customers are routed to queue 2 at station 2 after service and the rest are rejected (leave the system). After the service at queue 2 each customer returns to station 1 in queue 3 for another service before leaving the system.

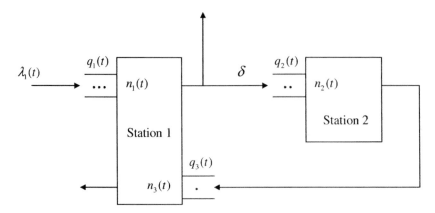

Figure 13: Two stations / three activities queueing network example.

This system could represents a service system such as a hospital where patients arrive at reception room (queue 1, station 1) and then are received by a nurse first and after examination of the problem the nurse can decide that the patient must see a physician (queue 2, station 2) or not. After the patient have seen a physician he or she returns to reception room to pay fees (queue 3, station 1). Let us suppose that each personnel at station 1 can serve 4 customers of queue 1 or 6 customers of queue 3 per time unit respectively whereas a personnel at station 2 serves 3 customers per time unit and it is known that in general 80% ($\delta = 0.8$) of customers arriving at queue 1 are routed to queue 2 and the purpose is to maintain queues length around 0 customer that is $q^*(t) = 0$. For simulation, arrival rate data are shown by Figure 14 where the nominal arrival rate (dotted line) is obtained as a piece wise linear approximation of the real arrival rate (solid line) and the estimated (intelligence) arrival rate is a three periods moving average of the real arrival rate.

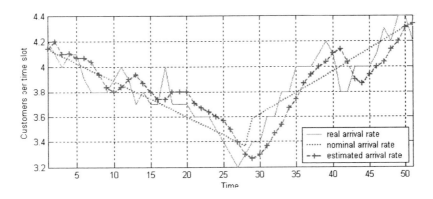

Figure 14: Arrival rate data for two stations / three activities queueing network.

For sake of simplicity, we treat resources as continuous. For constraint resources case with $n_1^{max} = 4$, $n_2^{max} = 3$ and $n_3^{max} = 2$, the behavior of queues length are given by Figure 15 (solid line when nominal controller is applied and dotted line when reactive controller is applied) with corresponding resources allocation given by Figure 16 (solid line = nominal controller and dotted line = reactive controller).

In the case of variable resources per station we obtain the allocations of Figure 17 (solid line= nominal controller and dotted line = reactive controller)

for activity 1, 2 & 3 respectively with corresponding queues length behavior (solid line = nominal controller and dotted line = reactive controller) shown by Figure 18 respectively.

We can see that reactive controller perform better as the nominal controller leads to instability of queues 2 & 3 as shown by Figures 15 & 18.

Simulation results 2: Time slot allocation   Let consider a traffic system consisting of a pivoting bridge that is used to route vehicles in an intersection as shown by Figure 19 (a). Let suppose that a proportion $\delta_1 = 0.1$ of vehicles coming from 1 join 2 and a proportion $\delta_2 = 0.1$ of vehicles coming from 2 join 1. Arrival rates from outside the system are $\lambda_1(t)$ and $\lambda_2(t)$ vehicles per time unit respectively. It is assumed a heavy traffic so that the flow is almost continuous. The equivalent queueing network is shown by Figure 19 (b) and the purpose of the control is to allocate during each time slot the proportion $\zeta_1(t)$ and $\zeta_2(t)$ to each queue respectively in order to minimize queues' length. It takes $\frac{1}{30}$ time unit for a vehicle to cross the intersection so that $\mu_1 = \mu_2 = 30$ vehicles per time slot and the desired

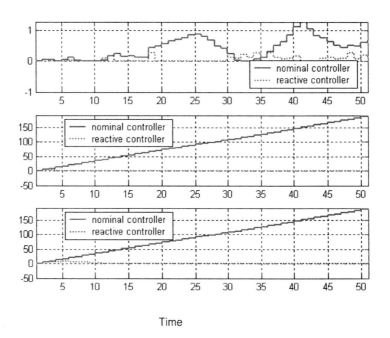

Figure 15: Queues 1, 2 & 3 behavior for two stations / three activities queueing net example.

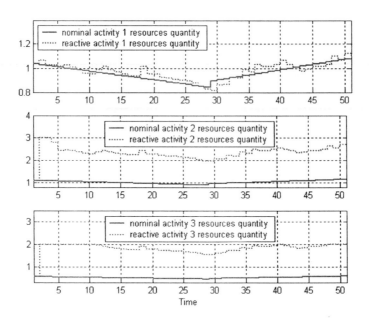

Figure 16: Quantity of resources allocated to activity 1, 2 & 3 respectively for two stations / three activities queueing net example.

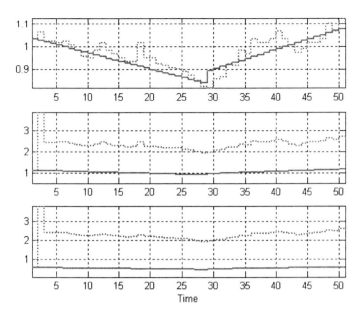

Figure 17: Quantity of resources allocated to activity 1, 2 & 3 respectively for two stations / three activities queueing net example in the case of variable resources per station.

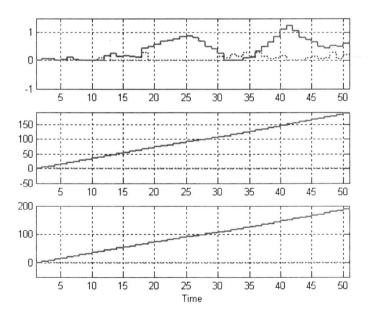

Figure 18: Queues 1, 2 & 3 behavior for two stations / three activities queueing net example in the case of variable resources per station.

queue length vector is $q^*(t) = \begin{bmatrix} 0 \\ 0 \end{bmatrix}$.

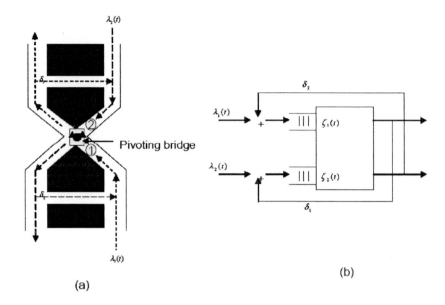

(a)   (b)

Figure 19: An example of a pivoting bridge for vehicules routing (a) and its queueing network representation (b)

The nominal arrival rates $\lambda_1^*(t)$ and $\lambda_2^*(t)$ behavior as well as real arrival rate behavior are shown on Figure 20 (solid line = nominal and dotted line = real) where one can see significant variation of $\lambda_2(t)$ around its nominal value of 10.

Let suppose that the nominal feasible time allocation is set to $\zeta_1^*(t) = 0.6$ and $\zeta_2^*(t) = 0.4$. The intelligence system consists in a 3 periods moving average of real arrival rate ; time allocation is then given by Figure 21 (solid line = nominal allocation and dotted line = reactive allocation)

leading to queues length behavior shown on Figure 22 (solid line = nominal allocation case and dotted line = reactive allocation case).

One can see that reactive controller perform better mainly with regards to queue 2.

3.3.2   Intelligent Real Time Control of Transformation Processes

A great number of transformation processes in practice (inventory system, supply chain, water pools, assembly line, etc.) can be described as an input-output system with controllable inflow, $u_{in}(t)$, and outflow, $u_{out}(t)$, material as well as uncontrollable inflow, $w_{in}(t)$, and outflow, $w_{out}(t)$, material as shown by Figure 23. Notice that depending on particular applications some of signals $u_{in}(t)$, $u_{out}(t)$, $w_{in}(t)$ and $w_{out}(t)$ may not exist.

A possible general behavior model of system described by Figure 23 is given by

$$x(t) = Ax(t-1) + B_{in}u_{in}(t-1) - B_{out}u_{out}(t-1) + C_{in}w_{in}(t-1) - C_{out}w_{out}(t-1), \quad (19)$$

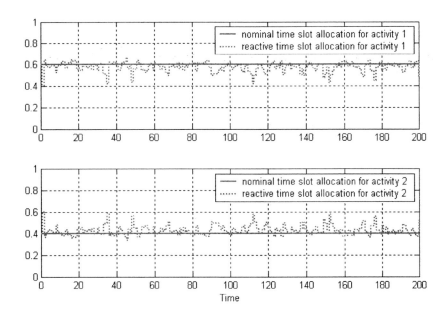

Figure 20: Nominal and real arrival rates for vehicles routing example.

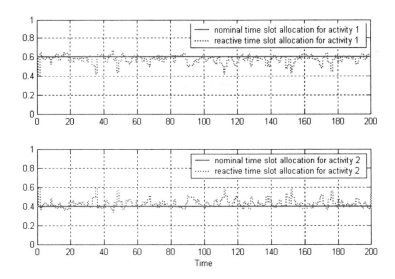

Figure 21: Nominal and reactive time slot allocation for activity 1 and 2 respectively for vehicules routing example.

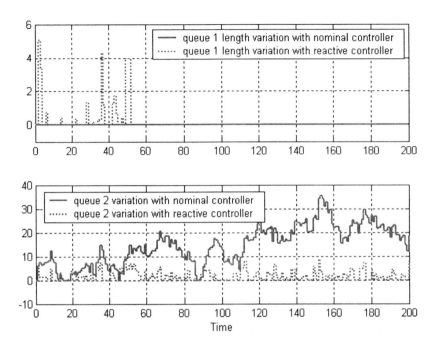

Figure 22: Queues 1 & 2 variation according to applied controller for vehicles routing example.

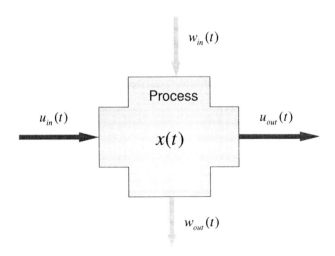

Figure 23: A typical representation of a transformation process

where

- $x(t) \in \mathcal{D}_x \subseteq \mathbb{R}^n_+$ represents the internal state of the system (the inventory level, water level in a pool, ..) at the end of period or time slot $(t-1,t)$; $x(0)$ is the initial state.

- $u_{in}(t-1) \in \mathcal{D}_{u_{in}} \subseteq \mathbb{R}^{m_{in}}_+$ and $u_{out}(t-1) \in \mathcal{D}_{u_{out}} \subseteq \mathbb{R}^{m_{out}}_+$ are control signals that represent incoming and outgoing material level respectively (production level, controlled incoming water flow, material level, ...) during period or time slot $(t-1,t)$.

- $w_{in}(t-1) \in \mathcal{D}_{w_{in}} \subseteq \mathbb{R}^{p_{in}}_+$ and $w_{out}(t-1) \in \mathcal{D}_{w_{out}} \subseteq \mathbb{R}^{p_{out}}_+$ are perturbation signals that represent uncontrollable incoming and outgoing material level respectively (demand level, incoming water flow due to rain, outgoing water flow for industrial and domestic usage, ...) during period or time slot $(t-1,t)$.

- Matrices A, B_{in}, B_{out}, C_{in} and C_{out} with appropriate dimensions are technical matrices that we suppose constant.

- $\mathcal{D}_x$, $\mathcal{D}_{u_{in}}$, $\mathcal{D}_{u_{out}}$, $\mathcal{D}_{w_{in}}$ and $\mathcal{D}_{w_{out}}$ are the admissible set for corresponding variables respectively.

Equation (19) can be compactly rewritten as (20)

$$x(t) = Ax(t-1) + Bu(t-1) + Cw(t-1) \qquad (20)$$

with

$$u(t) = \begin{bmatrix} u_{in}(t) \\ u_{out}(t) \end{bmatrix}, \ w(t) = \begin{bmatrix} w_{in}(t) \\ w_{out}(t) \end{bmatrix},$$
$$B = \begin{bmatrix} B_{in} & -B_{out} \end{bmatrix} \text{ and } C = \begin{bmatrix} C_{in} & -C_{out} \end{bmatrix}.$$

The purpose of control process is to manage $u_{in}(t)$ and $u_{out}(t)$ to compensate $w_{in}(t)$ and $w_{out}(t)$ in order to realize some objective function depending on the state of the system, $x(t)$. In normal conditions, it is supposed (and this is almost the case in practice) that a long term behavior $w^*(t)$ of the perturbation is known that leads to an admissible long term controller $u^*(t)$ and the objective of the intelligent reactive controller is to determine at each instant $\delta u(t)$ that will compensate an estimated deviation $\overline{\delta w}(t)$ around the long term perturbation trajectory according to the theory established in former paragraph. To have a concrete idea, we consider in the following paragraph a problem of production and row material planning problem driven by the demand of product.

Demand-driven production and row material planning Material requirements planning (MRP) facilities play an interesting role in manufacturing and services business. Effective production planning is essential in a manufacturing business and is the foundation on which the manufacturing organization operates. The manufacturing environment is one of constant change with conditions changing daily or hourly, customer emergencies, rush orders, machine breakdowns, quality problems, the list is endless. Production planning is a complex process that covers a wide range of activities that insures that material, capacity and human resources are available when needed. Material requirements planning involves getting material on hand when needed for production or sales. An effective production scheduling is closely related to material requirement planning. Each company has an overall goal and a strategy for achieving that goal, but within the company there are groups whose focus may seem to be in contradiction to the materials requirements planning process. Production planning and row material requirements processes are strongly dependent on demand that is supplied by sales forecast, or sales plan process. Sales forecasting is a key factor in any company's success and should include demand forecasting, trend forecasting and sales and marketing plans. Accurate sales forecasting (necessity of intelligence system) allows a company to effectively control inventory, production facilities, labor, inventory levels and logistics and is the base on which most of other operations within the company function. For many industries (clothes industry, energy industry, agriculture industry for instance), demand is time dependent and have a long term profile that can be estimated.

Here row material and production level planning problem for a given horizon is considered through a dynamic model with demand as an external perturbation that must be rejected in some sense. It is well known that in a deterministic unperturbed dynamic system, open loop control is as good as closed loop control. Indeed, one of the main objective of a closed loop control is to reject perturbation. A close loop type control scheme may be hard to be implemented because of difficulty to obtain a close loop control law or in the context of distributed systems because of delay in the reactivity of unit or in some cases because of the structure of decision processes (decisions made for an horizon for example). For instance, in the context of manufacturing, production maybe seasonal (cloths industry, toys industry, etc.) but decisions are made at the beginning of the season. In other hand for many cases in real world, perturbations profile can be estimated (demand level for an inventory management system, energy consumption level, water consumption level, etc.) using historical data or expert knowledge for a given period because of the fact that these perturbations are, most of time, random variables with known or good estimated statistics possibilities. Classical approach to dealing with randomness in optimal control system is to optimize the mean value of a performance index [3]. This approach suppose that one disposes of a prior probability distribution of the random variable (perturbation). But we argue that when a profile of the perturbation is known or can be estimated, it is possible to derive a long term controller to compensate the long term profile of the perturbation and use a reactive controller to compensate the deviation around the known profile.

We consider the problem of controlling a performance index of a production system described by following dynamic system

$$x_p(t) = x_p(t-1) + u_p(t-1) - w_p(t-1) \tag{21}$$

$$x_m(t) = x_m(t-1) - B_c u_p(t-1) + u_m(t-1) \tag{22}$$

where

- $x(t) = \begin{bmatrix} x_p(t) \\ x_m(t) \end{bmatrix} \in \mathcal{D} \subseteq \mathbb{R}^n_+$ is the state of the system ; the ith component of vector $x_p(t) \in \mathcal{D}_p \subseteq \mathbb{R}^p_+$ and the jth component of vector $x_m(t) \in \mathcal{D}_m \subseteq \mathbb{R}^m_+$ represents the stock level of product item i and the level of row material item j respectively at the *end* of period $(\ t,\ \ t+1\ )$; $x_p(0)$ and $x_m(0)$ are their initial values. This means that p different products are produced by using m different row material items.

- $u_p(t) \in \mathcal{D}_{u_p} \subseteq \mathbb{R}^p_+$ is the production control vector, its ith component represents the quantity of product item i to be produced *during* period $(\ t,\ \ t+1\ )$ and $u_m(t) \in \mathcal{D}_{u_m} \subseteq \mathbb{R}^m$ represents the row material requirement control vector, its ith component is the quantity of material i to be received at the *beginning* of period $(\ t,\ \ t+1\ )$.

- Sets $\mathcal{D}_p$, $\mathcal{D}_m$, $\mathcal{D}_{u_p}$, and $\mathcal{D}_{u_m}$ define some constraints ; they are most of time polyhedra with possible additional integrity constraints.

- $B_c \in \mathbb{R}^{m \times p}_+$ is the material consumption matrix ; $B_c(i,j)$ is the quantity of row material item i consumed to produce one unit of product item j.

- $w_p(t) \in \mathbb{R}^p_+$ is the vector of products demand for period $(\ t,\ \ t+1\ )$.

Let suppose that a long term feasible controller $u_p^*(t)$ and $u_m^*(t)$ given by

$$u_p^*(t) = x_p^*(t+1) - x_p^*(t) + w_p^*(t),$$
$$u_m^*(t) = x_m^*(t+1) - x_m^*(t) + B_c u_p^*(t)$$

exist for a known perturbation (demand) profile $w_p^*(t)$ and a desired product stock level $x_p^*(t)$ and material stock level $x_m^*(t)$; then the behaviors of actual $x_p(t)$ and $x_m(t)$ when the real perturbation is $w_p(t) = w_p^*(t) + \delta w_p(t)$ are given by

$$x_p(t) = \max\left(0,\ x_p^*(t) + \delta x_p(t-1) + \delta u_p(t-1) - \delta w_p(t-1)\right),$$
$$x_m(t) = x_m^*(t) + \delta x_m(t-1) - B_c \delta u_p(t-1) + \delta u_m(t-1).$$

As stated previously, at the time the adjustments $\delta u_p(t-1)$ and $\delta u_m(t-1)$ are to be made, only an estimate $\overline{\delta w_p}(t-1)$ of $\delta w_p(t-1)$ will be known and so these corrections are given by

$$\delta u_p(t-1) = -\delta x_p(t-1) + \overline{\delta w_p}(t-1),$$
$$\delta u_m(t-1) = -\delta x_m(t-1) + B_c \delta u_p(t-1),$$

and the implemented control signals are given by

$$u_p(t-1) = \max\left(0,\ u^*(t-1) + \delta u_p(t-1)\right), \tag{23}$$
$$u_m(t-1) = \max\left(0,\ u_m^*(t-1) + \delta u_m(t-1)\right). \tag{24}$$

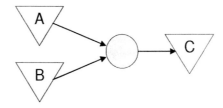

Figure 24: Example of an assembly system

Simulation results   Consider an assembly station depicted on Figure 24 where 2 parts of row material A are combined with 1 part of row material B to produce 1 part of product C (we suppose continuous materials and product such as in chemical or petrol processes).

We have then following variables, parameters and corresponding initial conditions

$$x_p(t) = x_C(t), \; x_m(t) = \begin{bmatrix} x_A(t) \\ x_B(t) \end{bmatrix}, \; x_p(0) = 0, \; x_m(0) = \begin{bmatrix} 0 \\ 0 \end{bmatrix},$$

$$u_p(t) = u_C(t), \; u_m(t) = \begin{bmatrix} u_A(t) \\ u_B(t) \end{bmatrix}, \text{ and } B_c = \begin{bmatrix} 2 \\ 1 \end{bmatrix}.$$

We suppose that the priority is to satisfy demand and keep stocks (product and row materials) as low as possible by purchasing supplementary raw material at any time if necessary. This means that we are attempting to ensure $x_p^*(t) = 0$ and $x_m^*(t) = \begin{bmatrix} 0 \\ 0 \end{bmatrix}$. Real demand, estimated demand (three periods moving average of real demand) and nominal demand (piece wise linear approximation of real demand) behaviors are shown on Figure 25.

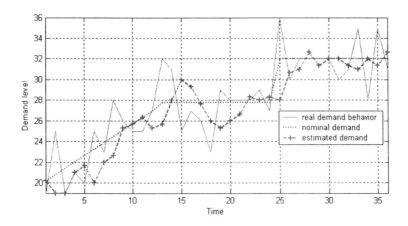

Figure 25: Real, nominal and estimated demands of product C for assembly problem.

Simulation shows that production of C must behave as given by Figure 26 where solid line corresponds to the case where the nominal controller is used without intelligent consideration of demand variation and the dotted line corresponds to the case where an intelligent reactive control is used. Figure 27 shows the corresponding stock of product C and we can see that the reactive control is performing better by keeping the stock level lower than that performed by nominal control. Finally Figure 28 shows the corresponding row materials A and B planning (solid line in the case of nominal control and dotted line in the case of intelligent reactive control). The stocks of row materials A and B are kept always equal to zero.

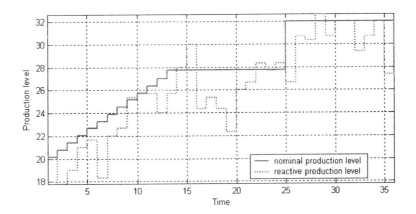

Figure 26: Nominal (solid line) and reactive (dotted line) production of produt C in the assemblt problem.

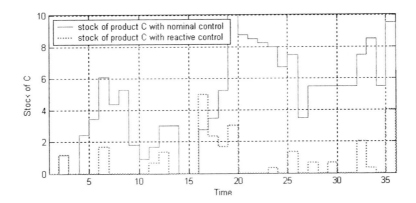

Figure 27: Stock of product C whith nominal production policy (solid line) and reactive production policy (dotted line).

Remark 3 *Though this system is not large, the intention here is to show the capability of established control method ; the extension to a real world large problem is straightforward.*

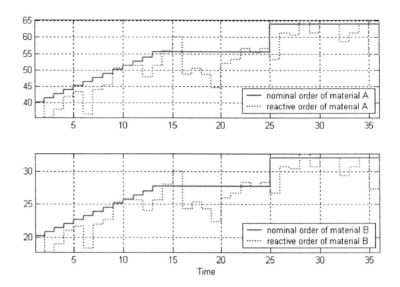

Figure 28: Planned orders of material *A* and *B* respectively when nominal production policy is considered (solid line) and reactive policy (dotted line).

4 Conclusion

An approach to build a system to support *reactive* decision making and control problems has been considered and formulated in this paper. Reactivity is related to changes of the control or decision process environment that is always uncertain and/or conflicting (state of the world model for decision and non cooperative game theory in classical formulation). Indeed, decision or control problem is solved in two stages: first of all, a nominal decision policy or control law is established (that realize some desired goal) in the case of nominal behavior of the environment ; in a second stage a reactive decision or control law is designed to compensate possible deviation of the environment around its nominal behavior. This is possible by setting up an *intelligence system* that collect data from environment and that process and transform them into a partial knowledge of the environment changes that will be used (modification of decision maker or controller belief about its environment) to react by eventually changing nominal decision or control. As that, decision maker or controller always dispose of a policy: if environment changes are not sensed accurately just apply nominal policy otherwise, apply reactive policy. Mathematical tools underlying the constructed decision and control support system are *influence diagrams* (graphical representation of discrete decision problems) in the case of decision making (discrete free decision variables) and *mathematical programming* in the case of control (continuous free variables). Applications considered in the field of maintenance, services (queueing networks) as well as manufacturing (row material and production planning) show the potentiality of this approach for real world applications.

References

[1] Bayes, T. (1958). An Essay Towards Solving a Problem in the Doctrine of Chances, *Biometrica*, **46**, p.293-298 (reprinted from an original paper of 1763).

[2] Becker, A. & Naïm, P.(1999). *Les reseaux bayesiens*, Eyrolles, Paris, France.

[3] Bertsekas, D. P. & Tsitsiklis, J. (1996). Neuro-dynamic programming, *Athena Scientific*.

[4] Cassandras, C. G. & Lafortune, S. (1999). *Introduction to Discrete Event Systems*, Kluwer Academic Publ.

[5] Dantzig, G. B. (1963). *Linear programming and extensions*, Princeton University Press, Princeton, NJ.

[6] Chen, M., Dubrawski, R. & Meyn, S. (2004). Management of Demand-Driven Production Systems, *IEEE Transactions on Automatic Control*, Vol 49, No. 5, pp. 686-698.

[7] Howard, R. A. & Matheson, J. E. (1984). Influence Diagrams, in Howard R. A. and Matheson J. E.(Eds), *The principles and Applications of Decision Analysis*, Vol. 2, Palo Alto, Strategic Decision Group, pp. 719-762.

[8] Jensen, F. V. (1999). *Lecture Notes on Bayesian Networks and Influence Diagrams*, Department of Computer Science, Aalborg University.

[9] Kaelbling, L. P., Littman, M. L. & Cassandra, A. R. (1998). Planning and Acting in Partially Observable Stochastic Domains, *Artificial Intelligence*, Vol. 101.

[10] Lewis F. L. and Syrmos V. L. (1995). *Optimal Control*, 2nd Edition, Wiley.

[11] Moïsséev, N. (1985). *Problèmes mathématiques d'analyse des systèmes*, Editions MIR, Moscou.

[12] Moré, J.J. & Wright, S.J. (1993). Optimization Software Guid, *Frontiers Appl. Math.* **14**, SIAM, Philadelphia.

[13] Nesterov, A.S. & Nemiroski, A. (1994). Interior Point Polynomial Algorithms in Convex Programming, *SIAM*, Philadelphia.

[14] http://www.norsys.com/

[15] Pearl, J. (1988). *Probabilistic Reasoning in Intelligent Systems*, Morgan Kaufmann.

[16] Puterman, M. L. (1994). *Decision Processes*, John Wiley & Sons, New-York., 1994.

[17] Sutton, R. and Barto, A. (1998). *Reinforcement Learning: An Introduction*, MIT Press.

[18] Tchangani, A. Ph. (2004). Decision-making with uncertain data: Bayesian linear programming approach, *Journal of Intelligent Manufacturing*, Vol. 15, No. 1, pp. 17-27.

[19] Tchangani, A. Ph. (2002). Decision Support System with Uncertain Data: Bayesian Networks Approach, *Studies in Informatics & Control Journal*, Vol. 11, No. 3, 233-241.

[20] Teghem, J. (1996). *Programmation linéaire*, Editions Ellipses/Editions de l'Université de Bruxelles.

[21] Winston, W. L. (1994). *Operations Research: Applications and Algorithms*, Third Edition, Duxbury Press.

In: Control and Learning in Robotic Systems
Editor: John X. Liu, pp. 59-85

ISBN 1-59454-356-9
©2005 Nova Science Publishers, Inc.

Chapter 3

Evolutionary Acquisition of Fuzzy Control Based Performance Skill Using a Three-Link Rings Gymnastic Robot

Takaaki Yamada[†,*] *Keigo Watanabe*[††], *and Kazuo Kiguchi*[††]
[†]Department of Mechanical Engineering, Kagoshima National College
of Technology, 1460-1 Shinko, Hayato-cho, Aira-gun,
Kagoshima 899-5193, Japan
[††] Department of Advanced Systems Control Engineering, Graduate
School of Science and Engineering, Saga University,
1 Honjomachi, Saga 840-8502, Japan

Abstract

The rings event is one of men's apparatus gymnastics. The ring exercises have the characteristics that the apparatus, namely rings, can move in all directions freely, i.e. free-floating characteristics at gripping point. This is a different point from other gymnastic events, therefore its motion property is totally unique. So far the rings have been studied mainly in the field of biomechanics. These studies mainly discuss the results obtained by motion measurements of actual gymnasts' performance.

On the other hand, we describe researches on "the rings gymnastic robot" using robot technology in this chapter. In an approach using the rings gymnastic robot, it is possible to analyze performances according to body type and characteristics for each gymnast. It is also capable of developing feats that nobody has ever done. These are not realized in the conventional approach based on the above-mentioned motion measurements. Purpose of studies on the rings gymnastic robot is to understand ring exercises through the robot and to apply the acquired skill to gymnastic coaching. In other words, it is to acquire the performance skill through the robot and to transfer the skill from the robot to gymnasts (to improve their motions) respectively. In these studies, fuzzy control is adopted. Because how to generate forces required in performances

[*]E-mail address:t-yamada@ieee.org, Phone&Fax: +81-995-42-9101

is useful for the gymnastic coaching as indicated in the area of sports biomechanics. In addition, if the control method is represented as "if-then" rules, it is easy for gymnasts to understand the knowledge to realize the performance as compared to reading the time-series data.

In this chapter, we discuss not only derivation of a three-link model of the robot but also a way of acquiring the performance skill, which is represented by fuzzy control rules. These three-link model-based analyses are necessary for our approach as a preliminary step toward analyses with simulations based on a three-dimensional model and experiments. In order to realize an objective performance, it is divided into basic exercises. A fuzzy controller is given for each basic exercise and one of them is selected according to the situation. Parameters of the fuzzy control rules are optimized by genetic algorithms (GAs). Simulation results show that the acquired skill is effective for realizing the objective performance.

1 Introduction

The rings event is one of men's apparatus gymnastics (see **Fig. 1**). It has been adopted as an event since the first Olympic Games in Athens in 1896. The ring exercises have the characteristics that the apparatus, namely rings, can move in all directions freely, i.e., free-floating characteristics at gripping point [1]. This is a different point from other gymnastic events and therefore its motion property is totally unique.

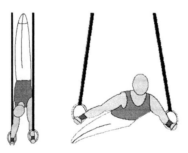

Figure 1: Rings event

The rings have been mainly studied in the field of biomechanics since the late twenty century. Chapman and Borchardt [2] have recorded the forces acting on gymnast's hands during a backward giant circle. Similarly these forces during a forward giant circle have been measured [3, 4, 5]. In recent years, Sprigings *et al.* [6] have studied a properly timed backward giant circle to remove swing from a handstand. Brewin *et al.* [7] have minimized peak forces at the shoulders during backward longswings. The aforementioned researches by Sprigings *et al.* and Brewin *et al.* using computer simulations as well as motion measurements have been recently reported with the advance of computer simulation techniques. However, the conventional studies on rings in the field of biomechanics as represented by references [2, 3, 4, 5] mainly discuss the results obtained by motion measurements of actual gymnasts' performances. On the other hand, we have already proposed "the rings gymnastic robot" [8] aiming at understanding rings and an application to gymnastic coaching.

In the area of robotics, the brachiation [9], aerial ski [10], trampoline [11], and horizontal bar gymnastic robots [12, 13, 14] have been developed. The purpose of these studies is to make a motion control method clear in skillful movements in human beings and animals, and to realize such motions using robots. Therefore, the purpose is different from our one. The use of an unformed object, namely a rope, in the rings gymnastic robot is also a big difference from those robots.

When gymnasts usually improve not only the ring exercises but also their motions, they first try to follow an expert's example. In such cases, they observe the expert's motions with the hope of imitation. However, from the visual information, they can obtain only the position and velocity. In other words, they cannot get the acceleration information related to the human motions. Therefore, it is difficult to estimate how to generate force from the outside intuitively. For this reason, the difference between the force which is estimated from the position and velocity based on the visual information, and the actual force is a problem in coaching. This problem is indicated in the field of sports biomechanics [15]. Hence, if we understand how to generate force, that is joint torque, required in a performance, it is useful for the coaching. In addition, if the control method is represented as "if-then" rules, it is easy for gymnasts to understand the knowledge to realize the performance as compared to reading time-series data. Consequently we adopt fuzzy control as a control strategy for the rings gymnastic robot.

In an approach based on the above motion measurements, the collected data are specific to the gymnast concerned and an expert gymnast is absolutely required to collect the data. On the other hand, in an approach using the rings gymnastic robot (including both simulations and experiments), it is possible to analyze not only performances according to body type and characteristics for each gymnast but also feats that nobody has ever done. The purpose of studies on the rings gymnastic robot is to understand ring exercises through the robot and to apply the acquired skill to gymnastic coaching. In other words, it is to acquire the performance skill through the robot and is finally to transfer the skill from the robot to gymnasts (to improve their motions). Also, the robot-based approach is capable of testing a limit of exercises that gymnasts can perform and developing new feats within the limit, so that it has potential to provide guidelines to realize them. In addition, gymnasts will be able to watch and feel the three-dimensional motions in practice by demonstration using the robot in the future. These experiences will be very helpful to them in the case of practicing a new feat that nobody has ever done. Thus, greater coaching effect can be expected by using both such experiences and fuzzy control rules on input torque required to perform the feat.

So far we have been mainly studying the skills acquisition, which is the first purpose of our research as mentioned above, of some performances. The skill of a performance including a backward giant circle and handstand has been already acquired by using two-link model [8, 16]. In order to apply the performance skills, which are acquired by simulations and experiments, to the gymnastic coaching as a final objective, we further need to acquire the skills of various performances by simulations based on the three-dimensional model and experiments using a prototype robot in stages. From this fact, in this chapter we acquire the performance skill, which is described by fuzzy control rules, of a handstand from backward giant circle using a three-link model as a preliminary step toward analyses in three-dimension and experiments. The rest of this chapter is organized as follows: In section 2, a dynamic equation of the three-link model of the robot is derived. Section 3 presents fuzzy control rules as the performance skill for realizing the objective performance. In section 4, how to obtain parameters of the fuzzy control rules with genetic algorithms (GAs) is described. Simulation results and the effectiveness of the obtained skill are shown in section 6.

2   Derivation of Three-Link Model

Although ring exercises are essentially complex three-dimensional motions, a three-link rings gymnastic robot that moves in two-dimensional plane is used in this chapter as the preliminary step toward motion analyses by a three-dimensional model and experiments. Note that Sprigings *et al.* [6] indicate that the performance skill to realize a backward giant circle appropriate for transition to a handstand does not become easier, though arm abduction is not considered in a two-dimensional model.

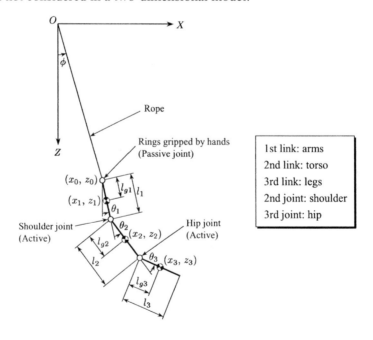

Figure 2: Three-link model of rings gymnastic robot

Assuming that the robot's arms, torso and legs are represented by one rigid link each leads to the three-link model of the robot, which corresponds to a triple physical pendulum suspended by a rope, as shown in **Fig. 2**. In this model, the second and third joints denote the shoulder and hip joints, which can be driven, respectively. In contrast, the robot's hands gripping the rings can revolute freely. In other words, the first joint is passive. The nomenclatures are as follows:

$\Sigma\,(O - XZ)$ absolute coordinate system;
(x_i, z_i) coordinates of mass center of link i;
θ_1 angle between Z-axis and link 1;
θ_2 angle of the shoulder joint, i.e., angle between links 1 and 2;
θ_3 angle of the hip joint, i.e., angle between links 2 and 3;
τ_j input torque to joint j;
(x_0, z_0) coordinates of the robot's end point, i.e., joint 1;
m_i mass of link i;
I_i moment of inertia about the gravity axis of link i;
l_i length of link i;
l_{gi} length between joint i and mass center of link i;
ϕ angle between Z-axis and the rope;
T_j torque caused by an impedance at joint j;
k_j spring constant at joint j;
b_j viscous damping coefficient at joint j;
F_r tension of the rope;
l_r natural length of the rope;
k_r spring constant of the rope;
b_r viscous damping coefficient of the rope;
r length between the robot's end point and the fixed point of the rope;
g gravitational acceleration;

where $i = 1, 2, 3$ and $j = 2, 3$.

In order to derive the dynamic equation of the three-link model of the robot, we use the Lagrangian formulation [17, 18]. Lagrange's equation of motion is expressed as

$$\frac{d}{dt}\left(\frac{\partial T}{\partial \dot{q}_i}\right) - \frac{\partial T}{\partial q_i} + \frac{\partial U}{\partial q_i} = Q_i, \tag{1}$$

where T is the total kinetic energy of the robot, U is the total potential energy stored in the robot, q_i is the generalized coordinate, and Q_i is the generalized force corresponding to q_i. Choosing $q_1 = x_1$, $q_2 = z_1$, $q_3 = \theta_1$, $q_4 = \theta_2$ and $q_5 = \theta_3$ as the generalized coordinates in the model, we have

$$\begin{aligned}
T &= \frac{1}{2}m_1(\dot{x}_1^2 + \dot{z}_1^2) + \frac{1}{2}m_2(\dot{x}_2^2 + \dot{z}_2^2) + \frac{1}{2}m_3(\dot{x}_3^2 + \dot{z}_3^2) \\
&\quad + \frac{1}{2}I_1\dot{\theta}_1^2 + \frac{1}{2}I_2(\dot{\theta}_1 + \dot{\theta}_2)^2 + \frac{1}{2}I_3(\dot{\theta}_1 + \dot{\theta}_2 + \dot{\theta}_3)^2,
\end{aligned} \tag{2}$$

$$U = -m_1 g z_1 - m_2 g z_2 - m_3 g z_3. \tag{3}$$

Using the generalized coordinates, the coordinates of mass center of links 2 and 3, (x_2, z_2) and (x_3, z_3), can be written as

$$x_2 = x_1 + (l_1 - l_{g1})\sin\theta_1 + l_{g2}\sin(\theta_1 + \theta_2), \tag{4}$$

$$z_2 = z_1 + (l_1 - l_{g1})\cos\theta_1 + l_{g2}\cos(\theta_1 + \theta_2), \tag{5}$$

$$x_3 = x_1 + (l_1 - l_{g1})\sin\theta_1 + l_2\sin(\theta_1 + \theta_2) + l_{g3}\sin(\theta_1 + \theta_2 + \theta_3), \tag{6}$$

$$z_3 = z_1 + (l_1 - l_{g1})\cos\theta_1 + l_2\cos(\theta_1 + \theta_2) + l_{g3}\cos(\theta_1 + \theta_2 + \theta_3). \tag{7}$$

Thus, from equations (1)–(7), the equations of motion for the robot can be given by

$$M(q)\ddot{q} + h(q,\dot{q}) + g(q) = Q, \tag{8}$$

where $q = [x_1\ z_1\ \theta_1\ \theta_2\ \theta_3]^T$ is a 5×1 vector of generalized coordinate terms, $M(q)$ is a 5×5 symmetric mass matrix, $h(q,\dot{q})$ is a 5×1 vector of centrifugal and Coriolis terms, $g(q)$ is a 5×1 vector of gravity terms, and Q is a 5×1 vector of generalized force terms. Elements, M_{ij}, h_i and g_i, of the matrix and vectors are represented as follows:

$$M_{11} = m_1 + m_2 + m_3, \tag{9}$$

$$M_{12} = 0, \tag{10}$$

$$M_{13} = (m_2 + m_3)(l_1 - l_{g1})\cos\theta_1 + (m_2 l_{g2} + m_3 l_2)\cos(\theta_1 + \theta_2)$$
$$+ m_3 l_{g3}\cos(\theta_1 + \theta_2 + \theta_3), \tag{11}$$

$$M_{14} = (m_2 l_{g2} + m_3 l_2)\cos(\theta_1 + \theta_2) + m_3 l_{g3}\cos(\theta_1 + \theta_2 + \theta_3), \tag{12}$$

$$M_{15} = m_3 l_{g3}\cos(\theta_1 + \theta_2 + \theta_3), \tag{13}$$

$$M_{22} = m_1 + m_2 + m_3, \tag{14}$$

$$M_{23} = -(m_2 + m_3)(l_1 - l_{g1})\sin\theta_1 - (m_2 l_{g2} + m_3 l_2)\sin(\theta_1 + \theta_2)$$
$$- m_3 l_{g3}\sin(\theta_1 + \theta_2 + \theta_3), \tag{15}$$

$$M_{24} = -(m_2 l_{g2} + m_3 l_2)\sin(\theta_1 + \theta_2) - m_3 l_{g3}\sin(\theta_1 + \theta_2 + \theta_3), \tag{16}$$

$$M_{25} = -m_3 l_{g3}\sin(\theta_1 + \theta_2 + \theta_3), \tag{17}$$

$$M_{33} = (m_2 + m_3)(l_1 - l_{g1})^2 + m_2 l_{g2}^2 + m_3 l_2^2 + m_3 l_{g3}^2$$
$$+ 2(m_2 l_{g2} + m_3 l_2)(l_1 - l_{g1})\cos\theta_2 + 2m_3(l_1 - l_{g1})l_{g3}\cos(\theta_2 + \theta_3)$$
$$+ 2m_3 l_2 l_{g3}\cos\theta_3 + I_1 + I_2 + I_3, \tag{18}$$

$$M_{34} = m_2 l_{g2}^2 + m_3 l_2^2 + m_3 l_{g3}^2 + (m_2 l_{g2} + m_3 l_2)(l_1 - l_{g1})\cos\theta_2$$
$$+ m_3(l_1 - l_{g1})l_{g3}\cos(\theta_2 + \theta_3) + 2m_3 l_2 l_{g3}\cos\theta_3 + I_2 + I_3, \tag{19}$$

$$M_{35} = m_3 l_{g3}^2 + m_3(l_1 - l_{g1})l_{g3}\cos(\theta_2 + \theta_3) + m_3 l_2 l_{g3}\cos\theta_3 + I_3, \tag{20}$$

$$M_{44} = m_2 l_{g2}^2 + m_3 l_2^2 + m_3 l_{g3}^2 + 2m_3 l_2 l_{g3}\cos\theta_3 + I_2 + I_3, \tag{21}$$

$$M_{45} = m_3 l_{g3}^2 + m_3 l_2 l_{g3}\cos\theta_3 + I_3, \tag{22}$$

$$M_{55} = m_3 l_{g3}^2 + I_3, \tag{23}$$

$$h_1 = -(m_2 + m_3)(l_1 - l_{g1})\dot{\theta}_1^2\sin\theta_1$$

$$-(m_2 l_{g2} + m_3 l_2)(\dot{\theta}_1 + \dot{\theta}_2)^2 \sin(\theta_1 + \theta_2)$$
$$-m_3 l_{g3}(\dot{\theta}_1 + \dot{\theta}_2 + \dot{\theta}_3)^2 \sin(\theta_1 + \theta_2 + \theta_3), \tag{24}$$

$$h_2 = -(m_2 + m_3)(l_1 - l_{g1})\dot{\theta}_1^2 \cos\theta_1$$
$$-(m_2 l_{g2} + m_3 l_2)(\dot{\theta}_1 + \dot{\theta}_2)^2 \cos(\theta_1 + \theta_2)$$
$$-m_3 l_{g3}(\dot{\theta}_1 + \dot{\theta}_2 + \dot{\theta}_3)^2 \cos(\theta_1 + \theta_2 + \theta_3), \tag{25}$$

$$h_3 = (m_2 l_{g2} + m_3 l_2)(l_1 - l_{g1})\{\dot{\theta}_1^2 - (\dot{\theta}_1 + \dot{\theta}_2)^2\}\sin\theta_2$$
$$+ m_3(l_1 - l_{g1})l_{g3}\{\dot{\theta}_1^2 - (\dot{\theta}_1 + \dot{\theta}_2 + \dot{\theta}_3)^2\}\sin(\theta_2 + \theta_3)$$
$$+ m_3 l_2 l_{g3}\{(\dot{\theta}_1 + \dot{\theta}_2)^2 - (\dot{\theta}_1 + \dot{\theta}_2 + \dot{\theta}_3)^2\}\sin\theta_3, \tag{26}$$

$$h_4 = (m_2 l_{g2} + m_3 l_2)(l_1 - l_{g1})\dot{\theta}_1^2 \sin\theta_2 + m_3(l_1 - l_{g1})l_{g3}\dot{\theta}_1^2 \sin(\theta_2 + \theta_3)$$
$$+ m_3 l_2 l_{g3}\{(\dot{\theta}_1 + \dot{\theta}_2)^2 - (\dot{\theta}_1 + \dot{\theta}_2 + \dot{\theta}_3)^2\}\sin\theta_3, \tag{27}$$

$$h_5 = m_3(l_1 - l_{g1})l_{g3}\dot{\theta}_1^2 \sin(\theta_2 + \theta_3) + m_3 l_2 l_{g3}(\dot{\theta}_1 + \dot{\theta}_2)^2 \sin\theta_3, \tag{28}$$

$$g_1 = 0, \tag{29}$$

$$g_2 = -(m_1 + m_2 + m_3)g, \tag{30}$$

$$g_3 = (m_2 + m_3)g(l_1 - l_{g1})\sin\theta_1 + (m_2 l_{g2} + m_3 l_2)g\sin(\theta_1 + \theta_2)$$
$$+ m_3 g l_{g3}\sin(\theta_1 + \theta_2 + \theta_3), \tag{31}$$

$$g_4 = (m_2 l_{g2} + m_3 l_2)g\sin(\theta_1 + \theta_2) + m_3 g l_{g3}\sin(\theta_1 + \theta_2 + \theta_3), \tag{32}$$

$$g_5 = m_3 g l_{g3}\sin(\theta_1 + \theta_2 + \theta_3), \tag{33}$$

where $i = 1–5$ and $j = 1–5$ denote their column and row respectively.

The generalized force vector Q consists of the input torques to joints, the torques caused by the joint impedance, and the forces and torques based on the rope tension acting on the robot's end point. Therefore, by statics we determine the vector $Q' = [Q'_1 \; Q'_2 \; Q'_3 \; Q'_4 \; Q'_5]^T$ whose elements, Q_i ($i=1–5$), represent the forces and torques caused by the rope tension applied to the robot's end point as shown in **Fig. 3**. Q'_1 and Q'_2 are the forces acting on the mass center (x_1, z_1) of link 1. Q'_3, Q'_4 and Q'_5 are the torques generated at joints 1, 2 and 3 respectively. The relationship between (x_0, z_0), the coordinates of the robot's end point, and the generalized coordinates can be described as

$$\begin{bmatrix} \dot{x}_0 \\ \dot{z}_0 \end{bmatrix} = \begin{bmatrix} 1 & 0 & -l_{g1}\cos\theta_1 & 0 & 0 \\ 0 & 1 & l_{g1}\sin\theta_1 & 0 & 0 \end{bmatrix} \begin{bmatrix} \dot{x}_1 \\ \dot{z}_1 \\ \dot{\theta}_1 \\ \dot{\theta}_2 \\ \dot{\theta}_3 \end{bmatrix}. \tag{34}$$

Hence, from the principle of virtual work, Q' is expressed as

$$Q' = \begin{bmatrix} Q'_1 \\ Q'_2 \\ Q'_3 \\ Q'_4 \\ Q'_5 \end{bmatrix} = \begin{bmatrix} 1 & 0 & -l_{g1}\cos\theta_1 & 0 & 0 \\ 0 & 1 & l_{g1}\sin\theta_1 & 0 & 0 \end{bmatrix}^T \begin{bmatrix} f_{rx} \\ f_{rz} \end{bmatrix}$$

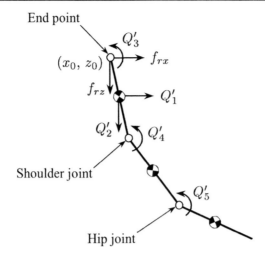

Figure 3: Forces and torques caused by rope tension

$$
= \begin{bmatrix}
1 & 0 \\
0 & 1 \\
-l_{g1}\cos\theta_1 & l_{g1}\sin\theta_1 \\
0 & 0 \\
0 & 0
\end{bmatrix}
\begin{bmatrix}
F_r\sin\phi \\
F_r\cos\phi
\end{bmatrix}
$$

$$
= \begin{bmatrix}
F_r\sin\phi \\
F_r\cos\phi \\
F_r l_{g1}\sin(\theta_1 - \phi) \\
0 \\
0
\end{bmatrix}.
\tag{35}
$$

Note here that f_{rx} and f_{rz} represent X- and Z-directional components of the rope tension F_r, which is given by

$$
F_r = \begin{cases}
-k_r(r - l_r) - b_r\dot{r} & \text{for } r > l_r \\
0 & \text{for } r \leq l_r
\end{cases},
\tag{36}
$$

where

$$
r = \sqrt{x_0^2 + z_0^2},
\tag{37}
$$

$$
\dot{r} = \dot{x}_0\sin\phi + \dot{z}_0\cos\phi,
\tag{38}
$$

$$
\phi = \mathrm{atan2}(x_0, z_0),
\tag{39}
$$

$$
x_0 = x_1 - l_{g1}\sin\theta_1,
\tag{40}
$$

$$
z_0 = z_1 - l_{g1}\cos\theta_1,
\tag{41}
$$

$$
\dot{x}_0 = \dot{x}_1 - l_{g1}\dot{\theta}_1\cos\theta_1,
\tag{42}
$$

$$
\dot{z}_0 = \dot{z}_1 + l_{g1}\dot{\theta}_1\sin\theta_1.
\tag{43}
$$

From equation (35), elements of the generalized force vector Q can be written as

$$
\begin{aligned}
Q_1 &= Q_1' = F_r \sin\phi, & (44)\\
Q_2 &= Q_2' = F_r \cos\phi, & (45)\\
Q_3 &= Q_3' = F_r l_{g1} \sin(\theta_1 - \phi), & (46)\\
Q_4 &= Q_4' + \tau_2 + T_2 = \tau_2 + T_2, & (47)\\
Q_5 &= Q_5' + \tau_3 + T_3 = \tau_3 + T_3, & (48)
\end{aligned}
$$

where the torques, T_j $(j = 2, 3)$, caused by the impedance at joint j are expressed as

$$
T_j = -k_j \theta_j - b_j \dot{\theta}_j. \tag{49}
$$

3 Fuzzy Control Rules

In order to realize a handstand from backward giant circle, its performance is divided into two basic exercises consisting of the backward giant circle and handstand, in which the skills to realize such basic exercises are represented by fuzzy control rules. One of these skills, i.e., a fuzzy controller is selected according to the situation. Note that the controller is constructed by using a simplified fuzzy reasoning.

3.1 Control of Backward Giant Circle

3.1.1 Fuzzy Control Rules on Input Torque

The fuzzy control rules on the input torques τ_i $(i = 2, 3)$ to the shoulder and hip joints are represented by

Rule 1: If ψ_g is A_i then $\tau_i = c_{ai}$,
Rule 2: If ψ_g is $\bar{A}_i$ then $\tau_i = 0$,

where A_i is an antecedent fuzzy set when the input torque is not zero, i.e., c_{ai} and $\bar{A}_i$ is its complement. The antecedent membership function for A_i is Gaussian type (see **Fig. 4**), which is expressed as

$$
\mu_{ai}(\psi_g) = \exp\{\ln(0.5)(\psi_g - \alpha_{ai})^2 \beta_{ai}^2\}. \tag{50}
$$

Here, α_{ai} and β_{ai} denote the center value and the reciprocal number of the standard deviation respectively. ψ_g, which is an input to the antecedent part, is defined by

$$
\psi_g = \text{atan2}(x_g - x_0, z_g - z_0), \tag{51}
$$

where (x_g, z_g) is the coordinates of robot's mass center.

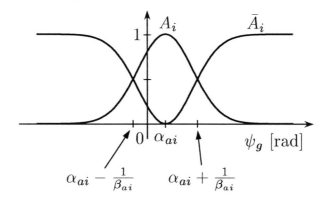

Figure 4: Antecedent membership functions in fuzzy control rules for τ_i to realize a backward giant circle

3.1.2 Fuzzy Control Rules on Joint Impedance

The joint impedance is adjusted by changing the spring constant k_i $(i = 2, 3)$ for the shoulder and hip joints according to the following equation:

$$k_i = \begin{cases} \gamma_i k_i, & \text{within range of motion} \\ k_i, & \text{otherwise} \end{cases}, \tag{52}$$

where γ_i is a changing rate for the spring constant k_i. The fuzzy control rules on γ_i are described as

Rule 1: If ψ_g is B_i then $\gamma_i = c_{bi}$,
Rule 2: If ψ_g is $\bar{B}_i$ then $\gamma_i = 1$,
Rule 3: If θ_1 is C_i then $\gamma_i = c_{ci}$,
Rule 4: If θ_1 is $\bar{C}_i$ then $\gamma_i = 1$,

in which the antecedent membership function for a fuzzy set B_i is the same Gaussian type as A_i. For a fuzzy set C_i, the following membership function represented by

$$\mu_{ci}(\theta_1) = \begin{cases} \exp\{\ln(0.5)(\theta_1 - \alpha_{ci})^2 \beta_{ci}^2\} & \text{for } \theta_1 \leq \alpha_{ci} \\ 1 & \text{for } \theta_1 > \alpha_{ci} \end{cases} \tag{53}$$

is adopted (see **Fig. 5**).

3.2 Control of Handstand

3.2.1 Realization Condition for Handstand

In order to control a handstand, we now consider a realization condition for handstand (see reference [8] about the condition for the two-link gymnastic robot). As shown in **Fig. 6**,

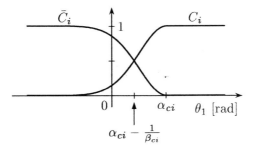

Figure 5: Antecedent membership functions in fuzzy control rules for γ_i to realize a backward giant circle

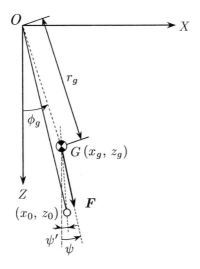

Figure 6: Definition of ψ and ψ' for handstand realization condition

assuming there is a lumped mass $m(= m_1 + m_2 + m_3)$ on the mass center $G\,(x_g, z_g)$ of the gymnastic robot, X- and Z-directional components of the force vector $\boldsymbol{F} = [F_x\ F_z]^T$ acting on the G are obtained by

$$F_x = -m\ddot{x}_g, \tag{54}$$
$$F_z = -m\ddot{z}_g + mg. \tag{55}$$

The angle ψ between the Z-axis and the vector $\boldsymbol{F}$ is expressed as

$$\psi = \mathrm{atan2}(F_x, F_z). \tag{56}$$

Also, the angle ψ' between the Z-axis and a straight line connecting (x_0, z_0) to (x_g, z_g) is determined by

$$\psi' = \mathrm{atan2}(x_0 - x_g, z_0 - z_g). \tag{57}$$

Furthermore, we define

$$s_h = \psi - \psi'. \tag{58}$$

Since the gymnastic robot can keep its balance if ψ and ψ' are equal, a handstand is controlled by making ψ and ψ' equal.

In simulations in the after-mentioned section 5, in order to calculate $\ddot{x}_g$ and $\ddot{z}_g$ in discrete-time, we will use discrete-time representations of equations (4)–(7) and the following equations:

$$x_g(k) = \frac{m_1 x_1(k) + m_2 x_2(k) + m_3 x_3(k)}{m_1 + m_2 + m_3}, \tag{59}$$

$$z_g(k) = \frac{m_1 z_1(k) + m_2 z_2(k) + m_3 z_3(k)}{m_1 + m_2 + m_3}, \tag{60}$$

$$\dot{x}_g(k) = \frac{x_g(k) - x_g(k-1)}{\Delta t}, \tag{61}$$

$$\dot{z}_g(k) = \frac{z_g(k) - z_g(k-1)}{\Delta t}, \tag{62}$$

$$\ddot{x}_g(k) = \frac{\dot{x}_g(k) - \dot{x}_g(k-1)}{\Delta t}, \tag{63}$$

$$\ddot{z}_g(k) = \frac{\dot{z}_g(k) - \dot{z}_g(k-1)}{\Delta t}, \tag{64}$$

where k is a discrete-time instant and Δt is a sampling period.

3.2.2 Fuzzy Control Rules for Handstand

From the consideration in the previous section, s_h is given as one of the inputs to the antecedent part of fuzzy reasoning. The input torques during a handstand depend on the angle ϕ between the Z-axis and the rope and its rate $\dot{\phi}$ as well as s_h, so that ϕ and $\dot{\phi}$ are also given as the inputs to the antecedent part.

In case of three labels for each input, the fuzzy control rules on the input torques τ_i ($i = 2, 3$) are described as follows:

> Rule 1: If s_h is ^1D_i and ϕ is ^1E_i and $\dot{\phi}$ is ^1F_i then $\tau_i = {}^1c_i$,
> $\vdots$
> Rule 27: If s_h is $^{27}D_i$ and ϕ is $^{27}E_i$ and $\dot{\phi}$ is $^{27}F_i$ then $\tau_i = {}^{27}c_i$,

where ^jD_i, ^jE_i and ^jF_i ($j=1$–27) are fuzzy sets for the inputs, s_h, ϕ and $\dot{\phi}$, to the antecedent part in the rule j respectively, and ^jc_i is a constant in the consequent part. When all the inputs are zero, a handstand is stable. Therefore the input torques have to be zero. To satisfy this condition, a triangular and two trapezoidal functions are adopted as the antecedent membership functions for each input as shown in **Fig. 7**. Using these functions, when the grade of triangular one with a label Z is 1, the grades of the other ones are 0. Thus, if the consequent constant of the rule that labels for all inputs are Z is set to 0, then the above condition is satisfied. **Table 1** presents the parameters of the antecedent membership

Table 1: Parameters of antecedent membership functions for a handstand

Inputs	a_N	a_P
s_h	$-\pi/90$	$\pi/90$
ϕ	$-\pi/18$	$\pi/18$
$\dot{\phi}$	-1	1

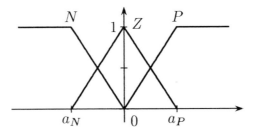

Figure 7: Antecedent membership functions for a handstand

functions for each input. Note here that the antecedent grade is given as the product of the grade for each membership function.

While doing a handstand, γ_i ($i=2, 3$) are kept at the output values from the controllers for backward giant circle at transition from the backward giant circle to the handstand.

4   Evolutionary Acquisition of Performance Skill

The parameters of the above-mentioned fuzzy control rules are searched by a two-stage approach using GAs as shown in **Fig. 8**. In case of the three-link model, setting of more evaluations is necessary to acquire the skill of a backward giant circle, because its motion is more complicated due to increase in the number of the active joints as compared to the two-link model. Therefore, in the first stage, the parameters for backward giant circle are obtained by the first GA using the subgroup-based evaluation method [16], which can set multiple fitness functions. In the second stage, those for handstand are determined with the second GA, in which the backward giant circle before handstand is performed by using the best controller obtained in the first stage. Design parameters used in GAs are shown in **Table 2**.

Table 2: Parameters used in genetic algorithms

Population size	60
Number of elites	6
Selection	Tournament selection with 3 individuals
Crossover	Uniform crossover (rate=0.6)
Mutation rate	0.01

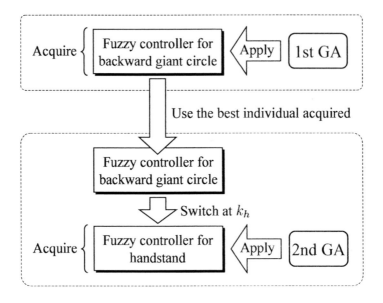

Figure 8: Procedure for acquiring parameters of fuzzy controllers by two-stage approach

Size of chromosome: $16\ bit \times 18 = 288\ bit$

16bit

| c_{a2} | α_{a2} | β_{a2} | c_{b2} | α_{b2} | β_{b2} | c_{c2} | α_{c2} | β_{c2} | c_{a3} | α_{a3} | β_{a3} | c_{b3} | α_{b3} | β_{b3} | c_{c3} | α_{c3} | β_{c3} |

Figure 9: Structure of chromosome in the first GA

4.1 The First GA for Backward Giant Circle

4.1.1 Parameters Acquired for Fuzzy Control Rules of Backward Giant Circle

In the fuzzy control rules for backward giant circle, 18 parameters, which are the parameters (α_{ai}, β_{ai}, α_{bi}, β_{bi}, α_{ci} and β_{ci}; $i = 2, 3$) of the antecedent membership functions and the consequent constants (c_{ai}, c_{bi} and c_{ci}) to be optimized, are acquired by the first GA. The search spaces are given in the following ranges:

$$\begin{aligned}
&c_{ai} : [0, 400], &&\alpha_{b2}: [-\pi/9, \pi/2], \\
&\alpha_{ai}: [-\pi/6, \pi/6], &&\alpha_{b3}: [-\pi/9, \pi/9], \\
&\beta_{ai}: [180/\pi, 6/\pi], &&c_{ci} : [1, 7], \\
&c_{bi}: [1, 7], &&\alpha_{ci} : [11\pi/12, \pi], \\
&\beta_{bi}: [180/\pi, 6/\pi], &&\beta_{ci} : [180/\pi, 12/\pi].
\end{aligned}$$

Figure 9 shows the structure of chromosome in the first GA.

4.1.2 Classification of Individuals into Five Subgroups in Evaluation

To acquire the parameters of fuzzy control rules for a backward giant circle that is appropriate for transition to a handstand, the subgroup-based evaluation method proposed in reference [16] is introduced into the first GA (see **Fig. 10**). The number of subgroups used in this study is five. Note that this evaluation method has the following properties: (1) a subgroup has an independent evaluation; (2) priority exists among subgroups; and (3) individuals are classified into subgroups based on judgment conditions established according to the applied problem. The details on the above (2) are as follows: the priority of evaluation is diminishing from the fifth subgroup to the first subgroup. With the alternation of generations, a percentage of the fifth subgroup is expected to be increased.

A flowchart of classifying individuals into five subgroups is presented in **Fig. 11**. The first condition:

$$r(k) > l_r$$

checks the slack of the rope during a backward giant circle by comparing the rope length $r(k)$ at the discrete-time instant k with its natural length l_r. Since the slack of rope is undesirable in a handstand, the individuals that do not pass through the first condition are assigned to the first subgroup. In order to shift to a handstand, it is required that there exists

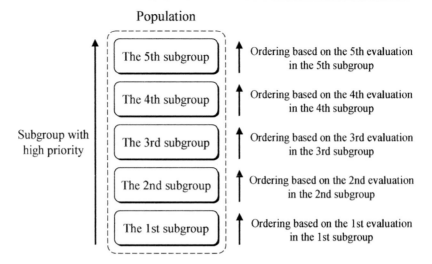

Figure 10: Ordering in population with five subgroups

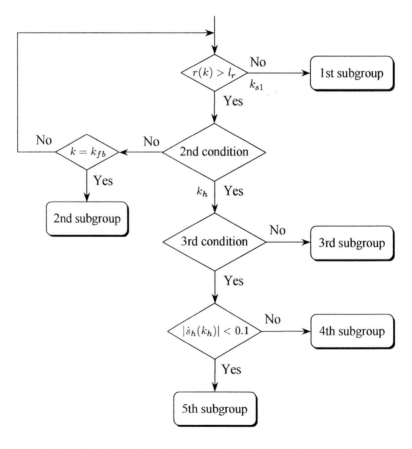

Figure 11: Classification of each individual into five subgroups in the first GA

a state satisfying the realization condition for handstand, i.e. $s_h(k) = 0$ [rad], when $\psi'(k)$ nearly equals zero [rad]. Hence, the second condition:

$$|\psi'(k)| < \tfrac{\pi}{12} \text{ and } s_h(k-1) > 0 \text{ and } s_h(k) < 0$$

is used to judge whether such a state exists or not at every discrete-time instant for 2.2 [s], i.e., until $k_{fb} = 2.2/\Delta t$. Note that k_{fb} is the preset final discrete-time instant in the first GA. If the second condition is not satisfied within the preset time, the corresponding individuals are classified into the second subgroup. Even if this condition is satisfied, it is undesirable to bend the shoulder and hip joints more than needs, from the viewpoint of both the skill and the appearance of performance. Accordingly the third condition:

$$|\theta_2(k_h)| < \tfrac{\pi}{60} \text{ and } |\theta_3(k_h)| < \tfrac{\pi}{60}$$

limits the angles of the shoulder and hip joints within the specified range at k_h. Here, k_h is the discrete-time instant at which the second condition is satisfied. If the individuals pass through the third condition, they go to the third subgroup. In addition, it is impossible to shift to a handstand unless the rate of change of $s_h(k_h)$ is small to some degree. For this reason, the magnitude of its value is restricted in the fourth condition such that

$$|\dot{s}_h(k_h)| < 0.1.$$

The individuals that satisfy the fourth condition are assigned to the fourth subgroup. The others are classified into the fifth subgroup. Note here that if the individual is classified into the first, third, fourth and fifth subgroups, then the corresponding simulation is stopped to save the computational time.

4.1.3 Independent Fitness Function in Each Subgroup

Each subgroup has an independent measure in the process of evaluation. As maximum θ_1 until k_{s1} is bigger, the evaluation in the first subgroup becomes better. Consequently,

$$f_{b1} = \max\{\theta_1(0), ..., \theta_1(k_{s1})\} \tag{65}$$

is maximized. Here, k_{s1} is the discrete-time instant at which an individual is assigned to the first subgroup because it does not satisfy $r(k) > l_r$. In the second subgroup, since a swing is not enough to do a backward giant circle, the following fitness function is maximized to make its swing bigger:

$$f_{b2} = \max\{\theta_1(0), ..., \theta_1(k_{fb})\}. \tag{66}$$

In the third subgroup, the joint angles are too big at which the realization condition for handstand is satisfied. Therefore, the third evaluation:

$$f_{b3} = |\theta_2(k_h)| + |\theta_3(k_h)| \tag{67}$$

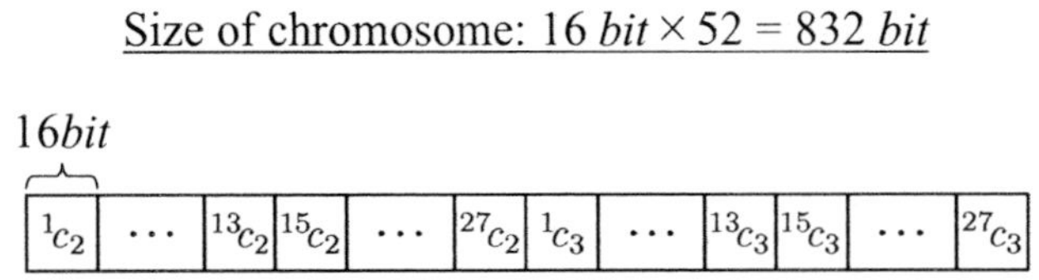

Size of chromosome: $16 \, bit \times 52 = 832 \, bit$

$16bit$

Figure 12: Structure of chromosome in the second GA

is minimized to make the joint angles smaller. In the fourth subgroup, it is difficult to shift from a backward giant circle to a handstand because of too big $\dot{s}_h(k_h)$. Hence, it is minimized by this fitness function:

$$f_{b4} = |\dot{s}_h(k_h)|. \tag{68}$$

In the fifth subgroup, in order to realize an effective performance, energy is minimized by the fifth evaluation:

$$f_{b5} = f'_{b5} + f''_{b5}, \tag{69}$$

where

$$f'_{b5} = \sum_{i=0}^{k_h-1} \{|T_2(i)| + |T_3(i)|\}, \tag{70}$$

$$f''_{b5} = \sum_{i=0}^{k_h-1} \{|\tau_2(i)| + |\tau_3(i)|\}. \tag{71}$$

Note here that f'_{b5} and f''_{b5} denote the sum of torques caused by the joint impedance and the sum of the joint torques, respectively. Also, k_{s1} in f_{b1} is the discrete-time instant at which an individual is assigned to the first subgroup.

4.2   The Second GA for Handstand

4.2.1   Parameters Acquired for Fuzzy Control Rules of Handstand

In the fuzzy control rules for handstand, the consequent constant $^{14}c_i$ in which all labels are Z in the fuzzy control rules to do a handstand is set to zero. The other 52 constants, ^1c_i to $^{13}c_i$ and $^{15}c_i$ to $^{27}c_i$, are obtained with the second GA. The search spaces of these consequent constants are set in the range of $[-200, 100]$. The structure of chromosome in the second GA is shown in **Fig. 12**.

4.2.2   Fitness Functions

In simulations in evaluation process of the second GA, the fuzzy controller for backward giant circle is selected at the beginning, and then it is switched to that for handstand at k_h.

In the evaluation of the second GA, the following fitness function:

$$f_h = f_{h1} + f_{h2} \tag{72}$$

is minimized. This function represents a stability of handstand. Note here that

$$f_{h1} = \sum_{i=k_h}^{k} |s_h(i)| \tag{73}$$

is used if the following condition is satisfied:

$$|s_h(k)| < \pi/90 \text{ and } |\theta_2(k)| < \pi/12 \text{ and } |\theta_3(k)| < \pi/12.$$

In this condition, the joint angles are limited during a handstand like the fourth condition in the individual classification of the first GA, because it is undesirable to bend the joints more than needs, from the viewpoints of the skill and the appearance. On the other hand, a handstand is treated as a failure, if all of the above inequalities are not satisfied. Then, a penalty:

$$f_{h2} = (k_{fh} + 1) - k \tag{74}$$

is given and the corresponding simulation using the individual is stopped to save the computational time. Here, k_{fh} is the predefined final discrete-time instant for a handstand from backward giant circle in the second GA. In this case, k_{fh} equals $5/\Delta t$.

5 Simulation

Physical parameters of the three-link robot [19], parameters of the rope [20], range of motion for each joint [21], and spring constants and viscous damping coefficients varied with the joint angles are shown in **Tables 3-6**, respectively. Note that the spring constants and viscous damping coefficients for each joint are adjusted by comparing the animation of the three-link rings gymnastic robot with the moving images of the actual gymnast's performance. In acquiring a backward giant circle by the first GA, the simulation time is 2.2 [s]. In case of determining a handstand from the backward giant circle with the second GA, the simulation time is 5 [s]. The sampling period is 1 [ms]. The initial states of the gymnastic robot are set as follows:

$$\begin{aligned}
\theta_1 &= -\frac{175}{180}\pi, \\
\theta_2 &= 0, \\
\theta_3 &= 0, \\
x_1 &= l_{g1} \sin\theta_1, \\
z_1 &= l_r + \frac{m_1 + m_2 + m_3}{k_r} g + l_{g1} \cos\theta_1, \\
\dot{q} &= [0\ 0\ 0\ 0\ 0]^T.
\end{aligned}$$

Table 3: Physical parameters of rings gymnastic robot

Link i	1	2	3
m_i [kg]	5.4	29.5	18.5
I_i [kgm^2]	0.15	1.93	1.03
l_i [m]	0.58	0.5	0.79
l_{gi} [m]	0.31	0.2	0.33

Table 4: Parameters of rope

Parameters		Values
Natural length	l_r [m]	2.95
Spring constant	k_r [N/m]	20000
Viscous damping coefficient b_r [Ns/m]		2000

The parameters of fuzzy control rules for backward giant circle are determined by genetic operations of 2000 generations in the first GA using the subgroup-based evaluation method. **Figures 13** and **14** show the simulation results using the fuzzy control rules obtained and the graphics sequence of the performance based on them, respectively. From these results, it is found that the backward giant circle is properly realized by the three-link rings gymnastic robot.

Furthermore, in order to realize a handstand from the backward giant circle, the parameters of fuzzy control rules for handstand are acquired with the second GA. Evolutionary histories of the best and mean fitness in the second GA are shown in **Fig. 15**. These graphs show that there appears better fitness with the alternation of generations. **Figures 16** and **17** show the simulation results using the fuzzy control rules acquired by genetic operations of 3000 generations in the second GA and the graphics sequence of the performance based on them, respectively. From these results, it is confirmed that the robot adequately realizes a series of exercises, i.e. a handstand form backward giant circle. The results also show that the backward giant circle determined by the first GA is appropriate for transition to handstand. Since the controller for backward giant circle is switched to that for handstand and

Table 5: Range of motion for each joint

Joint i	2	3
Upper limit θ_{iu} [deg]	230	90
Lower limit θ_{il} [deg]	0	-15

Table 6: Spring constant and viscous damping coefficient for each joint

Joint i	2		3	
Range	k_2	b_2	k_3	b_3
$\theta_{iu} < \theta_i$	400	50	500	50
$0 \leq \theta_i \leq \theta_{iu}$	40	4	50	5
$\theta_{il} \leq \theta_i < 0$	–	–	100	10
$\theta_i < \theta_{il}$	2000	50	1000	50

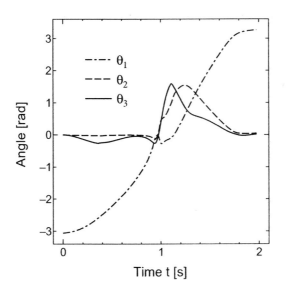

Figure 13: Simulation results of a backward giant circle acquired by the first GA

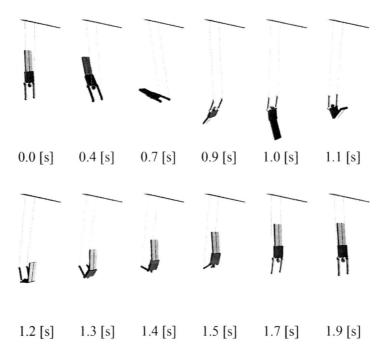

0.0 [s]   0.4 [s]   0.7 [s]   0.9 [s]   1.0 [s]   1.1 [s]

1.2 [s]   1.3 [s]   1.4 [s]   1.5 [s]   1.7 [s]   1.9 [s]

Figure 14: Graphics sequence of three-link rings gymnastic robot performing a backward giant circle

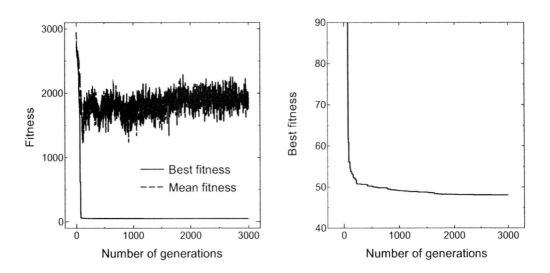

Figure 15: Evolutionary histories of fitness in the second GA

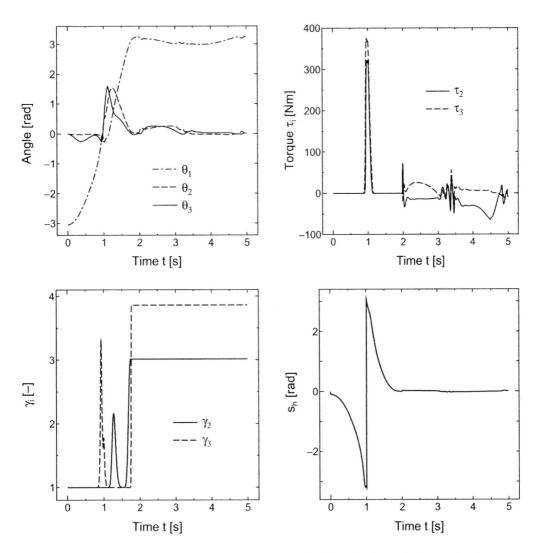

Figure 16: Simulation results of a handstand from backward giant circle acquired by two-stage approach

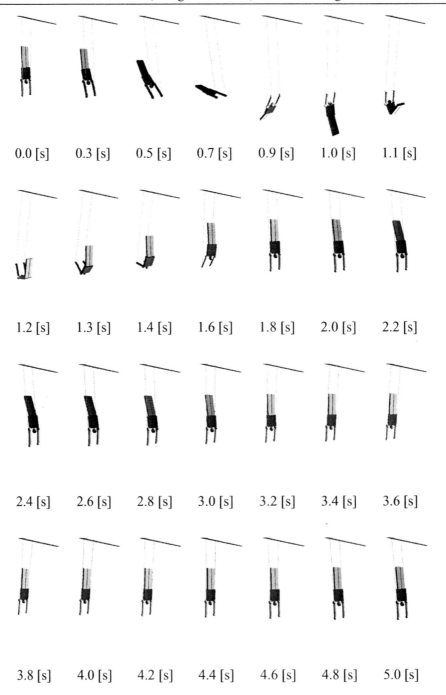

Figure 17: Graphics sequence of three-link rings gymnastic robot performing a handstand from backward giant circle

then s_h keeps nearly zero, it is found that the realization condition for handstand is effective in case of the three-link model as well as the two-link one.

6 Conclusion

In this chapter, we have acquired the performance skill, which is described by the fuzzy control rules, to realize the handstand from backward giant circle by the three-link rings gymnastic robot, as the preliminary step toward motion analyses by simulations of a three-dimensional model and experiments. The three-link model with two active joints was derived by the Lagrangian formulation to do computer simulations. In order to control the objective performance that is one of the compulsory requirements in the rings event, it was divided into two basic exercises, i.e. a backward giant circle and a handstand. One of the fuzzy controllers for the basic exercises was selected according to the situation. The two-stage approach using GAs was used for determining the parameters of the fuzzy control rules. In the first stage, the parameters for backward giant circle were obtained with the first GA using the subgroup-based evaluation method. By the second GA in the second stage, those for handstand were acquired. Finally simulation results showed the effectiveness of the performance skill obtained. The effectiveness of the realization condition for handstand in case of the three-link model was also confirmed.

In the future, we will acquire the performance skill by not only simulations of the three-dimensional model but also experiments as the next step. Also, under the present situation fifty-four fuzzy control rules in total are needed for the input torques to the shoulder and hip joints, in order to realize a handstand in case of the three-link model. For this reason, it may be difficult for gymnasts to understand the skill directly. Therefore, it will be necessary to reduce the number of the rules.

References

[1] A. Kaneko, Coaching of Gymnastics Competition. Tokyo, Japan: Taishukan-Shoten, 1974. (in Japanese)

[2] A. E. Chapman and W. Borchardt, "Biomechanical Factors Underlying the Dislocate on Still Rings," *J. of Human Movement Studies*, vol. 3, no. 4, pp. 221–231, 1977.

[3] D. G. Sale and R. L. Judd, "Dynamometric Instrumentation of the Rings for Analysis of Gymnastic Movements," *Medical Science in Sport*, vol. 6, no. 3, pp. 209–216, 1974.

[4] M. A. Nissinen, "Kinematic and Kinetic Analysis of the Giant Swing on Rings," *Biomechanics VIII-B: Proc. of the 8th Int. Congress of Biomechanics*, Nagoya, Japan: Human Kinetics Publishers, pp. 781–786, 1983.

[5] G. P. Brüggemann, "Biomechanics in Gymnastics," *Current Research in Sports Biomechanics*, (*Medicine and sport science*, vol. 25), Basel: Karger, pp. 142–176, 1987.

[6] E. J. Sprigings, J. L. Lanovaz, L. G. Watson, and K. W. Russell, "Removing swing from a handstand on rings using a properly timed backward giant circle: a simulation solution," *J. of Biomechanics*, vol. 31, no. 1, pp. 27–35, Nov. 1998.

[7] M. A. Brewin, M. R. Yeadon, and D. G. Kerwin, "Minimising peak forces at the shoulders during backward longswings on rings," *Human Movement Science*, vol. 19, no. 5, pp. 717–736, Nov. 2000.

[8] T. Yamada, K. Watanabe, K. Kiguchi, and K. Izumi, "Acquisition of Fuzzy Control Based Exercises of a Rings Gymnastic Robot," in *Proc. of IEEE Int. Conf. on Robotics and Automation*, 2001, pp. 2584–2589.

[9] T. Fukuda, F. Saito, and F. Arai, "A Study on the Brachiation Type of Mobile Robot (Heuristic Creation of Driving Input and Control Using CMAC)," in *Proc. of IEEE/RSJ Int. Workshop on Intelligent Robots and Systems*, vol. 2, 1991, pp. 478–483.

[10] K. Yoshida, M. Hirata, and H. Nagato, "Dynamics and Motion Control of a Somersault in Free-Style Skiing," *Trans. on the Japan Society of Mechanical Engineers*, C, vol. 57, no. 543, pp. 3558–3564, Nov. 1991. (in Japanese)

[11] S. Ohara, T. Imai, T. Shishikura, and S. Takashima, "Development of a Trampoline Gymnast Robot," in *Proc. of JSME Annual Conference on Robotics and Mechatronics*, vol. A, 1995, pp. 247–250. (in Japanese)

[12] S. Takashima, "Dynamic Modeling of a Gymnast on a High Bar," in *Proc. of IEEE/RSJ Int. Workshop on Intelligent Robots and Systems*, vol. 2, 1990, pp. 955–962.

[13] S. Takashima, "Control of Gymnast on a High Bar," in *Proc. of IEEE/RSJ Int. Workshop on Intelligent Robots and Systems*, vol. 3, 1991, pp. 1424–1429.

[14] M. W. Spong, "The Swing up Control Problem for the Acrobot," *IEEE Control Systems Magazine*, vol. 15, no. 1, pp. 49–55, Feb. 1995.

[15] S. Fukashiro, S. Sakurai, Y. Hirano, and M. Ae, Sports Biomechanics. Tokyo, Japan: Asakura-Shoten, 2000. (in Japanese)

[16] T. Yamada, K. Watanabe, K. Kiguchi, and K. Izumi, "Acquisition of Exercise Skill Represented by Fuzzy Control Rules for a Rings Gymnastic Robot," in *Proc. of IEEE Int. Symp. on Computational Intelligence in Robotics and Automation*, 2001, pp. 368–373.

[17] T. Yoshikawa, Foundations of Robotics: Analysis and Control. Cambridge, U.K.: MIT Press, 1990.

[18] J. J. Craig, *Introduction to Robotics: Mechanics and Control, 2nd ed. Reading*, MA: Addison-Wesley, 1989.

[19] K. Suzuki, N. Kawai, T. Miyamoto, H. Tsuchiya, and S. Kimura, "Mechanics of Kip Motion," *Trans. on the Japan Society of Mechanical Engineers*, C, vol. 62, no. 602, pp. 3979–3984, Oct. 1996. (in Japanese)

[20] Y. Shimouchi, Apparatus Exercises. Tokyo, Japan: Hitotsubashi-Shuppan Co. Ltd., 1998. (in Japanese)

[21] R. Nakamura and H. Saito, Basic Kinematics. Tokyo, Japan: Ishiyaku Publishers, Inc., 1976. (in Japanese)

In: Control and Learning in Robotic Systems
Editor: John X. Liu, pp. 87-121

ISBN 1-59454-356-9
© 2005 Nova Science Publishers, Inc.

Chapter 4

UNCERTAIN VARIABLES AND THEIR APPLICATIONS IN INTELLIGENT UNCERTAIN SYSTEMS

Z. Bubnicki[*]

Institute of Information Science and Engineering, Wroclaw University of Technology
Wyb. Wyspianskiego 27, 50-370 Wroclaw, POLAND,

Abstract

The chapter concerns a class of uncertain systems described by classical mathematical models or by relational knowledge representations with unknown parameters. In recent years the concept of so called uncertain variables has been introduced and developed as a tool for analysis and decision making problems in this class of uncertain knowledge-based systems. The formal descriptions and methods based on the uncertain variables may be used for the systems characterized by an expert presenting his/her knowledge on the unknown parameters and for the intelligent uncertain systems with learning algorithms.

The comprehensive description of the uncertain variables and their applications in different kinds of uncertain and intelligent systems with details and examples has been presented in the book: Z. Bubnicki, *Analysis and Decision Making in Uncertain Systems* (Springer Verlag, 2004). The purpose of this chapter is to describe shortly basic concepts and definitions concerning uncertain variables, and a brief review of new problems and results in this area. The definitions of uncertain logics and variables, and their applications to basic analysis and decision problems are presented in Secs 2, 3, 4. The Secs 5–10 are devoted to new problems and results, in particular, to stability and stabilization of uncertain systems and learning systems in which the learning process consists in *step by step* knowledge validation and updating. The chapter is completed with a list of other problems and practical applications (Sec. 11).

1 Introduction

Uncertainty is one of the main features of complex and intelligent decision making systems. Consequently, methods and algorithms of decision making based on an uncertain knowledge

---

[*]E-mail address: zdzislaw.bubnicki@pwr.wroc.pl, tel: +48 71 320 33 28, fax: +48 71 320 38 84

create now a large and intensively developing area in the field of knowledge-based decision support systems.

There exists a variety of definitions and formal models of uncertainties and uncertain systems, adequate for different types of the uncertainty and used for different problems concerning the systems. The idea of uncertain variables, introduced and developed in recent years, is specially oriented for analysis and decision problems in a class of uncertain systems described by traditional mathematical models and by relational knowledge representations (with number variables) which are treated as an extension of classical functional models [3-5, 9-11, 19]. The considerations are then directly related to respective problems and methods in traditional decision systems theory. The uncertain variable is described by a certainty distribution given by an expert and characterizing his/her knowledge concerning approximate values of the variable. The uncertain variables are related to random and fuzzy variables, but there are also essential differences discussed in the books cited above.

In the first part of this chapter (Secs 2-6), a short description of the uncertain variables, the basic analysis and decision problems and a generalization in the form of so called soft variables are presented. The second part (Secs 7-11) contains a brief review of the applications of the uncertain variables to selected analysis and decision problems in uncertain systems (including some practical applications). Details may be found in author's book *Analysis and Decision Making in Uncertain Systems* (Springer Verlag, 2004) and in other works cited in this text.

2   Uncertain Logics and Variables

Consider a universal set Ω, $\omega \in \Omega$, a set $\overline{X} \subset R^k$, a function $g : \Omega \to \overline{X}$, a crisp property (predicate) $P(\overline{x})$ and the crisp property $\Psi(\omega, P)$ generated by P and g: "For $\overline{x} = g(\omega) \stackrel{\Delta}{=} \overline{x}(\omega)$ assigned to ω the property P is satisfied". Let us introduce the property $G_\omega(x) =$ " $\overline{x}(\omega) \cong x$ " for $x \in X \subseteq \overline{X}$, which means: "$\overline{x}$ is approximately equal to x" or " x is the approximate value of $\overline{x}$". The properties P and G_ω generate the soft property $\overline{\Psi}(\omega, P)$ in Ω : "the approximate value of $\overline{x}(\omega)$ satisfies P ", i.e.

$$\overline{\Psi}(\omega, P) = "\overline{x}(\omega) \stackrel{\sim}{\in} D_x", \quad D_x = \{\overline{x} \in \overline{X} : P(\overline{x})\}, \tag{1}$$

which means: "$\overline{x}$ approximately belongs to D_x". Denote by $h_\omega(x)$ the logic value of $G_\omega(x)$:

$$w[\overline{x}(\omega) \cong x] \stackrel{\Delta}{=} h_\omega(x), \quad \bigwedge_{x \in X} h_\omega(x) \geq 0, \quad \max_{x \in X} h_\omega(x) = 1$$

Definition 1. The *uncertain logic L* is defined by Ω, $\overline{X}$, X, crisp predicates $P(\overline{x})$, the properties $G_\omega(x)$ and the corresponding functions $h_\omega(x)$ for $\omega \in \Omega$. In this logic we consider soft properties (1) generated by P and G_ω. The logic value of $\overline{\Psi}$ is

$$w[\overline{\Psi}(\omega,P)] \overset{\Delta}{=} v[\overline{\Psi}(\omega,P)] = \begin{cases} \max\limits_{x \in D_x} h_\omega(x) & \text{for } D_x \neq \varnothing \\ 0 & \text{for } D_x = \varnothing \end{cases}$$

and is called a *certainty index*. The operations are defined as follows:

$$v[\neg\overline{\Psi}(\omega,P)] = 1 - v[\overline{\Psi}(\omega,P)],$$

$$v[\Psi_1(\omega,P_1) \vee \Psi_2(\omega,P_2)] = \max\{v[\Psi_1(\omega,P_1)], v[\Psi_2(\omega,P_2)]\},$$

$$v[\Psi_1(\omega,P_1) \wedge \Psi_2(\omega,P_2)] = \begin{cases} 0 & \text{if for each } x\ w(P_1 \wedge P_2) = 0 \\ \min\{v[\Psi_1(\omega,P_1)], v[\Psi_2(\omega,P_2)]\} & \text{otherwise} \end{cases}$$

where Ψ_1 is $\overline{\Psi}$ or $\neg\overline{\Psi}$, and Ψ_2 is $\overline{\Psi}$ or $\neg\overline{\Psi}$ □

For the logic L one can prove the following statements:

$$v[\overline{\Psi}(\omega,P_1 \vee P_2)] = v[\overline{\Psi}(\omega,P_1)] \vee \overline{\Psi}(\omega,P_2)], \tag{2}$$

$$v[\overline{\Psi}(\omega,P_1 \wedge P_2)] \leq \min\{v[\overline{\Psi}(\omega,P_1)], v[\overline{\Psi}(\omega,P_2)]\}, \tag{3}$$

$$v[\overline{\Psi}(\omega,\neg P)] \geq v[\neg\overline{\Psi}(\omega,P)]. \tag{4}$$

The interpretation (semantics) of the uncertain logic L is the following: The uncertain logic operates with crisp predicates P, but for the given ω it is not possible to state if P is true or false because the function g and consequently the value $\overline{x}$ is unknown. The function $h_\omega(x)$ is given by an expert, who "looking at" ω obtains some information concerning $\overline{x}$ and uses it to evaluate his opinion that $\overline{x} \cong x$.

Definition 2 (the *uncertain logic C*). The first part is the same as in Def.1. The certainty index of $\overline{\Psi}$ and the operations are defined as follows:

$$v_c[\overline{\Psi}(\omega,P)] = \frac{1}{2}\{v[\overline{\Psi}(\omega,P)] + 1 - v[\overline{\Psi}(\omega,\neg P)]\},$$

$$\neg\overline{\Psi}(\omega,P) = \overline{\Psi}(\omega,\neg P), \tag{5}$$

$$\overline{\Psi}(\omega,P_1) \vee \overline{\Psi}(\omega,P_2) = \overline{\Psi}(\omega,P_1 \vee P_2),$$

$$\overline{\Psi}(\omega,P_1) \wedge \overline{\Psi}(\omega,P_2) = \overline{\Psi}(\omega,P_1 \wedge P_2) \qquad\qquad □$$

For the logic C one can prove the following statements:

$$v_c[\overline{\Psi}(\omega,P_1 \vee P_2)] \geq \max\{v_c[\overline{\Psi}(\omega,P_1)], v_c[\overline{\Psi}(\omega,P_2)]\}, \tag{6}$$

$$v_c[\overline{\Psi}(\omega, P_1 \wedge P_2)] \leq \min\{v_c[\overline{\Psi}(\omega, P_1)], v_c[\overline{\Psi}(\omega, P_2)]\},\qquad(7)$$

$$v_c[\neg \overline{\Psi}(\omega, P)] = 1 - v_c[\overline{\Psi}(\omega, P)].\qquad(8)$$

The variable $\overline{x}$ for a fixed ω will be called an uncertain variable. Two versions of uncertain variables will be defined by: $h(x)$ given by an expert and the definitions of certainty indexes $w(\overline{x} \,\widetilde{\in}\, D_x)$, $w(\overline{x} \,\widetilde{\notin}\, D_x)$, $w(\overline{x} \,\widetilde{\in}\, D_1 \vee \overline{x} \,\widetilde{\in}\, D_2)$, $w(\overline{x} \,\widetilde{\in}\, D_1 \wedge \overline{x} \,\widetilde{\in}\, D_2)$.

Definition 3. The *uncertain variable* $\overline{x}$ is defined by X, the function $h(x) = v(\overline{x} \cong x)$ given by an expert and the following definitions:

$$v(\overline{x} \,\widetilde{\in}\, D_x) = \max_{x \in D_x} h(x) \text{ for } D_x \neq \varnothing \text{ and } 0 \text{ for } D_x = \varnothing,$$

$$v(\overline{x} \,\widetilde{\notin}\, D_x) = 1 - v(\overline{x} \,\widetilde{\in}\, D_x),$$

$$v(\overline{x} \,\widetilde{\in}\, D_1 \vee \overline{x} \,\widetilde{\in}\, D_2) = \max\{v(\overline{x} \,\widetilde{\in}\, D_1), v(\overline{x} \,\widetilde{\in}\, D_2)\},$$

$$v(\overline{x} \,\widetilde{\in}\, D_1 \wedge \overline{x} \,\widetilde{\in}\, D_2) = \begin{cases} \min\{v(\overline{x} \,\widetilde{\in}\, D_1), v(\overline{x} \,\widetilde{\in}\, D_2)\} & \text{for } D_1 \cap D_2 \neq \varnothing \\ 0 & \text{for } D_1 \cap D_2 = \varnothing. \end{cases}$$

The function $h(x)$ will be called a *certainty distribution*   □

The definition of the uncertain variable is based on logic L. Then for (1) the properties (2), (3), (4) are satisfied. In particular, (4) becomes $v(\overline{x} \,\widetilde{\in}\, \overline{D}_x) \geq v(\overline{x} \,\widetilde{\notin}\, D_x) = 1 - v(\overline{x} \,\widetilde{\in}\, D_x)$ where $\overline{D}_x = X - D_x$.

Definition 4. *C-uncertain variable* is defined by X, $h(x) = v(\overline{x} \cong x)$ given by an expert and the following definitions:

$$v_c(\overline{x} \,\widetilde{\in}\, D_x) = \frac{1}{2}[\max_{x \in D_x} h(x) + 1 - \max_{x \in X - D_x} h(x)],\qquad(9)$$

$$v_c(\overline{x} \,\widetilde{\notin}\, D_x) = 1 - v_c(\overline{x} \,\widetilde{\in}\, D_x),$$
$$v_c(\overline{x} \,\widetilde{\in}\, D_1 \vee \overline{x} \,\widetilde{\in}\, D_2) = v_c(\overline{x} \,\widetilde{\in}\, D_1 \cup D_2),$$
$$v_c(\overline{x} \,\widetilde{\in}\, D_1 \wedge \overline{x} \,\widetilde{\in}\, D_2) = v_c(\overline{x} \,\widetilde{\in}\, D_1 \cap D_2)\qquad□$$

The definition of C-uncertain variable is based on logic C. Then for (1) the properties (6), (7), (8) are satisfied. According to (5) and (8) $v_c(\overline{x} \,\widetilde{\notin}\, D_x) = v_c(\overline{x} \,\widetilde{\in}\, X - D_x)$. The function

$v_c(\bar{x} \cong x) \stackrel{\Delta}{=} h_c(x)$ may be called *C-certainty distribution*. In the case of *C*-uncertain variable the expert's knowledge is used in a better way.

In a *continuous case* $h(x)$ is a continuous function in X and in a *discrete case* $X = \{x_1, x_2, ..., x_m\}$. The mean value of $\bar{x}$ is defined as follows:

$$M(\bar{x}) = \int_X xh(x)dx \cdot [\int_X h(x)dx]^{-1} \quad \text{or} \quad M(\bar{x}) = \sum_{i=1}^m x_i h(x_i)[\sum_{i=1}^m h(x_i)]^{-1}$$

in a continuous case or a discrete case, respectively. For a pair of variables $(\bar{x}, \bar{y})$ described by $h(x, y) = v[(\bar{x}, \bar{y}) \cong (x, y)]$ we may consider marginal and conditional distributions with the following relationships:

$$h_x(x) = v(\bar{x} \cong x) = \max_{y \in Y} h(x, y), \quad h_y(y) = v(\bar{y} \cong y) = \max_{x \in X} h(x, y),$$

$$h(x, y) = \min\{h_x(x), h_y(y \mid x)\} = \min\{h_y(y), h_x(x \mid y)\}$$

where $h_y(y \mid x) = v[\bar{x} = x \rightarrow \bar{y} \cong y]$, $h_x(x \mid y) = v[\bar{y} = y \rightarrow \bar{x} \cong x]$.

3   Analysis Problem

Let us consider an uncertain plant described by a relation $R(u, y; x) \subset U \times Y$ (*a relational knowledge representation*) where $u \in U$, $y \in Y$ are the input and output vectors, respectively, and $x \in X$ is an unknown vector parameter, which is assumed to be a value of an uncertain variable $\bar{x}$ described by the certainty distribution $h_x(x)$ given by an expert. The given value u determines the set of all possible outputs

$$D_y(u; x) = \{y \in Y : (u, y) \in R(u, y; x)\} . \tag{10}$$

The analysis may consist in evaluating the output with respect to a set $D_y \subset Y$ given by a user. One of the possible formulations of the **analysis problem** is as follows: For the given $R(u, y; x)$, $h_x(x)$, u and $D_y \subset Y$ one should determine

$$v[D_y \cong D_y(u; \bar{x})] \stackrel{\Delta}{=} g(D_y, u) . \tag{11}$$

The value (11) denotes the certainty index of the soft property: "the set of all possible outputs approximately contains the set D_y given by a user" or "the approximate value of $\bar{x}$ is such that $D_y \subseteq D_y(u;x)$" or "the approximate set of the possible outputs contains all values from the set D_y". Let us note that

$$v[D_y \tilde{\subseteq} D_y(u;\bar{x})] = v[\bar{x} \tilde{\in} D_x(D_y,u)]$$

where

$$D_x(D_y,u) = \{x \in X : \ D_y \subseteq D_y(u;x)\} .$$

Then

$$g(D_y,u) = \max_{x \in D_x(D_y,u)} h_x(x) .$$

When $\bar{x}$ is considered as a C-uncertain variable, it is necessary to determine

$$v[\bar{x} \tilde{\in} \overline{D}_x(D_y,u)] = \max_{x \in \overline{D}_x(D_y,u)} h_x(x)$$

where $\overline{D}_x(D_y,u) = X - D_x(D_y,u)$. Then, according to (9)

$$v_c[D_y \tilde{\subseteq} D_y(u;\bar{x})] = \frac{1}{2}\{v[\bar{x} \tilde{\in} D_x(D_y,u)] + 1 - v[\bar{x} \tilde{\in} \overline{D}_x(D_y,u)]\} .$$

The considerations may be extended for a plant described by a relation $R(u,y,z;x)$ where $z \in Z$ is the vector of disturbances which may be measured. For the given z

$$D_y(u,z;x) = \{y \in Y : \ (u,y,z) \in R(u,y,z;x)\}$$

and

$$v[D_y \tilde{\subseteq} D_y(u,z;\bar{x})] = \max_{x \in D_x(D_y,u,z)} h_x(x)$$

where

$$D_x(D_y,u,z) = \{x \in X : \ D_y \subseteq D_y(u,z;x)\} .$$

Consequently, the certainty index that the approximate set of the possible outputs contains all the values from the set D_y depends on z.

4   Parametric Decision Problem

Let us consider an uncertain plant described by a relation $R(u, y, z; x) \subset U \times Y \times Z$ where $x \in X$ is an unknown vector parameter, which is assumed to be a value of an uncertain variable described by $h_x(x)$ given by an expert. For the property $y \in D_y \subset Y$ required by a user, we can formulate the following **decision problem**: For the given $R, z, h_x(x)$ and D_y one should find the decision u^* maximizing the certainty index of the property: "the set of all possible outputs approximately belongs to D_y" or "the set of all possible outputs belongs to D_y for an approximate value of $\bar{x}$". Then

$$u^* = \arg\max_{u \in U} v[D_y(u, z; \bar{x}) \stackrel{\sim}{\subseteq} D_y] = \arg\max_{u \in U} \max_{x \in D_x(u,z)} h_x(x) \tag{12}$$

where $D_y(u, z; x) = \{y \in Y : (u, y, z) \in R\}$ and $D_x(u, z) = \{x \in X : D_y(u, z; x) \subseteq D_y\}$. If u^* is the unique result of the maximization then $u^* = \Psi(z)$ denotes a deterministic decision algorithm in the open-loop decision system. If $\bar{x}$ is considered as C-uncertain variable, then one should determine u_c^* maximizing $v_c[\bar{x} \stackrel{\sim}{\in} D_x(u, z)]$. In this case the calculations are more complicated.

Example 1.

Let $u, y, x \geq 0$ (onedimensional variables), the relation R is given by the inequality $xu \leq y \leq 2xu$ (there are no disturbances), $D_y = [y_1, y_2]$, $y_2 > 2y_1$. Assume that x is a value of an uncertain variable $\bar{x}$ with triangular certainty distribution presented in Fig.1.

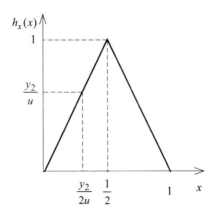

Figure 1. Example of certainty distribution

In our case $D_x(u) = [\dfrac{y_1}{u}, \dfrac{y_2}{2u}]$. According to (12) it is easy to obtain

$$v[D_y(\bar{x}) \tilde{\subseteq} D_y] = \begin{cases} \dfrac{y_2}{u} & \text{when} & y_2 \le u \\ 1 & \text{when} & 2y_1 \le u \text{ and } y_2 \ge u \\ 2(1 - \dfrac{y_1}{u}) & \text{when} & 2y_1 \ge u \text{ and } y_2 \le u \\ 0 & \text{when} & y_1 \ge u. \end{cases}$$

Thus, u^* is any value from $[2y_1, y_2]$ and in (12) $v(u^*) = 1$. In the case of a C-uncertain variable, using (9), we obtain

$$v_c(u) = \begin{cases} \dfrac{y_2}{2u} & \text{when} & u \ge y_1 + 0.5y_2 \\ 1 - \dfrac{y_1}{u} & \text{when} & y_1 \le u \le y_1 + 0.5y_2 \\ 0 & \text{when} & u \le y_1. \end{cases}$$

It is easy to see that $u_c^* = y_1 + 0.5y_2$ and $v_c(u_c^*) = \dfrac{y_2}{2y_1 + y_2}$. For example, for $y_1 = 2$, $y_2 = 12$ we obtain $u^* \in [4, 12]$ and $v = 1$, $u_c^* = 8$ and $v_c = 0.75$. The function $v_c(u)$ is illustrated in Fig. 2.

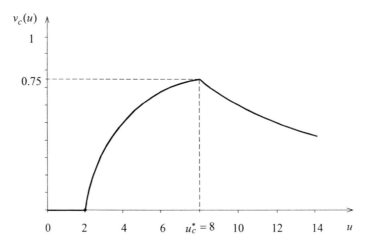

Figure 2. Example of the relationship between v_c and u

5 Nonparametric Decision Problem

The uncertain variables may be applied to a nonparametric description of the uncertain plant. Let us assume that (u, y, z) are values of uncertain variables $(\bar{u}, \bar{y}, \bar{z})$, and the knowledge of the plant (KP) has the form of the conditional certainty distribution $h_y(y \mid u, z)$ given by an expert. The decision problem may consist in finding a conditional distribution $h_u(u \mid z)$ for the given certainty distribution $h_y(y)$ required by a user. The **nonparametric decision problem** is then the following: For the given $h_y(y \mid u, z)$ and $h_y(y)$ one should determine $h_u(u \mid z)$. The determination of $h_u(u \mid z)$ may be decomposed into two steps. In the first step, one should find the distribution $h_{uz}(u, z)$ satisfying the equation

$$h_y(y) = \max_{u \in U, z \in Z} \min\{h_{uz}(u, z), h_y(y \mid u, z)\}, \tag{13}$$

and in the second step, one should determine the distribution $h_u(u \mid z)$ satisfying the equation

$$h_{uz}(u, z) = \min\{h_z(z), h_u(u \mid z)\} \tag{14}$$

where

$$h_z(z) = \max_{u \in U} h_{uz}(u, z). \tag{15}$$

The distribution $h_u(u \mid z)$ may be considered as a knowledge of the decision making (KD) or an *uncertain decision algorithm*. The deterministic decision algorithm Ψ may be obtained via a *determinization* of the uncertain algorithm, which may consist in using the mean value: $u_d = M(\bar{u} \mid z) = \Psi(z)$. It is worth noting that the analogous problem may be considered for the nonparametric descriptions of the uncertainty using random or fuzzy variables. If (u, y, z) are values of continuous random variables $(\tilde{u}, \tilde{y}, \tilde{z})$ then KP has the form of a conditional probability density $f_y(y \mid u, z)$ and the formulas analogous to (13), (14), (15) are as follows:

$$f_y(y) = \int_U \int_Z f_{uz}(u, z) f_y(y \mid u, z) du dz, \tag{16}$$

$$f_u(u \mid z) = f_{uz}(u, z) \cdot [\int_U f_{uz}(u, z) du]^{-1}. \tag{17}$$

In this case, the *random decision algorithm* (KD) has the form of $f_u(u \mid z)$ and the determinization may consist in using the expected value: $u_d = E(\bar{u} \mid z) = \Psi(z)$. If (u, y, z) are values of fuzzy variables $(\hat{u}, \hat{y}, \hat{z})$, the nonparametric description of the uncertainty is formulated by introducing three soft properties concerning u, y, z: $\varphi_u(u)$, $\varphi_y(y)$, $\varphi_z(z)$. Then KP given by an expert has the form of a conditional membership function

$$w[\varphi_u \wedge \varphi_z \to \varphi_y] \overset{\Delta}{=} \mu_y(y \mid u, z)$$

where $w \in [0,1]$ denotes the logic value. The formulas analogous to (13), (14), (15) are now as follows:

$$\mu_y(y) = \max_{u \in U, z \in Z} \min\{\mu_{uz}(u, z), \mu_y(y \mid u, z)\},$$

$$\mu_{uz}(u, z) = \min\{\mu_z(z), \mu_u(u \mid z)\},$$

$$\mu_z(z) = \max_{u \in U} \mu_{uz}(u, z)$$

where $\mu_y(y) = w[\varphi_y]$, $\mu_z(z) = w[\varphi_z]$, $\mu_{uz}(u, z) = w[\varphi_u \wedge \varphi_z]$, $\mu_u(u \mid z) = w[\varphi_z \to \varphi_u]$. In this case, the *fuzzy decision algorithm* (KD) has the form of $\mu_u(u \mid z)$ and the deterministic decision algorithm $u_d = \Psi(z)$ may be obtained via the determinization (defuzzification) using the mean value. The formulas for the fuzzy variables, similar to those for the uncertain variables, have different practical interpretations caused by the fact that they do not concern directly the *values* of the variables but are concerned with the *properties* φ. The differences between the descriptions using uncertain and fuzzy variables are discussed in [19].

6   Generalization. Soft Variables

The uncertain, random and fuzzy variables may be considered as special cases of a more general description of the uncertainty in the form of *soft variables* and *evaluating functions* [19], which may be introduced as a tool for a unification and generalization of nonparametric analysis and decision problems based on the uncertain knowledge representation. The definition of a soft variable should be completed with the determination of relationships for the pair of soft variables.

Definition 5 (*soft variable and the pair of soft variables*). A soft variable $\overset{\vee}{x} = \langle X, g(x) \rangle$ is defined by the set of values X (a real number vector space) and a bounded evaluating function $g : X \to R^+$, satisfying the following condition:

$$\int_X x g(x) < \infty$$

for the continuous case and

$$\sum_{i=1}^{\infty} x_i g(x_i) < \infty$$

for the discrete case.

Let us consider two soft variables $\overset{\vee}{x} = < X, g_x(x) >$, $\overset{\vee}{y} = < Y, g_y(y) >$ and the variable $(\overset{\vee}{x}, \overset{\vee}{y})$ described by $g_{xy}(x, y) : X \times Y \to R^+$. Denote by $g_y(y \mid x)$ the evaluating function of $\overset{\vee}{y}$ for the given value x (the conditional evaluating function). The pair $(\overset{\vee}{x}, \overset{\vee}{y})$ is defined by $g_{xy}(x, y)$ and two operations:

$$g_{xy}(x, y) = O_1[g_x(x), g_y(y \mid x)], \tag{18}$$

$$g_x(x) = O_2[g_{xy}(x, y)], \tag{19}$$

i.e.

$$O_1 : D_{gx} \times D_{gy} \to D_{g,xy}, \quad O_2 : D_{g,xy} \to D_{g,x}$$

where D_{gx}, $D_{gy}(x)$ and $D_{g,xy}$ are sets of the functions $g_x(x)$, $g_y(y \mid x)$ and $g_{xy}(x, y)$, respectively. The mean value $M(\overset{\vee}{x})$ is defined in the same way as for an uncertain variable.

□

The evaluating function may have different practical interpretations. In the random case, a soft variable is a random variable described by the probability density $g(x) = f(x)$ or by probabilities $g(x_i) = P(\tilde{x} = x_i)$. In the case of an uncertain variable, $g(x) = h(x)$ is the certainty distribution. In the case of the fuzzy description, a soft variable is a fuzzy variable described by the membership function $g(x) = \mu(x) = w[\varphi(x)]$ where w denotes a logic value of a given soft property $\varphi(x)$. In general, we can say that $g(x)$ describes an evaluation of the set of possible values X, characterizing for every value x its significance (importance or weight).

The nonparametric problems considered for random, uncertain and fuzzy variables may be generalized by using soft variables. For the plant with input $u \in U$ and output $y \in Y$ we assume that (u, y) are values of soft variables $(\overset{\vee}{u}, \overset{\vee}{y})$ and the knowledge of the plant has the form of a conditional evaluating function

$$KP = < g_y(y \mid u) > .$$

According to (18) and (19)

$$g_y(y) = O_2\{O_1[g_u(u), g_y(y \mid u)]\}. \tag{20}$$

In the analysis problem one should determine $g_y(y)$ for the given KP and $g_u(u)$, and the decision problem consists in finding $g_u(u)$ for the required evaluating function $g_y(y)$. To find the solution one should solve equation (20) with respect to the function $g_u(u)$.

For the plant with $u \in U$, $y \in Y$ and the external disturbance $z \in Z$, the knowledge of the plant has the form of a conditional evaluating function

$$KP = \, <g_y(y|u,z)> \, .$$

Analysis problem: For the given $KP = \, <g_y(y|u,z)> \,$, $g_u(u|z)$ and $g_z(z)$ find the evaluating function $g_y(y)$.

According to (18) and (19)

$$g_y(y) = O_2\{O_1[\, O_1(g_z(z), g_u(u|z)), g_y(y|u,z)]\,\}. \qquad (21)$$

Decision problem: For the given $KP = \, <g_y(y|u,z)> \,$ and $g_y(y)$ required by a user one should determine $g_u(u|z)$.

The determination of $g_u(u|z)$ may be decomposed into two steps. In the first step, one should find the evaluating function $g_{uz}(u,z)$ satisfying the equation

$$g_y(y) = O_2\{O_1[\, g_{uz}(u,z), g_y(y|u,z)]\,\}.$$

In the second step, one should determine the function $g_u(u|z)$ satisfying the equation

$$g_{uz}(u,z) = O_1[\, g_z(z), g_u(u|z)]$$

where

$$g_z(z) = O_2[\, g_{uz}(u,z)\,] \ .$$

The function $g_u(u|z)$ may be called a knowledge of the decision making $KD = \, <g_u(u|z)> \,$ or a *soft decision algorithm* (the description of a *soft controller* in the open-loop control system). Having $g_u(u|z)$ one can obtain the deterministic decision algorithm as a result of the determinization of the soft decision algorithm.

The general approach to the nonparametric decision problem, common for the different descriptions presented above, is illustrated in Fig. 3.

In general, for a variable $\overset{\vee}{x}$ it is possible to formulate an uncertain (imprecise) knowledge on its value $x \in X$ in two ways: by introducing monotonic nonadditive measure $m(D_x)$

where the value m evaluates the property $x \in D_x$ for the given $D_x \subset X$, or directly by introducing a distribution $g(x)$ where the value g evaluates the given value x. In the second description of the uncertainty (used in the definition of the soft variable and sufficient for nonparametric problems), it is not necessary to define $m(D_x)$. It is worth noting that for the uncertain variables $v(\overline{x} \; \tilde{\in} \; D_x) \stackrel{\Delta}{=} m(D_x)$ is a possibility measure with a specific interpretation, and $v_c(\overline{x} \; \tilde{\in} \; D_x) \stackrel{\Delta}{=} m_c(D_x)$ is neither a belief nor a plausibility measure.

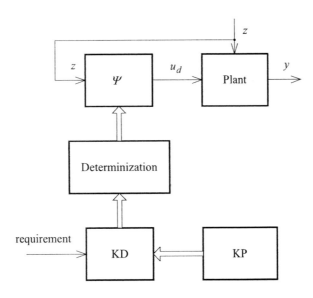

Figure 3. Open-loop knowledge-based decision system

7   Stability of Dynamical System with Uncertain Parameters

The uncertain variables may be used in the investigation of the stability of a dynamical system containing unknown parameter x which is assumed to be a value of an uncertain variable described by the certainty distribution $h(x)$. Consider a nonlinear and/or time-varying system with one equilibrium state equal to $\overline{0}$ (a vector with zero components), described by

$$s_{n+1} = A(s_n, c_n, x)s_n \qquad (22)$$

where $s_n \in S \in R^k$ is the state vector, $c_n \in C$ is the vector of time-varying parameters, and the uncertainty concerning c_n is formulated as follows

$$\bigwedge_{n \geq 0} c_n \in D_c \qquad (23)$$

where D_c is a given set in C. The system (22) is globally asymptotically stable (GAS) iff s_n converges to $\overline{0}$ for any s_0. For the fixed x, the uncertain system (22), (23) is GAS iff the system (22) is GAS for every sequence c_n satisfying (23). Let $W(x)$ and $P(x)$ denote properties concerning x such that $W(x)$ is a sufficient condition and $P(x)$ is a necessary condition of the global asymptotic stability for the system (22), (23). Then the certainty index v_s that the system (22), (23) is GAS may be estimated by the inequality $v_w \le v_s \le v_p$ where

$$v_w = \max_{x \in D_{xw}} h(x), \quad v_p = \max_{x \in D_{xp}} h(x),$$

$$D_{xw} = \{x \in X : W(x)\}, \quad D_{xp} = \{x \in X : P(x)\},$$

v_w is the certainty index that the sufficient condition is approximately satisfied, and v_p is the certainty index that the necessary condition is approximately satisfied. In general, $D_{xw} \subseteq D_{xp}$ and $D_{xp} - D_{xw}$ may be called "a grey zone" which is a result of an additional uncertainty caused by the fact that $W(x) \ne P(x)$. In particular, if it is possible to determine a sufficient and necessary condition $W(x) = P(x)$ then $v_w = v_p$ and the value v_s may be determined exactly. The condition $P(x)$ may be determined as a negation of a sufficient condition that the system is not GAS, i.e. such a property $P_{neg}(x)$ that

$$P_{neg}(x) \to \text{ there exists } c_n \text{ satisfying (23) such that (22) is not GAS.}$$

For the nonlinear and time-varying system under consideration we may use the following stability conditions presented in [1, 6, 14].

Theorem 1. The system (22), (23) where

$$D_c = \{c \in C : \bigwedge_{s \in S} [\underline{A}(x) \le A(s,c,x) \le \overline{A}(x)]\} \tag{24}$$

is GAS if all entries of the matrices $\underline{A}(x)$ and $\overline{A}(x)$ are nonnegative and

$$\| \overline{A}(x) \| < 1 \tag{25}$$

where

$$\| A \| = \max_{1 \le j \le k} \sum_{i=1}^{k} | aij | \qquad \square$$

The inequality in (24) denotes the inequalities for the entries:

$$\underline{a}_{ij}(x) \le a_{ij}(s,c,x) \le \overline{a}_{ij}(x).$$

Theorem 2. Assume that all entries of the matrix $\underline{A}(x)$ are nonnegative. If the system (22), (24) is GAS then

$$\bigvee_{j} \sum_{i=1}^{k} a_{ij}(x) < 1 \quad \square \tag{26}$$

From Theorems 1 and 2 we obtain

$$D_{xw} = \{x \in X : \bigwedge_{j} \sum_{i=1}^{k} \overline{a}_{ij}(x) < 1\}, \tag{27}$$

$$D_{xp} = X - D_{x,neg}$$

where

$$D_{x,neg} = \{x \in X : \bigwedge_{j} \sum_{i=1}^{k} \underline{a}_{ij}(x) \ge 1\}. \tag{28}$$

Example 2.
Consider an uncertain system (22) where $k = 2$ and

$$A(s_n, c_n, x) = \begin{bmatrix} a_{11}(s_n, c_n) + x & a_{12}(s_n, c_n) \\ a_{21}(s_n, c_n) & a_{22}(s_n, c_n) + x \end{bmatrix}$$

with the uncertainty (24), i.e. nonlinearities and the sequence c_n are such that

$$\bigwedge_{c \in D_c} \bigwedge_{s \in D_s} \underline{a}_{ij} \le a_{ij}(s,c) \le \overline{a}_{ij}, \quad i = 1,2 ; \quad j = 1,2 .$$

Assume that $x \ge 0$ and $\underline{a}_{ij} \ge 0$. Applying the condition (25) yields

$$\overline{a}_{11} + x + \overline{a}_{21} < 1, \quad \overline{a}_{12} + \overline{a}_{22} + x < 1$$

and D_{xw} in (27) is defined by: $x < 1 - \max(\overline{a}_{11} + \overline{a}_{21}, \overline{a}_{12} + \overline{a}_{22})$. Applying the negation of the condition (26) yields

$$\underline{a}_{11} + x + \underline{a}_{21} \ge 1, \quad \underline{a}_{12} + \underline{a}_{22} + x \ge 1 .$$

Then $D_{x,neg}$ in (28) is determined by

$$x \geq 1 - \min(\underline{a}_{11} + \underline{a}_{21}, \ \underline{a}_{12} + \underline{a}_{22})$$

and the necessary condition (26) defining the set $D_{xp} = X - D_{x,neg}$ is as follows

$$x < 1 - \min(\underline{a}_{11} + \underline{a}_{21}, \ \underline{a}_{12} + \underline{a}_{22}).$$

For the given certainty distribution $h_x(x)$ we can determine

$$v_w = \max_{0 \leq x \leq x_w} h(x), \quad v_p = \max_{0 \leq x \leq x_p} h(x) \qquad (29)$$

where

$$x_w = 1 - \max(\overline{a}_{11} + \overline{a}_{21}, \ \overline{a}_{12} + \overline{a}_{22}), \quad x_p = 1 - \min(\underline{a}_{11} + \underline{a}_{21}, \ \underline{a}_{12} + \underline{a}_{22}).$$

Assume that $h(x)$ has parabolic form

$$h_x(x) = -e^{-2}(x-d)^2 + 1 \quad \text{for} \quad d-e \leq x \leq d+e$$

and

$$h_x(x) = 0 \quad \text{otherwise,} \ 0 < e < d.$$

The results obtained from (29) for the different cases are as follows.

1. For $x_w \geq d$, $v_w = v_p = 1$.

2. For $d - e \leq x_w \leq d$

$$v_w = -e^{-2}(x_w - d)^2 + 1$$

$$v_p = \begin{cases} 1 & \text{for} \quad x_p \geq d \\ -e^{-2}(x_p - d)^2 + 1 & \text{for} \quad x_p \leq d. \end{cases}$$

3. For $x_w \leq d - e$, $v_w = 0$,

$$v_p = \begin{cases} 1 & \text{for} & x_p \ge d \\ -e^{-2}(x_p - d)^2 + 1 & \text{for} & d - e \le x_p \le d \\ 0 & \text{for} & x_p \le d - e. \end{cases}$$

For example, if $d - e \le x_w \le d$ and $x_p \le d$ then the certainty index v_s that the system is globally asymptotically stable satisfies the inequality

$$-e^{-2}(x_w - d)^2 + 1 \le v_s \le -e^{-2}(x_p - d)^2 + 1.$$

For the numerical data $\bar{a}_{11} = 0.2$, $\underline{a}_{11} = 0.1$, $\bar{a}_{21} = 0.3$, $\underline{a}_{21} = 0.2$, $\bar{a}_{12} = 0.2$, $\underline{a}_{12} = 0.1$, $\bar{a}_{22} = 0.2$, $\underline{a}_{22} = 0.1$, $d = 0.9$, $e^{-2} = 3$, we obtain $v_w = 0.52$, $v_p = 0.97$ and $0.52 \le v_s \le 0.97$. $\qquad\qquad\qquad\qquad\qquad\qquad\qquad\qquad\qquad\qquad\qquad\blacksquare$

The considerations may be extended to C-uncertain variables with v_c instead of v, and to a stabilization problem. Consider a system described by

$$s_{n+1} = A(s_n, c_n, x, e) s_n \qquad\qquad (30)$$

where $e \in E$ is a vector of parameters which may be chosen by a designer. Assume that x is a value of C-uncertain variable $\bar{x}$ described by the certainty distribution $h(x)$ given by an expert, and denote by $W(x,e)$, $P(x,e)$ the sufficient and necessary stability conditions, respectively. Then the certainty index $v_s(e)$ that the system (30), (23) is GAS may be estimated by the inequality $v_w(e) \le v_s(e) \le v_p(e)$. According to (9)

$$v_{cw}(e) = \frac{1}{2}[\max_{D_{xw}} h(x) + 1 - \max_{X - D_{xw}} h(x)], \qquad\qquad (31)$$

$$v_{cp}(e) = \frac{1}{2}[\max_{D_{xp}} h(x) + 1 - \max_{X - D_{xp}} h(x)] \qquad\qquad (32)$$

where $D_{xw} = \{x \in X : M(x,e)\}$, $D_{xp} = \{x \in X : G(x,e)\}$.

The stabilization consists here in a proper choosing of the stabilizing parameter e by a designer who in this way may have an influence on the values $v_{cw}(e)$ and $v_{cp}(e)$. Let us introduce the index of the grey zone $\delta(e) = v_{cp}(e) - v_{cw}(e)$ and take into account that usually there is a constraint $e \in D_e \subset E$ where D_e is determined by a requirement concerning a quality of the system. The stabilization problem may be formulated in the following ways:

1   Choose e maximizing $v_{cw}(e)$ subject to the constraint $e \in D_e$.

2   Choose e maximizing $v_{cp}(e)$ subject to the constraint $e \in D_e$.

3   Choose e maximizing $v_{cp}(e)$ subject to the constraints $e \in D_e$ and $v_{cw}(e) \geq \bar{v}$ where $0 < \bar{v} < 1$ is given.

4   Choose e maximizing $v_{cp}(e)$ subject to the constraints $e \in D_e$ and $\delta(e) \leq \bar{\delta}$ where $0 < \bar{\delta} < 1$ is given.

In the cases 3. and 4. the grey zone is included into the optimization problem in two different ways. Let us consider a special case where x and e are one-dimensional positive parameters and the conditions $W(x,e)$, $P(x,e)$ are reduced to inequalities $xe \leq b_w$, $xe \leq b_p$, respectively ($b_p \geq b_w$). In a typical case x denotes an unknown amplification factor of a control plant and e denotes an amplification factor of a controller in a closed-loop control system. Assume that $h(x) = 0$ for $x \leq \alpha$ or $x \geq \beta$ ($\alpha, \beta > 0$), $h(x) = 1$ for $x = z$ and $h(x)$ is an increasing (a decreasing) function for $\alpha \leq x \leq z$ ($z \leq x \leq \beta$). It is easy to show that (31) is then reduced to

$$
v_{cw}(e) = \begin{cases}
0 & \text{for} & e \geq \dfrac{b_w}{\alpha} \\[2mm]
\dfrac{1}{2}h(\dfrac{b_w}{e}) & \text{for} & \dfrac{b_w}{z} \leq e \leq \dfrac{b_w}{\alpha} \\[2mm]
1 - \dfrac{1}{2}h(\dfrac{b_w}{e}) & \text{for} & e \leq \dfrac{b_w}{z}.
\end{cases}
\tag{33}
$$

The function $v_{cp}(e)$ has an analogous form with b_p instead of bw. Introduce the constraint $e \geq \bar{e}$ and denote the solutions of the problems 1. and 2. (maximization of v_{cw} and v_{cp}) by e_w^* and e_p^*, respectively.

Theorem 3. Under the assumptions introduced above:

a. For $\dfrac{b_w}{\beta} \leq \bar{e} \leq \dfrac{b_p}{\beta}$, $e_w^* = \bar{e}$, e_p^* is any value from $[\bar{e}, \dfrac{b_p}{\beta}]$.

b. For $\dfrac{b_p}{\beta} \leq \bar{e} \leq \dfrac{b_w}{\alpha}$, $e_w^* = e_p^* = \bar{e}$.

c. For $\dfrac{b_w}{\alpha} \leq \bar{e} \leq \dfrac{b_p}{\alpha}$ e_w^* is any value from $[\dfrac{b_w}{\alpha}, \bar{e}]$, $e_p^* = \bar{e}$.   □

Proof. According to the assumptions concerning $h(x)$, it follows from (33) and from the analogous formula for $v_{cp}(e)$ that $v_{cw}(e)$ is a decreasing function of e for $\dfrac{b_w}{\beta} \leq e \leq \dfrac{b_w}{\alpha}$ and

$v_{cp}(e)$ is a decreasing function of e for $\dfrac{b_p}{\beta} \le e \le \dfrac{b_p}{\alpha}$. Then for $\dfrac{b_p}{\beta} \le e \le \dfrac{b_w}{\alpha}$ the both functions

are decreasing what proves the case b. of the Theorem. The cases a. and c. follow from the

fact that $v_{cp}(e) = 1$ for $\dfrac{b_w}{\beta} \le e \le \dfrac{b_p}{\beta}$ and $v_{cw}(e) = 0$ for $\dfrac{b_w}{\alpha} \le e \le \dfrac{b_p}{\alpha}$.   □

Example 3.

Consider the system (22) where $k = 2$,

$$A(s_n, c_n, x, e) = \begin{bmatrix} a_{11}(s_n, c_n) + xe & a_{12}(s_n, c_n) \\ a_{21}(s_n, c_n) & a_{22}(s_n, c_n) + xe \end{bmatrix}$$

with the uncertainty the same as in Example 2, i.e. nonlinearities and the sequence c_n are

such that $\forall c \in D_c$, $\forall s \in S$ $\underline{a}_{ij} \le a_{ij}(s, c) \le \overline{a}_{ij}$ and $\underline{a}_{ij} \ge 0$. Applying the results in Example 2

yields $b_w = 1 - \max(\overline{a}_{11} + \overline{a}_{21}, \overline{a}_{12} + \overline{a}_{22})$ and $b_p = 1 - \min(\underline{a}_{11} + \underline{a}_{21}, \underline{a}_{12} + \underline{a}_{22})$. Assume that x is a

value of C-uncertain variable $\overline{x}$ described by a triangular $h(x)$ presented in Fig. 4. In this

case, according to (33)

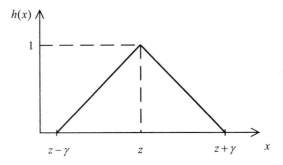

Figure 4. Certainty distribution

$$v_{cw}(e) = \begin{cases} 0 & \text{for} & e \ge \dfrac{b_w}{z - \gamma} \\ \dfrac{b_w}{2\gamma e} - \dfrac{z - \gamma}{2\gamma} & \text{for} & \dfrac{b_w}{z + \gamma} \le e \le \dfrac{b_w}{z - \gamma} \\ 1 & \text{for} & e \le \dfrac{b_w}{z + \gamma}, \end{cases} \tag{34}$$

and $v_{cp}(e)$ has the same form with b_p in the place of b_w. The functions $v_{cw}(e)$, $v_{cp}(e)$ and

$\delta(e)$ are illustrated in Fig. 5, for $b_w = 0.2$, $b_p = 0.5$, $z = 0.4$.

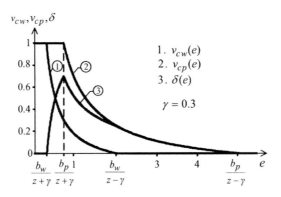

Figure 5. Illustration of the results

The solutions of the problems 1. and 2. are such as in Theorem 3, with $\alpha = z - \gamma$, $\beta = z + \gamma$. In case a. $v_{cp} = 1$, $v_{cw} = v_{cw}(\bar{e})$. In case b. $v_{cp} = v_{cp}(\bar{e})$, $v_{cw} = v_{cw}(\bar{e})$ and according to (34) $\delta = (b_p - b_w)(2\gamma \bar{e})^{-1}$. In case c. $v_{cp} = v_{cp}(\bar{e})$, $v_{cw} = 0$. For the numerical data presented above, choosing $e_w^* = e_p^* = \bar{e}$, we obtain the following estimation of the certainty index v_s that the system is GAS: $0.17 \leq v_s \leq 0.67$ for $\bar{e} = 1$, $0.25 \leq v_s \leq 0.86$ for $\bar{e} = 0.8$, $0.66 \leq v_s \leq 1$ for $\bar{e} = 0.4$. The solutions of the problems 3. and 4. are the same as in the problems 1. and 2., under the conditions $\bar{v} \leq v_{cp}(e)$ and $\delta(\bar{e}) \leq \bar{\delta}$, respectively.

8 Allocation Problems and Project Management

The uncertain variables may be applied to allocation problems consisting in the proper task or resource distribution in a complex of operations described by a relational knowledge representation with unknown parameters [19]. The parts of the complex may denote manufacturing operations [12], computational operations in a computer system [18] or operations in a project to be managed [15]. Let us consider a complex of k parallel operations described by a set of inequalities

$$T_i \leq \varphi_i(u_i, x_i), \quad i = 1, 2, ..., k \tag{35}$$

where T_i is the execution time of the i-th operation, u_i is the size of a task in the problem of task allocation or the amount of a resource in the problem of resource allocation, an unknown parameter $x_i \in R^1$ is a value of an uncertain variable $\bar{x}_i$ described by a certainty distribution $h_i(x_i)$ given by an expert, and $\bar{x}_1, ..., \bar{x}_k$ are independent variables. The complex may be considered as a decision plant described in Sec. 4 where y is the execution time of the whole complex $T = \max\{T_1, ..., T_k\}$, $x = (x_1, ..., x_k)$, $u = (u_1, ..., u_k) \in \bar{U}$. The set $\bar{U} \subset R^k$ is determined by the constraints: $u_i \geq 0$ for each i and $u_1 + ... + u_k = U$ where U is the total size of the task or the total amount of the resource to be distributed among the operations (Fig. 6).

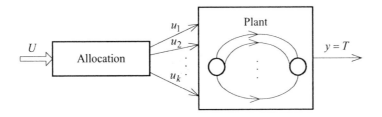

Figure 6. Complex of parallel operations as a decision plant

According to the general formulation of the decision problem presented in Sec. 4, the allocation problem may be formulated as an optimization problem consisting in finding the optimal allocation u^* that maximizes the certainty index of the soft property: "the set of possible values T approximately belongs to $[0, \alpha]$" (i.e. belongs to $[0, \alpha]$ for an approximate value of $\bar{x}$). **Optimal allocation problem**: For the given φ_i, h_i $(i \in \overline{1,k})$, U and α find

$$u^* = \arg\max_{u \in \bar{U}} v(u)$$

where

$$v(u) = v\{D_T(u;\bar{x}) \tilde{\subseteq} [0, \alpha]\} = v(T(u,\bar{x}) \tilde{\leq} \alpha).$$

The soft property "$D_T(u;\bar{x}) \tilde{\subseteq} [0, \alpha]$" is denoted here by "$T(u,\bar{x}) \tilde{\leq} \alpha$", and $D_T(u;x)$ denotes the set of possible values T for the fixed u, determined by the inequality

$$T \leq \max_i \varphi_i(u_i, x_i).$$

According to (35)

$$v(u) = v\{[T_1(u_1,\bar{x}_1) \tilde{\leq} \alpha)] \wedge [T_2(u_2,\bar{x}_2) \tilde{\leq} \alpha)] \wedge ... \wedge [T_k(u_k,\bar{x}_k) \tilde{\leq} \alpha)]\}.$$

Then

$$u^* = \arg\max_{u \in \bar{U}} \min_i v_i(u_i) \qquad (36)$$

where

$$v_i(u_i) = v[T_i(u_i,\bar{x}_i) \tilde{\leq} \alpha)] = v[\varphi_i(u_i,\bar{x}_i) \tilde{\leq} \alpha)] = v[\bar{x}_i \tilde{\in} D_i(u_i)],$$

$$D_i(u_i) = \{x_i \in R^1 : \varphi_i(u_i, x_i) \leq \alpha\}.$$

Finally

$$v_i(u_i) = \max_{x_i \in D_i(u_i)} h_i(x_i) \qquad (37)$$

and

$$u^* = \arg\max_{u \in \overline{U}} \min_i \max_{x_i \in D_i(u_i)} h_i(x_i).$$

In many cases an expert gives the value x_i^* and the interval of the approximate values of $\overline{x}_i$: $x_i^* - d_i \le x_i \le x_i^* + d_i$. Then we assume that $h_{xi}(x_i)$ has a triangular form presented in Fig. 7 where $d_i \le x_i^*$. Let us consider the relation (35) in the form $T_i \le x_i u_i$ where $x_i > 0$ and u_i denotes the size of a task. In this case, using (37) it is easy to obtain the following formula for the function $v_i(u_i)$:

$$v_i(u_i) = \begin{cases} 1 & \text{for} & u_i \le \dfrac{\alpha}{x_i^*} \\[2mm] \dfrac{1}{d_i}\left(\dfrac{\alpha}{u_i} - x_i^*\right) + 1 & \text{for} & \dfrac{\alpha}{x_i^*} \le u_i \le \dfrac{\alpha}{x_i^* - d_i} \\[2mm] 0 & \text{for} & u_i \ge \dfrac{\alpha}{x_i^* - d_i} . \end{cases}$$

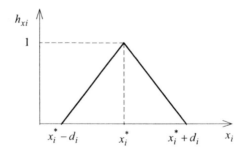

Figure 7. Example of the certainty distribution

For the relations $T_i \le x_i u_i^{-1}$ where u_i denotes the size of a resource, the function $v_i(u_i)$ has an analogous form with u_i^{-1} in the place of u_i :

$$v_i(u_i) = \begin{cases} 0 & \text{for} & u_i \le \dfrac{x_i^* - d_i}{\alpha} \\[2mm] \dfrac{1}{d_i}(\alpha u_i - x_i^*) + 1 & \text{for} & \dfrac{x_i^* - d_i}{\alpha} \le u_i \le \dfrac{x_i^*}{\alpha} \\[2mm] 1 & \text{for} & u_i \ge \dfrac{x_i^*}{\alpha} . \end{cases} \qquad (38)$$

Example 4.

Let us consider the resource allocation for two operations $(k = 2)$. Now in the maximization problem (36) the decision u_1^* may be found by solving the equation $v_1(u_1) = v_2(U - u_1)$ and $u_2^* = U - u_1^*$. Using (38) we obtain the following result:

1. For

$$\alpha \leq \frac{x_1^* - d_1 + x_2^* - d_2}{U} \tag{39}$$

$v(u) = 0$ for any u_1.

2. For

$$\frac{x_1^* - d_1 + x_2^* - d_2}{U} \leq \alpha \leq \frac{x_1^* + x_2^*}{U} \tag{40}$$

we obtain

$$u_1^* = \frac{\alpha d_1 U + x_1^* d_2 - x_2^* d_1}{\alpha(d_1 + d_2)}, \tag{41}$$

$$v(u^*) = \frac{1}{d_1}[\alpha u_1^* - x_1^*] + 1. \tag{42}$$

3. For

$$\alpha \geq \frac{x_1^* + x_2^*}{U} \tag{}$$

we obtain $v(u^*) = 1$ for any u_1 satisfying the condition

$$\frac{x_1^*}{\alpha} \leq u_1 \leq U - \frac{x_2^*}{\alpha}.$$

In the case (39) α is too small (the requirement is too strong) and it is not possible to find the allocation for which $v(u)$ is greater than 0. In the case (40) we obtain one solution maximizing $v(u)$. For the numerical data $U = 9$, $\alpha = 0.5$, $x_1^* = 2$, $x_2^* = 3$, $d_1 = d_2 = 1$, using (41) and (42) we obtain $u_1^* = 3.5$, $u_2^* = 5.5$ and $v = 0.75$ which means that the requirement $T \leq \alpha$ will be approximately satisfied with the certainty index 0.75   □

The determination of the decision u^* may be difficult for $k > 2$ because of the great computational difficulties. To decrease these difficulties we can apply the decomposition of the complex into subcomplexes, which leads to two-level allocation system [19]. The considerations may be extended to more complicated structures of the complex of operations and applied to a knowledge-based project management with cascade-parallel structure [15].

9   Systems with Uncertain and Random Parameters

It is interesting and useful to consider a decision plant containing two kinds of unknown parameters in the relational knowledge representation: uncertain parameters described by certainty distributions and random parameters described by probability distributions. The different versions of problem formulations in this case are presented in [17, 19]. Let us consider shortly selected versions. Assume that the plant is described by a relation $R(u, y, z; x, w)$ where x is a value of an uncertain variable described by $h(x)$ and $w \in W$ is a value of a random variable described by a probability density $f(w)$. Now the certainty index in (12) depends on w, i.e.

$$v[D_y(u, z; \overline{x}, w) \widetilde{\subseteq} D_y] \overset{\Delta}{=} v(u, z, w)$$

and u^* may be obtained as a result of the maximization of the expected value of v. In another version (case a), the plant is described by $R(u, y, z; x)$ with the uncertain parameter described by $h(x; w)$ where w is a value of a random variable described by $f(w)$.

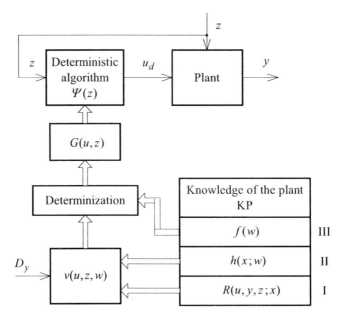

Figure 8. Decision system with three-level uncertainty in the case a; I – relational level, II – uncertain level, III – random level

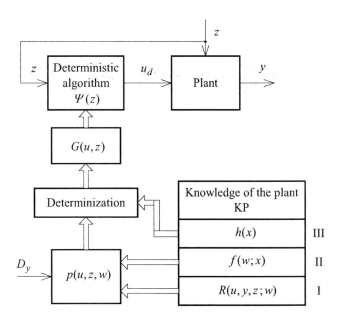

Figure 9. Decision system with three-level uncertainty in the case b; I – relational level, II – random level, III – uncertain level

Then $v(u, z, w)$ depends on w and the deterministic algorithm $u^* = u_d = \Psi(z)$ may be obtained by the maximization of the expected value of v, as in the first version. It is possible to invert the order of uncertain and random parameters (case b): The plant is described by $R(u, y, z; w)$, the random parameter w has the density $f(w; x)$ and x is a value of $\overline{x}$ with the distribution $h(x)$. Now $u_d = \Psi(z)$ may be obtained via the maximization of the mean value

(with respect to $\overline{x}$) of the probability $P[D_y(u, z; w) \subseteq D_y] \triangleq p(u, z, x)$. The knowledge-based decision systems with three levels of uncertainty in cases a and b are illustrated in Figs. 8 and 9, respectively. The result of the determinization $G(u, z)$ denotes the expected value of $v(u, z, w)$ in case a and the mean value of $p(u, z, x)$ in case b.

10 Learning Systems

For a class of uncertain decision systems described by a relational knowledge representation with unknown parameters, learning algorithms have been elaborated. The learning process consists here in *step by step* knowledge evaluation and updating, and the successive decisions are based on the current knowledge [4, 7, 21, 19]. This approach may be considered as an extension of the known idea of adaptation via identification (see e.g. [2]). The combination of the knowledge updating with the approach based on the uncertain variables leads to the learning system based on the current expert's knowledge. In such a system, at each step of the learning process the expert's knowledge in the form of the certainty distribution is modified according to the current result of the learning [20, 22].

Let us consider an uncertain plant described by a relation $R(u, y; x) \subset U \times Y$ (see Sec. 3). For the requirement $y \in D_y \subset Y$ and the given value x one may determine the largest set of decisions $D_u(x) \subset U$ such that the implication $u \in D_u(x) \to y \in D_y$ is satisfied:

$$D_u(x) = \{u \in U : \ D_y(u; x) \subseteq D_y\}$$

where $D_y(u; x)$ is defined by (10). Assume now that the parameter x in the relation R has the value $x = a$ and a is unknown. Then, for the fixed value u it is not known if u is a correct decision, i.e. if $u \in D_u(a)$ and consequently $y \in D_y$. Our problem may be considered as a classification problem with two classes. The point u should be classified to class $j = 1$ if $u \in D_u(a)$ and to class $j = 2$ if $u \notin D_u(a)$. Assume that we can use the learning sequence $(u_1, j_1), (u_2, j_2), \ldots, (u_n, j_n) \overset{\Delta}{=} S_n$ where $j_i \in \{1, 2\}$ are the results of the correct classification given by an external trainer.

Let us denote by $\bar{u}_i$ the subsequence for which $j_i = 1$, i.e. $\bar{u}_i \in D_u(a)$ and by $\hat{u}_i$ the subsequence for which $j_i = 2$, and introduce the following sets in X:

$$\bar{D}_x(n) = \{x \in X : \bar{u}_i \in D_u(x) \text{ for every } \bar{u}_i \text{ in } S_n\}, \tag{43}$$

$$\hat{D}_x(n) = \{x \in X : \hat{u}_i \in U - D_u(x) \text{ for every } \hat{u}_i \text{ in } S_n\}. \tag{44}$$

The set

$$\bar{D}_x(n) \cap \hat{D}_x(n) \overset{\Delta}{=} \Delta_x(n)$$

may be proposed as the estimation of a. For example, if $D_u(a)$ is described by the inequality $u^T u \le a$ then

$$\bar{D}_x(n) = [x_{\min,n}, \infty), \quad \hat{D}_x(n) = [0, x_{\max,n}),$$

$$\Delta_x(n) = [x_{\min,n}, x_{\max,n})$$

where

$$x_{\min,n} = \max_i \bar{u}_i^T \bar{u}_i, \quad x_{\max,n} = \min_i \hat{u}_i^T \hat{u}_i.$$

The determination of $\Delta_x(n)$ may be presented in the form of the following recursive algorithm:

$$\text{If } j_n = 1 \ (u_n = \overline{u}_n).$$

1. **Knowledge validation** for $\overline{u}_n$. Prove if

$$\bigwedge_{x \in \overline{D}_x(n-1)} [u_n \in D_u(x)].$$

If yes then $\overline{D}_x(n) = \overline{D}_x(n-1)$. If not then one should determine the new $\overline{D}_x(n)$, i.e. update the knowledge.

2. **Knowledge updating** for $\overline{u}_n$

$$\overline{D}_x(n) = \{x \in \overline{D}_x(n-1) : u_n \in D_u(x)\}.$$

Put $\hat{D}_x(n) = \hat{D}_x(n-1)$.
If $j_n = 2 \ (u_n = \hat{u}_n)$.

3. **Knowledge validation** for $\hat{u}_n$. Prove if

$$\bigwedge_{x \in \hat{D}_x(n-1)} [u_n \in U - D_u(x)].$$

If yes then $\hat{D}_x(n) = \hat{D}_x(n-1)$. If not then one should determine the new $\hat{D}_x(n)$, i.e. update the knowledge.

4. **Knowledge updating** for $\hat{u}_n$

$$\hat{D}_x(n) = \{x \in \hat{D}_x(n-1) : u_n \in U - D_u(x)\}.$$

Put $\overline{D}_x(n) = \overline{D}_x(n-1)$ and $\varDelta_x(n) = \overline{D}_x(n) \cap \hat{D}_x(n)$.

The successive estimation of a may be performed in a closed-loop learning system where u_i is the sequence of the decisions. The decision making algorithm is as follows:

1. Put u_n at the input of the plant and introduce y_n.

2. Determine $\varDelta_x(n)$ using the estimation algorithm with knowledge validation and updating.

3. Choose randomly x_n from $\varDelta_x(n)$, put x_n into $D_u(x)$ and choose randomly u_{n+1} from $D_u(x_n)$.

At the n-th step, the result of the learning process in the form of a set $\Delta_x(n)$ may be used to present an expert's knowledge in the form of a certainty distribution $h_n(x)$ such that $h_n(x) = 0$ for every $x \notin \Delta_x(n)$. Thus, the expert formulates his / her current knowledge, using his / her experience and the current result of the learning process based on the knowledge of the external trainer. In particular, $h_n(x) = h(x, b_n)$, i.e. the form of the certainty distribution is fixed, but the parameter b_n (in general, b_n is a vector of parameters) is currently adjusted. For example, if in one-dimensional case $\Delta_x(n) = [x_{\min,n}, x_{\max,n})$ and $h_n(x) = h(x, x_n^*, d_n)$ has a triangular form presented in Fig. 7 with n instead of i, then $b_n = (x_n^*, d_n)$ and

$$x_n^* = \frac{x_{\min,n} + x_{\max,n}}{2}, \quad d_n = \frac{x_{\max,n} - x_{\min,n}}{2}. \tag{45}$$

For $h_n(x)$ the next decision u_{n+1} may be determined in the way presented in Sec. 4. For a plant without z, according to (12)

$$u_{n+1} = \arg \max_{u \in U} v_n(u) \tag{46}$$

where

$$v_n(u) = v[D_y(u; x) \widetilde{\subseteq} D_y] = v[u \widetilde{\in} D_u(\overline{x})] = v[\overline{x} \widetilde{\in} D_x(u)] = \max_{x \in D_x(u)} h_n(x), \tag{47}$$

and $D_x(u) = \{x \in X : u \in D_u(x)\}$. In general, as a result of the maximization (46) one may obtain a set of decisions $D_{u,n+1}$. For $h(x, b_n)$ we obtain the fixed form of the function $v(u, b_n)$:

$$v_n(u) = \max_{x \in D_x(u)} h(x, b_n) \overset{\Delta}{=} v(u, b_n)$$

and consequently, the fixed form of the final result, i.e. one decision $u_{n+1} = u(b_n)$ or the set of decisions $D_{u,n+1} = D_u(b_n)$. The same approach may be presented in the case of C-uncertain variables, with v_c instead of v.

The block scheme of the learning decision making system under consideration is presented in Fig. 10 where G is a generator of random variables for the random choosing of u_{n+1} from $D_{u,n+1}$. The blocks in the figure illustrate parts of the computer decision system or parts of the program which has been used for the computer simulations.

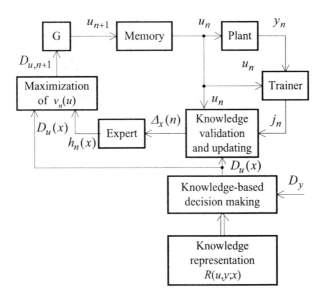

Figure 10. Learning system

The following remark is important for the evaluation of the presented approach: If for every $u_{n+1} \in D_{u,n+1}$ it is not necessary to update the knowledge (i.e. $\Delta_x(n+1) = \Delta_x(n)$), then

for $m \geq n+1$ the set $D_u(m) = D_u(n+1) \overset{\Delta}{=} D_u = \text{const}$. On the other hand, the set D_u may be much smaller than the set of all correct decisions $D_u(a)$ where $x = a$ is a real but unknown value of x. This property may be considered as a disadvantage of the presented approach.

Example 5.

Consider a decision plant with the input vector $\tilde{u}$, one-dimensional output y and one-dimensional unknown parameter x, described by a relation

$$y \leq (\tilde{u}^T \tilde{u}) x^{-1}, \quad x > 0.$$

The components $\tilde{u}^{(l)}$ of $\tilde{u}$ may denote some features of a raw material in a manufacturing process, which may be chosen as decisions, and y may denote a cost of the process. For the requirement $y \leq \bar{y}^2$ we obtain the set $D_u(x)$ described by inequality

$$u^T u \leq x \tag{48}$$

where $u^{(l)} = \tilde{u}^{(l)} \cdot (\bar{y})^{-1}$. In this case, according to (43) and (44)

$$\overline{D}_x(n) = [x_{\min,n}, \infty), \quad \hat{D}_x(n) = [0, x_{\max,n}),$$

$$\Delta_x(n) = [x_{\min,n}, x_{\max,n})$$

where

$$x_{\min,n} = \max_i \bar{u}_i^T \bar{u}_i, \quad x_{\max,n} = \min_i \hat{u}_i^T \hat{u}_i.$$

The estimation algorithm with the knowledge validation and updating is then as follows:

1. Put u_n at the input and introduce j_n.

2. For $j_n = 1$ $(u_n = \bar{u}_n)$, prove if

$$u_n^T u_n \leq x_{\min,n-1}.$$

If yes then $x_{\min,n} = x_{\min,n-1}$. If not $x_{\min,n} = u_n^T u_n$.

Put $x_{\max,n} = x_{\max,n-1}$.

3. For $j_n = 2$ $(u_n = \hat{u}_n)$, prove if

$$u_n^T u_n \geq x_{\max,n-1}.$$

If yes then $x_{\max,n} = x_{\max,n-1}$. If not $x_{\max,n} = u_n^T u_n$.

Put $x_{\min,n} = x_{\min,n-1}$, $\Delta_x(n) = [x_{\min,n}, x_{\max,n})$.

Let us assume that $h_n(x)$ has a triangular form where x_n^* and d_n are determined by (45). Using (47) it is easy to obtain the certainty index that u "approximately" belongs to the set $D_u(x)$ determined by (48): $v_n(\alpha) = 1$ for $\alpha \leq x_n^*$, $v_n(\alpha) = -d_n^{-1}(d_n - x_n^*) + 1$ for $x_n^* \leq \alpha \leq x_n^* + d_n$, $v_n(\alpha) = 0$ for $\alpha \geq x_n^* + d_n$ where $\alpha = u^T u$. Then

$$D_{u,n+1} = \{u_{n+1} \in U : u_{n+1}^T u_{n+1} \leq x_n^*\} \tag{49}$$

and for every u_{n+1} from this set $v_n(u_{n+1}) = 1$. Consequently, the decision making algorithm in the learning system is the following:

1. Put u_n at the input and introduce j_n.

2. Determine $x_{\min,n}$ and $x_{\max,n}$ using the estimation algorithm with the knowledge validation and updating.

3. Choose randomly u_{n+1} from the set (49).

Assume that in (48) $x = a$, i.e. a is the unknown value of x. From (49) it is easy to note that x_n^* may be considered as an estimation of a. Under some conditions, in the same way as that in [19], it may be proved that x_n^* converges to a with probability 1. Figure 11 presents the result of simulations for the following data: $a = 5$, $\alpha_0 = u_0^T u_0 = 20$, $\alpha_{n+1} = u_{n+1}^T u_{n+1}$ is chosen randomly from the interval $[0, x_n^*]$ with the rectangular probability density.

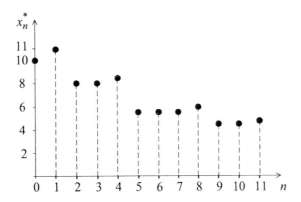

Figure 11. Result of simulations

11 Other Problems and Practical Applications

1. Uncertain dynamical systems [19]. The considerations presented here for static (memoryless) plants have been extended to relational dynamical plants described by a set of relations

$$R_\mathrm{I}(u_n, s_n, s_{n+1}\,; x) \subseteq U \times S \times S, \quad R_\mathrm{II}(s_n, y_n\,; w) \subseteq S \times Y$$

where n denotes the discrete time; $s_n \in S$, $u_n \in U$, $y_n \in Y$ are the state, the input and the output vectors, respectively; $x \in X$ and $w \in W$ are unknown vector parameters which are assumed to be values of uncertain variables $(\bar{x}, \bar{w})$ with the joint certainty distribution $h(x, w)$.

2. Parametric optimization of closed-loop control systems [19]. The uncertain variables have been applied to the optimization of an *uncertain controller* in the closed-loop system with a dynamical plant. The main idea is the following: for the given model of the plant with an uncertain parameter x characterized by the certainty distribution, and the given model of the controller with a parameter b – one should find $b(x)$ minimizing a quality index $Q(x, b)$. The controller with the uncertain parameter $b(x)$ may be called an uncertain controller, and via the determinization the deterministic control algorithm may be obtained.

3. Quality of decisions based on uncertain variables. For the known deterministic models of the decision plants, it is possible to evaluate the quality of the decisions based on the certainty distributions given by an expert. The quality index may be used to compare different values of parameters in the certainty distributions and consequently, to compare different experts. It may be also used in an adaptive system with *step by step* adjusting of the parameters in the certainty distributions, based on the current quality evaluation [19, 25].

4. Application to pattern recognition. Let an object to be recognized or classified be characterized by a vector of features u and the index of a class j to which the object belongs. The set of the objects may be described by a relational knowledge representation $R(u, j; x)$ where the vector of unknown parameters x is assumed to be a value of an uncertain variable described by a certainty distribution. Optimal recognition problem may consist in the determination of a class j maximizing the certainty index that j belongs to the set of all possible classes for the known u. Different versions of this problem and the corresponding results are presented in [19, 16].

5. Complex systems with distributed knowledge. Interesting results in this area concern the complex structures of uncertain systems and the knowledge-based computer systems with the distributed knowledge. Each part of the complex system is described by a relational knowledge representation with unknown parameters characterized by an expert in the form of certainty distributions [7, 13, 22].

The description of the uncertainty based on the uncertain variables has been used in different **practical applications**. As the selected examples, one can list the following applications:

a. Analysis and design of a **congestion control system in computer networks**. In particular, the general method of the stability analysis presented in Sec. 7 has been applied to the estimation of v_s for the congestion control system with an uncertain parameter characterized by an expert [27].

b. Intelligent control of an uncertain **assembly process** and other **manufacturing processes** [12, 19]. A class of knowledge-based assembly systems may be described by relations between successive operations, states and variables characterizing the current effect of the assembly process. Then the problem of choosing assembly operations from the set given at each stage is considered as a specific multistage decisions process for a relational dynamical plant with uncertain parameters.

c. Knowledge-based control of an **industrial transportation system** [23, 26]. In the simplest case, the system consists of three parts: the transport of a raw material, the set of parallel production units processing the raw material, and the transport of a product. For the system described by relations with uncertain parameters, the decision problem consisting in the proper distribution of the raw material, may be considered as a specific allocation problem presented in Sec. 8.

d. The approach presented in Sec. 8 may be applied to the **allocation of computational tasks in a group of parallel processors** with uncertain execution times [18]. In this case (35) is reduced to $T_i \le x_i n_i$ where n_i is a number of elementary tasks (programs or parts of programs) and x_i is the upper bound of the execution time for the elementary task, estimated by an expert.

e. **Resource allocation in a group of research units** [24]. A specific knowledge-based allocation problem may consider a resource allocation in a group of research units for which the relation between the amount of the resource and a numerical effect of the research activity for the particular unit is described by the help of uncertain variables characterized by an expert.

12 Conclusion

The uncertain variables have been proved to be a convenient tool for handling the decision problems based on an uncertain knowledge given by an expert in the form of certainty distributions. The problems and methods described in this chapter for static (memoryless) systems have been extended to systems described by a dynamical knowledge representation. The formalism of the uncertain variables and its application in a wide spectrum of uncertain systems (complexes of operations, systems with distributed knowledge, pattern recognition and various practical systems) have been used as a basis for a uniform description of analysis and decision making in uncertain systems, presented in the book [19], where one can find the indications of new problems and perspectives in this area.

General ideas and new directions in the field of uncertain systems may be connected with two different concepts concerning the determinization, and two different approaches concerning the role of an expert. In the first case, one can consider the concept illustrated in Fig. 3 and another concept consisting in the determinization of KP (i.e. replacing the uncertain knowledge of the plant by its deterministic representation). Then the algorithm Ψ is obtained as a solution of a deterministic decision problem. In the second case, except the approach presented in Fig. 3 where KD is obtained from KP given by an expert, one can consider another approach where KD is given directly by an expert [8].

References

[1] Bubnicki, Z. On the stability condition of nonlinear sampled-data systems. *IEEE Trans. AC* 1964, 9, 280–281.

[2] Bubnicki, Z. Identification of Control Plants; Elsevier: Oxford, Amsterdam, New York, 1980.

[3] Bubnicki, Z. Uncertain logics, variables and systems. In *Proceedings of the 3rd Workshop of International Institute for General Systems Studies;* Tianjin People's Publishing House: Tianjin, 1998; pp 7–14.

[4] Bubnicki, Z. Uncertain variables and learning algorithms in knowledge-based control systems. *Artificial Life and Robotics* 1999, 3, 155–159.

[5] Bubnicki, Z. Uncertain variables in the computer aided analysis of uncertain systems. In *Computer Aided Systems Theory;* Lecture Notes in Computer Science; Springer Verlag: Berlin, 2000; Vol. 1798, pp 528–542.

[6] Bubnicki, Z. General approach to stability and stabilization for a class of uncertain discrete nonlinear systems. *Int. J. Control* 2000, 73, 1298–1306.

[7] Bubnicki, Z. Knowledge validation and updating in a class of uncertain distributed knowledge systems. In *Proceedings of the 16th IFIP World Computer Congress;* Publishing House of Electronics Industry: Beijing, 2000; Vol. "Intelligent Information Processing", pp 516–523.

[8]   Bubnicki, Z. (2001). A unified approach to descriptive and prescriptive concepts in uncertain decision systems [pdf]. *Proceedings of the European Control Conference*; Porto, Portugal, 2001; pp 2458–2463.

[9]   Bubnicki, Z. Uncertain variables and their applications for a class of uncertain systems. *Int. J. Systems Science* 2001, 32, 651–659.

[10] Bubnicki, Z. Uncertain variables and their application to decision making. *IEEE Trans. SMC, Part A: Systems and Humans* 2001, 31, 587–596.

[11] Bubnicki, Z. Uncertain Logics, Variables and Systems; Springer Verlag: Berlin London, N. York, 2002.

[12] Bubnicki, Z. Learning process in a class of computer integrated manufacturing systems with parametric uncertainties. *J. Intelligent Manufacturing* 2002, 13, 409–415.

[13] Bubnicki, Z. Application of uncertain variables to decision making in a class of distributed computer systems. In *Proceedings of the 17th IFIP World Computer Congress;* Kluwer Academic Publishers: Norwell, MA, 2002; Vol. "Intelligent Information Processing", pp 261–264.

[14] Bubnicki, Z. Stability and stabilization of discrete systems with random and uncertain parameters. In *Proceedings of the 15th IFAC World Congress;* Pergamon: Oxford, 2003; Vol. E, pp 193–198.

[15] Bubnicki, Z. Application of uncertain variables to a project management under uncertainty. *Systems Science* 2003, 29, 65–79.

[16] Bubnicki, Z.; Szala, M. Application of uncertain and random variables to knowledge-based pattern recognition. In *Proceedings of the 16th International Conference on Systems Engineering;* Coventry University: Coventry, UK, 2003; Vol. 1, pp 100–105.

[17] Bubnicki, Z. (2003). Application of uncertain variables in a class of control systems with uncertain and random parameters [pdf]. Proceedings of the European Control Conference; Cambridge, UK, 2003.

[18] Bubnicki, Z. Application of uncertain variables to task and resource distribution in complex computer systems. In *Computer Aided Systems Theory;* Lecture Notes in Computer Science; Springer Verlag: Berlin, Heidelberg, 2003; Vol. 2809, pp 38–49.

[19] Bubnicki, Z. *Analysis and Decision Making in Uncertain Systems;* Springer Verlag: Berlin, London, N. York, 2004.

[20] Bubnicki, Z. Application of uncertain variables to learning process in knowledge-based decision systems. In *Proceedings of the 9th International Symposium on Artificial Life and Robotics;* Oita, Japan, 2004; Vol. 2, pp 396–399.

[21] Bubnicki, Z. (2004). Knowledge-based and learning control systems [html]. http://www.eolss.net, Control Systems, Robotics and Automation, *Encyclopedia of Life Support Systems* (EOLSS), Developed under the auspices of the UNESCO.

[22] Bubnicki, Z. Application of uncertain variables and learning algorithms in a class of distributed knowledge systems. In *Proceedings of the 18th IFIP World Computer Congress;* Laboratoire d'Automatique et d'Analyse des Systemes: Toulouse, 2004; Vol. "The Symposium on Professional Practice in AI", pp 111-120.

[23] Bubnicki, Z. (2004). Uncertain variables and learning process in an intelligent transportation system with production units [pdf]. *Proceedings of the 5th IFAC Symposium on Intelligent Autonomous Vehicles*; Lisbon, Portugal, 2004.

[24] Bubnicki, Z.; Orski D. Application of uncertain variables to knowledge-based resource allocation in a group of research units. *Systems Science* 2004, 30.

[25] Orski, D. Quality analysis and adaptation in control systems with a controller based on uncertain variables. In *Proceedings of the 16th International Conference on Systems Engineering;* Coventry University: Coventry, UK, 2003; Vol. 2, pp 525–530.

[26] Toledo, R. A study of a simple transportation problem applying uncertain variables. In *Proceedings of the 15th International Conference on Systems Science;* Oficyna Wydawnicza PWr: Wroclaw, Poland, 2004; Vol. II, pp 339–345.

[27] Turowska, M. (2004). Application of uncertain variables to stability analysis and stabilization for ATM ABR congestion control systems [pdf]. *Proceedings of the International Conference on Enterprise Information Systems*; Porto, Portugal, 2004.

In: Control and Learning in Robotic Systems
Editor: John X. Liu, pp. 123-148

ISBN 1-59454-356-9
©2005 Nova Science Publishers, Inc.

Chapter 5

PERCEPTIVE PLANNING AND CONTROL OF ROBOTIC SYSTEM IN A HYBRID SYSTEM FRAMEWORK[*]

Yu Sun and Ning Xi
Department of Electrical and Computer Engineering
Michigan State University, East Lansing, MI 48824, USA

Abstract

In this chapter, a new planning and control scheme for intelligent robotic systems is presented. Robotic systems obtain environmental information from perceptive sensors and respond to the perceptions to execute tasks through decision and control process. In perceptive frame, the evolution of the robotic system is driven by perceptive references, which is directly related to sensory measurement of the system output, instead of the reference of time. The hybrid hierarchical perceptive framework of robotic systems has continuous and discrete layers, and it also has continuous and discrete perceptive references. The discrete layers enable the robot systems to operate at higher levels for improving the robotic intelligence, namely, planning and modifying original tasks and actions based on the perceptions through switching the low level perceptive controllers.

Using hybrid automata and hybrid formal language theory, a hybrid perceptive control theory for modeling and analyzing the motion planning and control of mobile robots has been developed. Hybrid and perceptive automata are able to exhibit the robot behaviors both continuous and discrete in perceptive frame. The discrete expressions of hybrid language can be accepted by discrete part of the automata, while the continuous part of the language performs the control of the continuous physical system. Stability can be guaranteed for the switched systems based on the dynamical properties. In contrast to a continuous perceptive reference, the hybrid perceptive

---

[*]Research partially supported under NSF Grants IIS-9796300, IIS-9796287 and EIA-9911077, and under DARPA Contract DABT63-99-1-0014.

framework is able to deal with unexpected events which can block the evolution of the continuous or discrete reference, and the reference keeps evolving during the unexpected events. At the discrete levels, the hierarchical architecture of the hybrid automata improves the formal language so that the automata are able to treat the unexpected events as the disturbances of a tracking control system, it can return back to execute the original task after processing the unexpected events; at the continuous level, the system can make a discrete switch to prevent the references from stagnation.

The experimental results, given by a tele-operation system consisting of a phantom joystick and a mobile manipulator, show the effectiveness of the proposed perceptive control theory. The controller is stable and able to process unexpected events and return back to the desired task. In the autonomous control mode, the hybrid perceptive references can not be stopped by blocking the continuous reference, including obstacles, in the tele-operation mode, the control performance is not affected by the block of discrete reference due to communication time delay.

1   Introduction

In recent years, there has been increasing interest in robot intelligence. As applications are becoming more and more complicated, robot systems need to have a certain amount of autonomy. One of the most important aspects of autonomous systems is the ability to deal with unexpected events in a changing environment. The robot processes environmental information obtained from onboard sensors, and responds to the information by changing the original path planning and control schemes. Based on the perceptive-based approach [1], a hybrid perceptive reference model is proposed, which enables the mobile robot system to autonomously respond to unexpected events.

A continuous perceptive planning and control approach [1] has been applied to the path planning and control for a single manipulator, and multiple manipulators coordination [2]. Song et al. [3] developed an perceptive scheme for integration of task scheduling, action planning and control in robotic manufacturing systems. The basic idea of the perceptive planning and control theory is to introduce the concept of a motion reference s, a parameter that is directly relevant to measured sensory outputs and the task. Instead of time, the control input is parameterized by the motion reference. since the action reference is a function of the real time measurement, the values of the desired vehicle states are functions of the measured data. This creates a mechanism to adjust or modify the plan based on the measurements. Thus, the planning becomes a closed loop real-time process. The planner generates the desired values of the system, according to the on-line computed action reference parameter s. Figure 1 indicates the perceptive control in comparison with the traditional control concepts.

The perceptive frame approach guarantees stability of the robot system in presence of unexpected events. However the robot system cannot avoid the obstacle based on the environmental sensor information. The hybrid perceptive reference model has three layers, one continuous layer, and two discrete layers. The continuous layer guarantees the system stability in the presence of unexpected events based on the perceptive approach. The discrete layers enable the robot system to make decisions and modify original path plans.

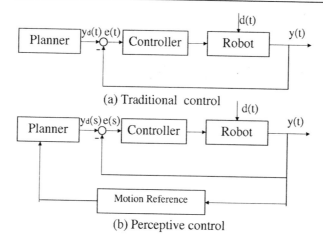

(a) Traditional control

(b) Perceptive control

Figure 1: Perceptive Control

Hybrid systems have been extensively investigated recently [4] [5] [6] [7] and a variety of hybrid system models and frameworks have been proposed[8] [9] [10] [11]. Linguistic method is an efficient approach proposed for motion planing and control of mobile robots.

Linguistic/automata approach is the method applying the mathematical tools, formal languages and automata theory, to describe the dynamics of the control system at discrete levels. Hybrid control system behaviors, thus, can be discussed by hybrid automata, exhibiting both continuous and discrete evolutions. Brockett abstracted the concepts of the lattice language based on the analysis of the robot motion[12]. A motion description language for kinetic state machines, which are the continuous analog of finite automata, is proposed in [11]. Furthermore, Manikonda at al. [14] defined atoms in the language alphabet that describes motion behaviors. The functions of the language are improved to deal with multiple interruptions. In Egerstedt's research, a complexity problem of a multiple obstacle avoidance is analyzed by using motion description language of Brockett's hybrid model [13]. However, both the continuous part and the discrete part of the system use time as the reference.

By nature of robotic systems, the motion planning and control of the mobile robot system can be modeled and analyzed in perceptive frame. In this chapter, using a hybrid framework, the hybrid perceptive references enable both the continuous part and the discrete part of the system to deal with unexpected events. The discrete part of the system governs the system based on perceptive information such that the system can keep the perceptive reference evolving through modifying original path plans. The discrete part and the continuous part are integrated by the hybrid perceptive motion reference.

2 Automata Theory

Automata theory is the key technology in hybrid system modeling and system analysis. It provides a mathematical tool to describe the properties of the system in a discrete fash-

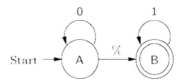

Figure 2: An Example of Finite Automata.

ion. Finite Automata and formal languages are useful models for many important kinds of hardware and software. First of all, some basic concepts about finite automata and formal languages [15] are introduced.

A *symbol* is the abstract entity of automata theory. Examples are letters and digits.

An *alphabet* is a finite, nonempty set of symbols. Conventionally, we use the symbol Σ for an alphabet. Common alphabets include:

1. $\Sigma = 0, 1$, the binary alphabet.
2. $\Sigma = a, b, ..., z$, the set of all lower-case letters.
3. The set of all ASCII characters.

A *string* (or word) is a finite sequence of symbols chosen from some alphabet. The empty string, denoted ε, is the string consisting of zero symbols.

Concatenation of strings: Let x and y be strings. Then xy denotes the concatenation of x and y, that is, the string formed by making a copy of x and following it by a copy of y.

A set of strings all of which are chosen from some Σ^*, where Σ is a particular alphabet, is called a *language*.

If Σ is an alphabet, and $L \subseteq \Sigma^*$, then L is a language over Σ. Notice that a language over Σ need not include strings with all the symbols of Σ, so once L is a language over Σ, it is a language over any alphabet that is a superset of Σ.

A finite automaton(FA) is a system $(Q, \Sigma, \delta, q_0, M)$ where Q is a finite set of states, Σ is the alphabet, $q_0 \in Q$ is the initial state, $M \subset Q$ is the accepting states.

A finite automaton , M is a "tuple", $(Q, \Sigma, \Delta, q_0, F)$, where:

–Q is a finite set of states

–Σ is a set of input symbols, an alphabet, call the input alphabet

–$\Delta \subset Q \times \Sigma \rightarrow Q$ is a set of the transition mappings $Q \times \Sigma$ into Q.

–$F \subseteq Q$ is a set of final states.

Σ^* set of string of finite length of elements of Σ. Define string of length 0 to be ϵ.

M accepts a string $s \in \Sigma^*$, with $|s|$ = n, if there exists a sequence of states $q \in Q^*$, with $|q| = n + 1$ such that:

–$q[0] = q_0$ –For $i = 0, 1, ...n$, $(q[i], s[i], q[i + 1]) \in \Delta$ –$q[n + 1] \in F$.

The language accepted by M, L(M) is the set of all string s accepted by M.

Two automata, M1 and M2, are equivalent if L(M1)=L(M2).

An example of finite automata, automaton M is shown in Fig. 2. The state set consists of A and B . A is the starting state, B is the final state. The alphabet of the input symbols includes "0, 1",and "%". The arrows in the figure denoote the state transition. For automaton M, "0%" and "%1" are the words in the language accepted by M.

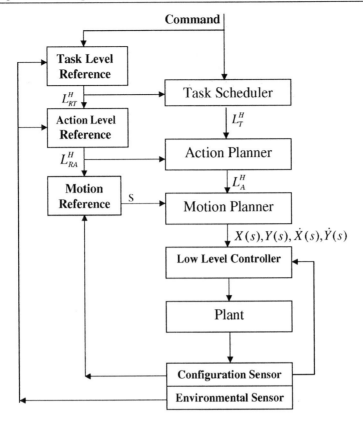

Figure 3: A Perceptive Framework for Robot Systems.

3   Hybrid Perceptive Framework

A perceptive framework represented as a hybrid system model, as shown in Figure 3, is developed for robot system to process both continuous and discrete information. The perceptive framework is composed of two major parts, a hierarchical command execution block and a hybrid perceptive reference block. All the units of the model are modeled by hybrid automata and hybrid languages.

3.1   Perceptive Automata

In the perceptive framework, an action set consists of all the action primitives the robot can execute, for example, moving a robot, picking up/placing a part, open a gripper, and so on. The action set can be notated as $A = \{A_1, A_2...A_{n1}\}$ where $A_i, i = 1, 2, ...n_1$ are action primitives.

A task set is a set of all the task primitives for a robot system, each of which may include a few actions. The task set can be notated as $T = \{T_1, T_2...T_{n2}\}$, where $T_i, i = 1, 2, ...n_2$ are task primitives. Usually, a task consists of several actions. For example, the task " pick up" consists of the actions that include "Go To" and "Close gripper".

An Action Reference Set refers to a subset of the action set. The set can be written in the form of $R_{Action} = \{r_{a1}, r_{a2}, ...r_{an3}\}$, where $r_{ai}, i = 1, 2, ...n_3$ are the references which govern the evolution of corresponding tasks.

Similarly, the Task Reference Set is a subset of the task set and related to motions. It is described as $R_{Task} = \{r_{t1}, r_{t2}, ...r_{tn4}\}$, where $r_{ti}, i = 1, 2, ...n_4$ are the references which govern the evolution of corresponding tasks.

The Motion Reference s is a continuous reference chosen for lower level path planning and control. For example, the distance the robot traveled, can be chosen as the motion reference. Examples of design and analysis of the mobile robot motion planning and control based on the continuous motion reference can be seen in [2].

The formal languages L_A, L_T, L_{RA} and L_{RT} can be built by the alphabets $\Sigma_A, \Sigma_T, \Sigma_{RA}$, and Σ_{RT} that are generated by sets $A, T, R_{Action}, R_{Task}$, respectively.

A hybrid language is generated from an original alphabet Σ, consisting of discrete atoms, and numbers $N \in R^n$. The alphabet of the hybrid language is an extended alphabet of a formal language. A hybrid language is defined as the set of all the strings over the extended alphabet Σ^H. a string is concatenation of the elements lying in the product set of the original alphabet and the Euclidean spaces. A hybrid language derived from the extended alphabet Σ^H is denoted L^H. A word in a hybrid language L_T^H can be "$T_1(x, y)$" meaning "going to point(x,y)". The alphabet gives the qualitative description, the numeric part is the quantization of the linguistic expressions.

A perceptive automaton can be defined as a tuple

$$M_e = (Q, \Sigma, \Sigma_R, \delta, Q_0, \eta, O). \tag{3.1}$$

where Q is the set of the discrete states of the automaton and Σ is the discrete control input set. Σ_R is the discrete perceptive reference set. Q_0 is a set of starting states and δ is the evolutions of states. η is the output function. O is output, it can be a vector.

The following properties describe the state transitions and the output action trigged by the control command input and the reference inputs.

$$q_j = \delta(q_i, \sigma_j), O_k = \eta(q_i, \sigma_k) \tag{3.2}$$

where $q_i, q_j \in Q, O_k \in O, \sigma_j \in \Sigma, \sigma_k \in \Sigma_R$.

Corresponding different inputs the automaton can have different responses including state switching and command issuing. A hybrid automaton is capable of bridging the two kinds of variables. Based on [4][16], it can be considered to be a tuple

$$M^H = (Q, X, \Sigma, \delta, Q_0, \eta, 0), \tag{3.3}$$

where X is the set of continuous variables. The automaton can be described as

$$q_j = \delta(q_i, \sigma_j, X), \sigma_j \in \Sigma \tag{3.4}$$

$$O_k = \eta(q_i, \sigma_i, X) \tag{3.5}$$

A hierarchical automaton consists of several nodes, some of which are other automata. The first automaton is an upper automaton, the other automata denoting the nodes of the upper automaton are called embedded automata. This architecture can be described as a tuple,

$$M_h = (Q_U, Q_{EM}, \Sigma_U, \Sigma_{EM}, \delta_U, \delta_{EM}, Q_{U0}, Q_{EM0}, O),$$

Σ_U is the input set for the upper automaton, Σ_{EM} is the input set of embedded automaton, Q_U is a set of the states(the nodes of the upper automaton), Q_{EM} is a set of the states of the automata embedded into nodes of Q_U, Q_{U0} is the set of the starting states of Q_U, Q_{EM0} is the set of the starting states of Q_{EM}. δ_U and δ_{EM} describe the transition functions of Q_U and Q_{EM}, respectively.

Perceptive, hybrid and hierarchical automata refer to three basic prototypes of automata in the framework.

3.2   Hierarchical Task Execution Architecture

As shown in Figure 1, The task execution includes three parts: task scheduler, action planner and motion planner. which can be modeled using automata.

The task scheduler sequentially generates tasks according to the commands and discrete task level references. It is a perceptive automaton hybrid variables.

$M_{TaskSch} = (Q, \Sigma_T^H, \Sigma_{RT}^H, X, \delta, Q_0, \eta, O_{TaskSch})$, where the unexpected event input and task level reference input are in language L_{RT}^H, and L_T^H, respectively. The inputs from the Task Level Reference automaton trigger the task output in hybrid language L_T^H. Logically, it chooses the higher priority task from the inputs, e.g. obstacle avoidance.

The function of the planner is to generate a sequence of actions according to the task input from the task level. According to the discussion above, the action planner is modeled as a hybrid perceptive automaton which can be described as a tuple,
$M_{ActPlan} = (Q, \Sigma_T^H, \Sigma_{RA}^H, X, \delta, Q_0, \eta, O_{ActPlan})$. It has perceptive reference inputs in hybrid language L_{RA}^H and task inputs in hybrid language L_T^H denoted as:

$$I_{ActPlan1} = L_T^H \tag{3.6}$$

$$I_{ActPlan2} = L_{RA}^H \tag{3.7}$$

The output $O_{ActPlan}$ is in the language L_A^H.

The continuous variables in set X carried out by hybrid language inputs give the automaton continuous evolutions and can also be used to generate the outputs in the hybrid language.

The transition function and the output functions of the Action Planner are as follows

$$q_{ActPlanj} = \delta(q_{ActPlani}, \sigma_j), \sigma_j \in \Sigma_T^H \tag{3.8}$$

$$O_{ActPlank} = \eta(q_{ActPlani}, \sigma_k), \sigma_k \in \Sigma_{RA}^H \tag{3.9}$$

The task inputs cause the state transition of the automaton. The perceptive reference input triggers the output, The output can be a vector comprising of several commands.

The motion planner in Fig. 3 can generate a continuous trajectory for the desired motion, based on the actions from the action planner. The motion planner is a hybrid automaton triggered by the perceptive reference. We describe it as a tuple: $M_{MotPlan} = (Q, \Sigma_A^H, s, X, \delta, Q_0, \eta, O_{MotPlan})$ The output is parameterized by reference s. Another input of the automaton is in hybrid language L_A^H from the alphabet Σ_A^H. The discrete part of the input σ in L_A^H make the automaton switch to the appropriate node. Each node of the automaton is a dynamic system (or a controller) which can issue the planned continuous trajectory.

Therefore, it can be formulated as:

$$q_{MotPlanj} = \delta(q_{MotPlani}, \sigma_j), \sigma_j \in \Sigma_A \tag{3.10}$$

$$O_{MotPlank} = \eta(q_{MotPlani}, s) \tag{3.11}$$

It can be seen that the state transition and output are triggered separately by two different inputs. For motion actions, the output to the robot controller can be a vector, $O_{MotPlank} = [X(s), Y(s), \dot{X}(s), \dot{Y}(s)]^T$, for a complete motion trajectory.

3.3   Reference Generation

The system has configuration sensors and environmental sensors. The environmental sensory measurements are used by the higher levels for discrete perceptive reference generation, and the motion reference can be issued based on the configuration sensor measurements. Reference generation can be described as automata.

The Action Reference automaton has a hierarchical embedded structure consisting of the upper automaton and the embedded automata that are nodes of the upper automaton(in Fig.4). This automaton receives the inputs from the task level reference automaton and the environmental sensors. The upper automaton and the embedded automata respond to the task inputs and sensor inputs respectively. The upper automaton has an architecture as shown in Figure 4, Each task input can change the states of the automaton which can be described as:

$$q_i = \delta(q_j, \sigma_i), j \neq i, \sigma_i \in \Sigma_T^H \tag{3.12}$$

where T_i is a word defined in the task set T. In this case, each node of the upper automaton is an automaton embedded in the node. These embedded automata may be different in state transitions and output generations. Figure 4 shows the architecture of an embedded automaton.

The states and the outputs can be changed based on environmental sensor measurements, i.e. Environmental Conditions(EnvCon), as shown in Figure 4.

The properties of such an embedded automaton can be described as:

$$o_{ik} = \eta(q_{ij}, EnvCon_k = TRUE), j \neq k \tag{3.13}$$

$$q_{ik} = \delta(q_{ij}, EnvCon_k = TRUE), j \neq k \tag{3.14}$$

Where o_k, a word in L, is the output to the motion reference automaton and the task scheduler, which resets the motion reference and triggers the next action.

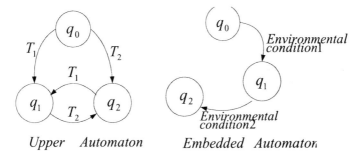

Figure 4: The Upper Automaton of the Action Level Reference automaton.

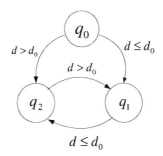

Figure 5: An Automaton of the Task Level Reference.

The Task Level Reference can be generated by automaton whose state transitions and output are triggered by change of the environment. As an example, Figure 5 shows how the automaton works while a mobile robot is approaching an unexpected obstacle. The automaton operates by issuing a proper command response to the event.

The transition function and the output function of the automaton are:

$$o_i = \eta(q_1 : \sigma_i) = AO(.), \sigma_i \in \Sigma_T^H \tag{3.15}$$

$$o_i = \eta(q_2 : \sigma_i) = \sigma_i, \sigma_i \in \Sigma_T^H \tag{3.16}$$

Where AO denotes Obstacle Avoidance, d is the distance between the mobile robot and its nearest obstacle, and d_0 is the minimum safe distance between them. The outputs in language L_{RT}^H go to Task Scheduler and the Action Level Reference automaton, respectively.

4   Hybrid Peceptive Control for Mobile Manipulation and Tele-operation

A typical task is to pick up an object and transport it to a desired destination with a mobile manipulator. In the route of transporting the object, an unexpected obstacle occurs. In order to complete the task, the mobile manipulator will execute an obstacle avoidance task. The mobile manipulator will then resume its original task and reach the destination.

To build the hybrid languages from task set and action set, some continuous variables can be involved to the atoms. In the extended task set, the $T1(m1, n1)(m2, n2)$ denotes "transporting an object from point(m1,n1) to point(m2,n2)." $T2(m, n, l_O)$ denotes "to avoid an obstacle at(m, n) at the distance l_O." In the extended action set, A1(m,n,l) denotes "following a straight line with directional cosine (m,n) and length l." A2(m,n,r) denotes "go through an arc centered at (m, n) with length r." A3 and A4 denote "close the gripper" and "open the gripper", respectively.

4.1   Task Execution in Mobile Manipulation

Task Scheduler: The output of the task scheduler is $T1(m1, n1)(m2, n2)$ until the obstacle has been detected.

   Action Planner: Based on $T1(m1, n1)(m2, n2)$, the transition and the outputs are $q_{AP1} = (q_0, T1)$ and $o_{AP1} = \eta(q_{AP1}, reset) = A1$. When the planner has the reference input A1 meaning that the A1 action is finished, the output action triggered by A1 is a vector $o_{AP1} = \eta(q_{AP1}, A1) = [A2, A3]^T$. The output triggered by action reference A11 is another action $o_{AP1} = \eta(q_{AP2}, A11) = A4$. The directional cosines and the length in A1 and A11 can be found with the parameters in T1. **Motion Planner:** The inputs $A1(m1, n1, l1)$, $A11(m11, n11, l11) \in \Sigma_A^H$ cause the states transition of the motion planner. $q_{MotPlan1} = \delta(q_0, A1)$, $q_{MotPlan11} = \delta(q_0, A11)$. The states denote different vector fields, i.e., trajectories, namely, straight line path and arc path.

   Corresponding to the continuous perceptive reference input s, the robot's motion can be described by the following equations:
for a straight line path

$$\dot{x} = mV(s), \dot{y} = nV(s) \tag{4.17}$$

for a circular Path(an arc)

$$x = m + rcos(s/r), y = n + rsin(s/r) \tag{4.18}$$
$$\dot{x} = -V(s)(y/r), \dot{y} = V(s)(x/r) \tag{4.19}$$

This trajectory is made up of two pieces of straight lines $A1$ and $A11$.

   Action Level Reference: The nodes of the upper automaton denote different paths, namely two segments of straight lines managed by the same procedure. The state transition is $q_{AR1} = \delta_{AR}(q_{AR0}, T1(m1, n1)(m2, n2))$. The output function and state transition of the embedded automaton M_{AR1} can be described as $O_{AR11} = \eta(q_{AR10}, L \geq l_1)$ and $q_{AR11} = \delta_{AR1}(q_{AR10}, L \geq l_1)$, respectively. L is the distance traveled by the robot since the new action started.

4.2   Task Execution in Tele-operation

In the Internet based tele-operation, the tele-operator sends the operation command. The commands can be in position or velocity. Therefore, the proposed hybrid system control approach can be implemented from the action level. The position command sent to the Puma controller is in the hybrid language word $A1(m,n,l) \in L_A^H$. Then the motion planner can generate the perceptive trajectory for the Puma arm, the arm can remotely execute the position or velocity commands during tele-operation by the proposed hybrid perceptive approach.

5   Analysis on Linguistic Description of Hybrid Perceptive Reference

5.1   Linguistic Description for Hybrid Perceptive Reference

In order to study the properties of the proposed model, the hierarchical architecture is considered as an integrated system. to describe the robotic manipulation system in the perceptive frame, Hybrid Perceptive Reference space is introduced in this chapter.

5.1.1   Hybrid Perceptive Reference Space of Robotic Manipulation System

Considering the traditional time based systems, the reference has two significant features. First, it is a variable in a metric space. Second, it evolves itself forward during the operation of the system.

The system illustrated in Figure 3, the hybrid system model, can be described with perceptive references and states, which can be expressed in a hybrid fashion. A hybrid perceptive trajectory is a finite or infinite sequence of reference interval $e = \{S_{TRi}S_{ARj}S_K\}$, where $S_K \in [0, s'_k] = I_{sk}, S_{TRi} \in \Sigma_{RT}^H, S_{ARj} \in \Sigma_{RA}^H$. Although the first two parts of $e = \{S_{TRi}S_{ARj}S_K\}$ have the continuous components, these components are stationary. The perceptive references at the task level and the action level can be still thought of as the discrete parts. The item s_K is a continuous variable.

5.1.2   Hybrid Metric Space

For the discrete topology, the hybrid metric can be generated in the following fashion:

$d(S_{TRi1}S_{ARj1}, S_{TRi2}S_{ARj2}) = 2$ if $S_{TRi1} \neq S_{TRi2}$,

$d(S_{TRi1}S_{ARj1}, S_{TRi2}S_{ARj2}) = 1$ if $S_{TRi1} = S_{TRi2}S_{ARj2}, S_{ARj1} \neq S_{ARj2}$

For the continuous item s_K, the metric is $d(s_{Ki}, s_{Kj}) = \parallel s_{Ki} - s_{Kj} \parallel$

Claim: the product space $S_{TR} \times S_{AR} \times S$ can be generated by the matric:

$d_D(S_{TRi1}S_{ARj1}S_{K1}, S_{TRi2}S_{ARj2}S_{K2}) = d(S_{TRi1}S_{ARj1}, S_{TRi2}S_{ARj2}) + \parallel s_{Ki} - s_{Kj} \parallel$.

Proof:

(1) "d" can satisfy the metric symmetric axiom ;

(2) $d_D(S_{TRi}S_{ARj}S_K, S_{TRi}S_{ARj}S_K) =$
$d(S_{TRi}S_{ARj}, S_{TRi}S_{ARj}) + \parallel s_K - s_K \parallel = 0$;

(3) It satisfies the triangle inequality.

For the discrete parts:

Let $d_D(1,2)$ denote $d_D(S_{TRi1}S_{ARj1}, S_{TRi2}S_{ARj2})$,
$d_D(2,3)$ denote $d_D(S_{TRi2}S_{ARj2}, S_{TRi3}S_{ARj3})$,
and $d_D(1,3)$ denote $d_D(S_{TRi1}S_{ARj1}, S_{TRi3}S_{ARj3})$.
If $d_D(1,2) = 2$, $d_D(2,3) = 2$. Then $d_D(1,3) \leq 2$, $d_D(1,3) \leq d_D(1,2) + d_D(2,3) = 4$;
If $d_D(1,2) = 1$, $d_D(2,3) = 2$. Then $d_D(1,3) = 2$, $d_D(1,3) \leq d_D(1,2) + d_D(2,3) = 3$;
If $d_D(1,2) = 1$, $d_D(2,3) = 1$. Then $d_D(1,3) \leq 1$, $d_D(1,3) \leq d_D(1,2) + d_D(2,3) = 2$;
If $d_D(1,2) = 0$, $d_D(2,3) = 1$. Then $d_D(1,3) = 1$, $d_D(1,3) \leq d_D(1,2) + d_D(2,3) = 1$;
If $d_D(1,2) = 0$, $d_D(2,3) = 0$. Then $d_D(1,3) = 0$, $d_D(1,3) \leq d_D(1,2) + d_D(2,3) = 0$;
For the continuous parts, the triangle inequality can be satisfied. Thus, d is a metric.

5.1.3 Independent Evolution

The perceptive reference is working as a "clock" or a "driver", to trigger the planning and control activities. The triggering reference should evolve independently along a chain, an evolving chain, which plays the role of the time in the traditional time-based systems.

The "time" can be described as an ordered set. There exists a subset of the partially ordered set which is a totally ordered set.

We define the binary relation "$\preceq$", for perceptive reference value a and b, we say $a \preceq b$ if the time at which event a happens precedes to the time at which b happens. It can be proven that the perceptive reference trajectory is ordered by the binary relation "$\preceq$".

In an execution of a given task, the continuous reference s is a monotonic increasing function of time. Then the continuous part of the reference is ordered. Based on the sensor information and the task level linguistic input, the output of task level reference is ordered by the task sequence.

From the model of the action level reference $q_i = \delta(q_j, \sigma_i)$, it can be seen that the output of this level follows the order of the order of the state transitions on the upper automaton.

It means that if $S_{TRi1} \preceq S_{TRi2}$, then $S_{TR1}S_{ARj}S_{Kj} \preceq S_{TR2}S_{ARi}S_{Ki}$ for any i and j.

Given $a, b \in \{S_{TRi}S_{ARj}S_K\}$, the reflexivity and antisymmetry can be satisfied.

The transitivity can be stated by the ordering characteristics of the three level references. Finally, for any $a, b \in \{S_{TRi}S_{ARj}S_K\}$, satisfy the relation $a \preceq b$ or $b \preceq a$.

A perceptive reference trajectory is $\{S_{TRi}S_{ARj}S_K\}$, a totally ordered set with the binary relation $\preceq$.

Therefore, the perceptive reference can evolve independently over time.

5.2 Hybrid State Space Model for Robotic Manipulations

We denote the state of the integrated system in a hybrid way:

Let $X^h = \{q_T q_A q_M X\}$ be a hybrid state, where q_T is the state of task scheduler, q_A is the state of action planner, q_M is the state of motion planner. q_T, q_T, and q_T are discrete variables, X is the continuous state of the low level control system. Similar to the perceptive reference description, the metric of the state space can be defined by

$$d_D(q_{T1}q_{A1}q_{M1}, q_{T2}q_{A2}q_{M1}) = 3 \text{ if } q_{T1} \neq q_{T2},$$
$$d_D(q_{T1}q_{A1}q_{M1}, q_{T2}q_{A2}q_{M1}) = 2 \text{ if } q_{T1} = q_{T2}, q_{A1} \neq q_{A2}$$
$$d_D(q_{T1}q_{A1}q_{M1}, q_{T2}q_{A2}q_{M1}) = 1 \text{ if } q_{T1}q_{A1} = q_{T2}q_{A2}, q_{M1} \neq q_{M2}$$
$$d(q_{H1}, q_{H2}) = d_D(q_{T1}q_{A1}q_{M1}, q_{T2}q_{A2}q_{M1}) + \parallel X_1 - X_2 \parallel$$

We denote the hybrid system as $dX^h/ds^h = f(X^h, u(s^h))$. Therefore, the entire system can be described in the hybrid space.

6   Dynamical Properties of Switching System in Perceptive Frame

During switching between the tasks, which can be described in subspaces, stability is a very crucial issue. The values of the vector in each subspace represent the parameter disturbance of the low level controllers.

6.1   Description of Task Space and Action Space

Task space can be defined as a linear space. According to the definition in the linguistic control model. Task Space is a N-dimensional vector space.

$$[T_1 T_2 ... T_n]^T \tag{6.20}$$
$$T_i = diag[0...I_i...0] \tag{6.21}$$
$$[T_1 T_2 ... T_n]^T = [0...x_{i1}, x_{i2}...x_{ip}...0]^T$$

Where I_i is a diagonal matrix with order p. Therefore, for each task, we have a multidimensional vector space. For several independent tasks, the combined task space can be generated and, each task is defined in a subspace. For example, the T_i is a task in p dimensional space. The dimension of the subspace depends on the specific task. Similarly, the action space can be defined as:

$$[A_1 A_2 ... A_m]^T \tag{6.22}$$

$$\begin{aligned} A_j &= diag[0...D_j...0][A_1 A_2 ... A_m]^T \\ &= [0...x_{j1}, x_{j2}...x_{jq}...0]^T \end{aligned} \tag{6.23}$$

Where D_j is a diagonal matrix with order q. The transform would result in mapping the task from a task subspace into an action subspace. The mapping $T- > A$, therefore, can be described as a linear mapping.

6.2   Mapping from Task to Low Level Controllers and Regions

According to the control model of action planning,

$$O_k^A = L(q_i, \sigma_i, \sigma_k) \tag{6.24}$$

$$
\begin{aligned}
&= L((q_i, \sigma_k), \sigma_i) \\
&= L((q_i, \sigma_k), diag[0...D_i...0][T]) \\
&= L'(q_i, \sigma_k, T)
\end{aligned}
\tag{6.25}
$$

Where the D_i is a diagonal matrix. The switch between the task level inputs is equivelent to the switch of the Linear mappings. Given a different task level input, there exist a linear mapping to generate the the vectors in Action Space.

$$
\begin{aligned}
O^M &= L_M(\eta(q_i^M, \sigma_i, s)) \\
&= L_M''((q_i, s), \sigma_i) \\
&= L''((q_i, s), diag[0...D_j...0][A]) \\
&= L''(q_i, s, A)
\end{aligned}
\tag{6.26}
$$

Where the D_j is a diagonal matrix. L' and L'' are linear mappings.

6.3 Perceptive Reference

The system model, shown in Figure 1, can be described with perceptive references and states, which can be expressed in a hybrid fashion. A hybrid perceptive trajectory is a finite or infinite sequence of reference interval $e = \{S_{TRi}S_{ARj}S_K\}$, where $S_K \in [S_k, s_k']$, $S_{TRi} \in \Sigma_{RT}^H$, $S_{ARj} \in \Sigma_{RA}^H$. Furthermore, An execution of a hybrid perceptive automaton describes a collection $\chi = \{e, q, x\}$, where e is a hybrid perceptive reference trajectory. q is a map from $S_{TRi}S_{ARj}$ to q, x maps the continuous reference s to the continuous state space.

6.4 Switching Conditions for Low Level Controllers

For the system $dx/ds = f(x)$, we say that $V(s)$ is a candidate Lyapunov function if $V(s)$ is a continuous positive definite function (about the origin) with continuous partial derivatives. Note this assumes $V(0) = 0$.

Considering switched hybrid systems, the systems operate in a hybrid metric space. The stability property of such systems can be described in a hybrid state space of the perceptive frame. Therefore, multiple Lyapunov function can be used to discuss the stability issue.

Given a dynamical system in the perceptive frame, if there exist Lyapunov functions $V_1(s^h)$, $V_2(s^h)$... with hybrid perceptive references, corresponding to the hybrid perceptive reference trajectory and different segments of executions, for a given strictly increasing sequence of $S_{TRi}S_{ARj}$ (discrete reference values), denoted by $I = S_{TRi}S_{ARj}$ in perceptive reference, we say that V is a *Lyapunov-like* function for function f and execution $\chi = \{e, q, x\}$, if

1. Given $s \in (s_i, s_{i+1})$ where $dV(x(s^h))/ds \leq 0$
2. V is monotonically non-increasing on ordered set I.

Theorem 1 (Stability): *Suppose we have candidate Lyapunov functions $V_i, i = 1...N$, and vector field $dx/ds = f_i(x)$ with $f_i(0) = 0$, Let E be the set of all switching sequences* *associated with the system. If for each $s^h \in E$ we have for all i, V_i is Lyapunov-like for f_i. Then the system is stable in the sense of Lyapunov.*

Proof: Given a strictly increasing sequence of $S_{TRi}S_{ARj}$, we denote the time at which the reference happen as $\tau(S_{TRb}S_{ARb})$. Since the perceptive reference is an ordered set by relation $\preceq$, i.e., $\tau(s_a^h) \leq \tau(s_b^h)$, for any $s_a^h \preceq s_b^h$. Based on the stability theorem for continuous perceptive system, the corresponding time based functions have the identical monotony. According to [6] for the case if $N = 2$, or $N > 2$, we always can pick a small neighborhood of the origin with ρ from that the trajectory will stay in $B(r)$, which is the minimum ball around the origin over all the possible Lyapunov functions.

Theorem 2(Exponential Stability): *Suppose we have candidate Lyapunov functions $V_i, i = 1...N$, and vector field $dx/ds = f_i(x)$ with $f_i(0) = 0$, Let E be the set of all switching sequences associated with the system. In the perceptive frame, if for each $s^h \in E$ we have for all i, V_i is Lyapunov-like for f_i. Then the system is exponentialy stable in the sense of Lyapunov. If There exist constants $\alpha_g, \beta_g, \gamma_g, g = 1, ..., l$, such that:*

$$-- \quad \forall x \in \Omega, \alpha_g(\| x \|) \leq V_g(x) \leq \beta_g(\| x \|) \tag{6.27}$$

$$-- \quad \forall x \in \Omega,$$

$$-- \quad \forall x \in E_{gr}, V_r(x) \leq V_g(x). \tag{6.28}$$

for switching from q to r.

Proof: Since the perceptive reference can be introduced in to the hybrid system. The Lyapunov function V_r has the reference segment, which evolves from 0, after discrete transitions. Then in the given case, $V_r(x) \leq V_r(x(t_1))/(-\gamma_r/\beta_r(s))$, since $\forall x \in E_{gr}, V_r(x) \leq V_g(x)$ for switching from g to r.

$$\| x(t) \| \leq \delta e^{-\delta(s-s^h)} \| x(s^h) \| . \tag{6.29}$$

Where perceptive reference s^h is a discrete transition. Therefore the system is exponentially stable in sense of Lyapunov.

6.5   Continuity of Input-Output Mapping

The offset on the continuous variables of the task level linguistic inputs. For the tasks, which have the same discrete task but different continuous variables, i.e., two task controls in the same task subspace, if the continuous parts approximate sufficiently, then the states of the system are sufficiently close to each other.

It can be formally described by followings:

For a given task subspace T_i, the mapping L is referred to as a continuous mapping, if there always exists a ϵ, for given δ such that $\| \Delta A \| < \epsilon$, when $\| \Delta T \| < \delta$. The intuitive meaning is that both the task and the action will evolve continuously. The norm $\| \Delta A \|$ is referred to as the sum of all the action offset. I.e., $\| \Delta A \| = \Sigma \| \Delta A_i \|$. Thus, the input and output of the task controller in the linear space T and A are continuous dependent.

6.6   Stability Conditions for Switched Systems

The stability conditions include two parts. First, the switching between the task subspace with different dimension is stable under certain conditions of Lyapunov functions. Second, the continuous variable of the task input do not spoil the stability, if for a given value, the system is stable.

Assumption: The input and the output of the task controllers in linear space are continuous dependent.

Claim 3: If the vector fields corresponding to the switching task T1 and T2 in Ω_1 and Ω_2 are exponentially stable. Then given any continous part T1 and T2, the switching tasks will be stable.

Proof: The control input of the low level controller is generated by linear mappings $L : T-> A$ and $L' : A-> \Omega$, so the mapping $L(L')$ follows lipschitz condition, such that $\| L(L')(x) - L(L')(y) \le l \| x - y \|$.

For the conditions of the theorem 2 still hold, for a close loop controller, we have:

$$\forall x \in \Omega_i, \partial V_i(x)/\partial s = \partial V_i(x)/\partial x \cdot f(x, s), \qquad (6.30)$$

The from different vectors T in the same task subspace, Δu is defined as the pertubation the controller caused by the offset of the vector T1 and T2. Substitute linear mappings $L : T-> A$ and $L' : A-> \Omega$ and lipschitz coefficient l.

$$\| x(t) \| \le 1/l\delta \cdot (e^{-\delta(s-s_i^h)/l}) \| x(s_i^h) \| . \qquad (6.31)$$

The system is still stable in sense of Lyapunov during the task switches.

7   Modeling of Unexpected Event Processing

The ability to deal with unexpected events is crucial for a perceptive control system. Q represents the set of the vector fields, $(q_1, ..., q_n)$, for a m-dimensional space, at least it has m m-dimensional orthogonal vector fileds. $f_1...f_m$. Therefore, for any 2 points $x1$ and $x2$, the system can start from $x1$, $x2$ within finite time of switches over the vector fields.

In task execution, the continuous reference s should satisfies: $L_f S(q, x) > 0$, it guarantees that the continuous reference is an increasing function of the time t.

7.1   Unexpected Events(UE)

Unexpected events can be described as follows:

For a given $e = \{S_{TRi} S_{ARj} S_K\}$, where $S_K \in [S_k, s_k']$, $S_{TRi} \in \Sigma_{RT}^H$, $S_{ARj} \in \Sigma_{RA}^H$. and an execution of a hybrid perceptive automaton $\chi = \{e, q, x\}$, A unexpected event happends, when the following two condition hold:

$$ds/dt \mid_{s_u^h} = 0, s_u^h \in (s_0^h, s_f^h), \qquad (7.32)$$

$$s_u^h \neq S_{TRi}, S_{ARj}. \qquad (7.33)$$

Furthermore, an unexpected event represents a convex region U_A that is an open set of states, in which the above conditions hold, the boundary is a scalar function $\omega(x) = 0$, and $\omega(x) > 0$ for $x \in U_A$, when UE happen, using "Lie Derivative", we have $L_f \omega(x) > 0, \omega(x) = 0$.

It is a local blocking event of Task T if we can find a subset of the trajectory $\chi_1 = \{s_t, q, x\}$, where $s_t = (s_u^h, s_f^h) \notin U_A$.

7.2   Unexpected Event Processing

For hybrid automata, the unexpected events can be described as a disturbance for a system. The system is desired to be able to return to the designed trajectory after the disturbance. Or, the unexpected events can be treated as a task executed before finishing the current task. The new task is to switch out of the blocking state then return to and resume the previous task. For the given model, the unexpected event will affect the states, causes a discrete state transition. A new vector field will be applied on the continuous states of the system. which is an orthogonal vector field to the old one, $< f_i(x) \cdot f_{i+1}(x) > = 0$, then it can satisfy the following conditions: $L_{f_{i+1}} \omega(x) < 0$, $L_{f_{i+1}} S(q, x) > 0$, the former makes the system leave domain U_A from the boundary $\omega(x) = 0$, the latter guarantees the evolution of the continuous reference.

The following theorem shows the existence of the solution at unexpected events in the perceptive frame.

Theorem 4: *if the unexpected event e_u is a local blocking event, there exists another finite automaton and with the same alphabet, corresponding to the new automaton, the automaton can go back to the original task trajectory $\chi = \{e, q, x\}$.*

The following is the sketch of the proof of the Theorem 4. Since the given UE is a local blocking event, $B - U_A$ is a path connected domain. There exists a sequence of vectors connecting x_{e_u} and $x_{s_{return}}$, $s_{return} \in s_t$, which is $l_1...l_p$, p is a finite number. Since the vectors can be described as the trajectory in the given orthogonal vector fields though a linear mapping M, then the trajectory can be decomposed as a finite number of the connected segments in the switching vector fields. We denote the segments with the action level alphabet, as $w_1...w_m$, m is a finite number.

Based on the sequence of the segments, a new automaton can be built for UE processing and switching back to the original task. The string of the corresponding automaton is $w_1...w_m$, which is embedded in the string of original task execution.

7.3   Unexpected Event Processing in Mobile Manipulation

In mobile manipulation, after an unexpected event that blocks the evolution of continuous reference is detected, the task reference generates the UE with the obstacle info $T2(m_o, n_o, l_o)$, the output of the task scheduler is $T2(m_o, n_o, l_o)$. It causes a discrete switch over discrete variables, an obstacle avoidance on the task execution to switch and resume the evolution of the continuous perceptive reference. When the effect of the unexpected event is removed, the robot will resume the previous task to reach the destination. The perceptive reference is never blocked.

The transition function and the output function are $q_{AP2} = \delta(q_{AP1}, T2(m_o, n_o))$ and $o_{AP2} = \eta(q_{AP2}, reset) = A2(m_{a2}, n_{a2}, r_{a2})$, respectively.

The motion planner receives the action $A2(m_{a2}, n_{a2}, r_{a2})$ from the action planner to generate the trajectory.

This automaton has a state transition on the upper automaton due to the unexpected event and returns to the previous state after processing the unexpected event. The triggered procedure is the same as in the task execution process.

In real time tele-operated systems, it is important to synchronize the command from the operator to the arm controller. The time delay of the communication between the operator and controller will significantly affect the control performance. If the controller can not receive the next position or velocity command before it finishes executing the current one. It has to stop or just keep going on the current velocity, the former reduces the smoothness of execution while the later causes large control errors. In the hybrid perceptive model, the delayed command is thought of as an unexpected event on action level, that can prevent the discrete reference from evolving.

The routing for the unexpected event is to switch to another action which decelerates the motion to make the trajectory smooth until the next control command is received then switch back to the previous state and the path plan is modified. As the result, the hybrid perceptive reference can not be blocked.

8   Experimental Implementation and Results

The proposed perceptive control model has been tested using a Mobile Manipulator-Phantom Joystick tele-operation System consisting of a Nomadic XR4000 mobile robot, a puma560 robot arm mounted on the mobile robot and a phantom joystick controller. There are two PCs on the mobile platform for the control of the mobile robot and the Puma, respectively. Another PC is driving the Phantom joystick and communicate with the Puma controller through the Internet.

8.1   Experimental System Setup

The experimental system, as shown in Figure 6, is an integrated control system. The hardware structure and the software structure of the system will be introduced.

8.1.1   Hardware Structure

The robot controller is a computer control system. The mobile manipulator has two PCs for the Puma 560 control and the mobile robot. Both PCs in the mobile manipulator are communicating with the workstation or PC whose interface can have humans involved into the control process. Human intervention is involved through joystick controller. Two kinds of

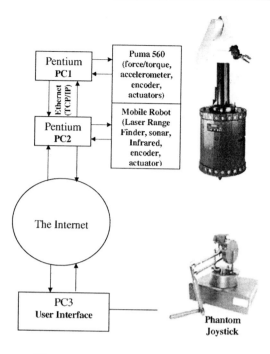

Figure 6: The Experimental System

force-feedback joysticks are employed in the experimental system: the MicroSoft(MS) joystick, which can send the human motion command in velocities, and the Phantom joystick, which issues the motion command in positions.

The mobile manipulator has a laser range finder, which can measure the distance of the object ahead with a uniformly distributed bundle of laser lights. The encoders of the mobile manipulator can measure the position of gripper.

8.1.2   Software Structure

The control software of the experimental system consists of:

- **Robot Arm Control program**: A program for controlling the Puma 560.

- **Mobile Base Control program:** This is designed for motion control of the base and collection of sensor measurements.

- **Mobile Base Control Proxy:** It is a program to receive the command from the shared memory and send the commands to Mobile Base Controller in PC2.

- **Hybrid Linguistic Control program:** This is the high level controller of the experimental system, running in the perceptive frame.

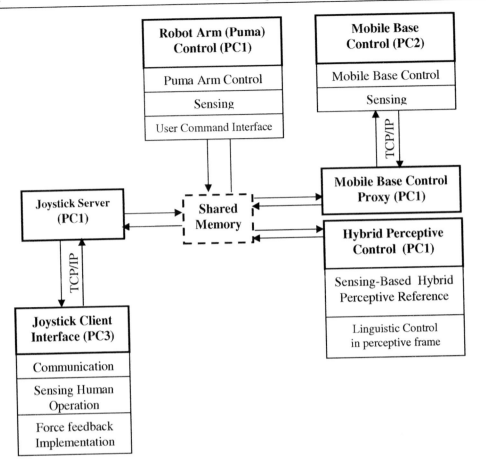

Figure 7: The Software Structure of the Experimental System

- **Joystick Client User Interface:** The interface software converts the measurements of human operation on the joystick into motion control signals and sends the control signals to a remote motion controller.

- **Joystick Control Server:** This program is running in PC1 and designed to receive the commands from Joystick Client User Interface, then send the commands to the shared memory in PC1

Figure 7 shows the modules in each program and the relations between the control programs, based on the hardware structure shown in Fig. 6.

8.2   Robot Controller

The mobile manipulator can be controlled remotely by an operator through the joysticks or can run autonomously by a computer program.

In perceptive frame, the robot controller can be described as

$$\tau = D(q)J_h^{-1}[\ddot{Y}^d(s) + K_v\dot{e}(s) + K_Pe(s)$$
$$- \dot{J}_h\dot{q}] + C(q, \dot{q}) + G(q)$$

(8.34)

The dynamics of the robot is described as

$$\frac{dw}{ds} = 2a$$

(8.35)

$$\frac{da}{ds} = u$$

(8.36)

Where $W = v^2$, s is the continuous perceptive reference.

8.3   Experimental Results

The first task is designed as picking up an object followed by transporting it to a destination autonomously, i.e., without teleoperation. An unexpected obstacle blocks the mobile robot and prevents the continuous event from evolving while the robot is executing the transporting task. In this implementation, environmental information is involved in the reference processing. The experimental results show that the proposed model is able to keep the perceptive reference evolving and trigger multiple events.

Figure 8 shows the trajectory of the experiment result. The movement started from $P0$ and went straight to point $P1$, and picked up an object at this point. During the object transportation phase to $P4$, it modified the designed trajectory and avoided the obstacle at $P2$, From there, it went through a segment of an arc with a length of π and from point $P3$ resumed the previously designed trajectory, From Figure 9b, it can be seen that the hybrid perceptive reference always evolves, The stagnation of the hybrid perceptive reference can be avoided by a discrete transition at $P2$.

As designed, the action reference has triggered multiple events including the moving path segment and action"Close Gripper"and"Open Gripper" at $P1$ and $P4$.

The second task is designed on a Phantom Joystick-Puma560 arm joint tele-operation to show unexpected event processing at the action level. As shown in Figure10, a phantom joystick is used to control a Puma 560 robot arm through the Internet. Phantom controller can measure and send out the position of the joystick in the phantom work space with a fixed rate of 10 Hz. The Puma arm controller receives the information as motion commands and maps it to the puma arm work space. The control frequency of the arm controller is 500Hz. Therefore, the arm runs 50 control cycles for one phantom command. The task is to implement position tracking from Phantom space to Puma arm space in real-time. This experiment is designed for processing the expected events caused by the time delay of the Internet communication between the joystick and controller. If the phantom controller can not receive the next position command within 50 cycles of the current command, it is an unexpected event which can block the discrete perceptive reference.

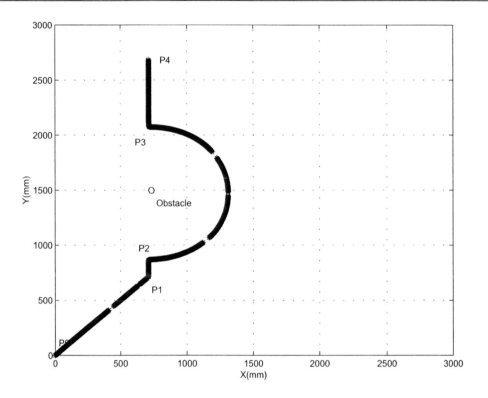

Figure 8: The motion trajectory in the experiment.

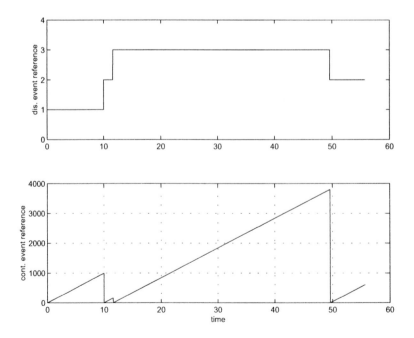

Figure 9: Discrete and Continuous perceptive references w.r.t time During the Motion.

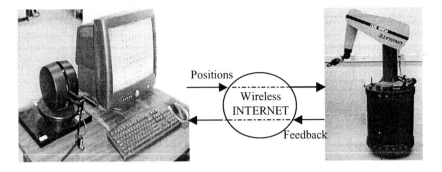

Figure 10: The phantom joystick teleoperation.

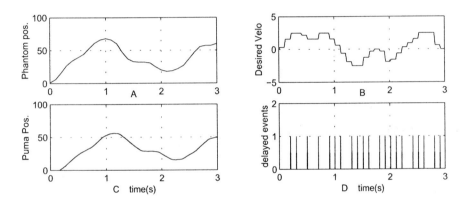

Figure 11: Unexpected Events in Phantom Manipulation

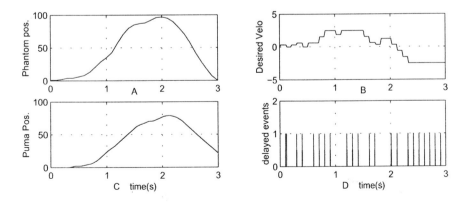

Figure 12: Unexpected Events in Phantom Manipulation along y-axis

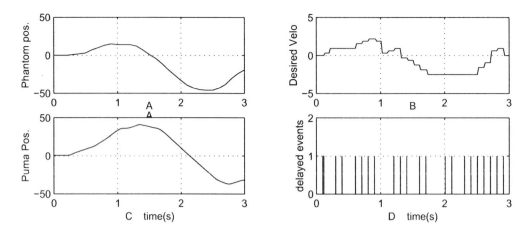

Figure 13: Unexpected Events in Phantom Manipulation along z-axis

Figure 11, 12, and 13 shows the Phantom-Puma position tracking control result on z axis. Fig.6 A illustrates the trajectory from phantom position data. Fig.6 C shows that the Puma position tracking is smooth with unexpected event processing. As shown in D, it can be found that the unexpected events (delayed command) occurred during the task execution. Corresponding to Fig.6 D, Fig.6 B shows that the perceptive velocity control signal is continuously evolving, i.e., the proposed hybrid reference can not be blocked at the occurrence of the unexpected events that may block the evolution of the single discrete perceptive reference.

9   Conclusion

A hybrid model for the perceptive robotic plan and control systems has been proposed. The model integrates the continuous perceptive reference and discrete references based on hybrid automata and offers improvements on continuous perceptive control models for planning and control of robotic systems. The hybrid perceptive reference keeps evolving despite the occurrence of an unexpected event, which could block the continuous perceptive reference or the discrete perceptive reference. It can be shown that the system is stable during the switches. The experimental results show the effectiveness of the proposed model to keep the hybrid perceptive reference evolving through the modification of the designed plan.

References

[1] N. Xi, "Event-Based Motion Planning and Control for Robotic Systems". Ph.D thesis, Washington University in St. Louis, 1993.

[2] N. Xi, T. J. Tarn and A. K. Bejczy "Intelligent Planning and Ctrl for Multirobot Coordination: An Event-Based Approach", *IEEE Transactions on Robotics and Automation* Vol.12, June, 1996.

[3] M. Song, T. J. Tarn, and N. Xi "Integration of Task Scheduling, Sensing, Planning and Contron in Robotic Manufacturing system", *Proceedings of the IEEE*, Vol. 88, July, 2000.

[4] J. Lygeros, K. H. Johansson, S. N. Simic, J. Zhang, and S. S. Sastry "Dynamical Properties of Hybrid System", *IEEE Transactions On Automatic Control.* Vol.48, No.1, January, 2003.

[5] H. Ye, A. N. Michel and L. Hou "Stability Theory for Hybrid Dynamical System", *IEEE Transactions On Automatic Control* Vol.43, No.43, pp461-pp474, April, 1998.

[6] M.S Braniky, "Studies in Hybrid Systems: Modeling, Analysis, and Control", Ph.D Thesis of M.I.T. 1995.

[7] Stefan Pettersson and Bengt Lennartson, Controller Design of Hybrid Systems, in"Hybrid Systems", volume 1201 of *Lecture Notes in Computer Science,* Springer-Verlag, New York, 1993.

[8] R. L. Grossman, Anil Nerode, A. P. Ravn, and H. Rischel. editors. "Hybrid Systems", volume 736 of *Lecture Notes in Computer Science,* Springer-Verlag, New York, 1993.

[9] P. J. Antsaklis, M. D. Lemmon and M. D. Lemmon. "Hybrid Systems Modeling and Autonomous Control Systems"volume 736 of *Lecture Notes in Computer Science,* Pages 366-392, Springer-Verlag, 1993.

[10] M. S. Braniky, V. S. Borkar, and S. K. Mitter "A Unified Framework for Hybrid Control" *Proceedings of the 33rd Conference on Decision and Control,* Lake Buena Vista, FL, Dec. 1994.

[11] R. W. Brockett."Hybrid Model for Motion Control Systems."in *Perspectives in Control.* Eds. H. Trantelman and J. C. Willems, pp.29-54, Birkhauser, Boston, 1993.

[12] R. Brockett, "Language Driven Hybrid Systems" *Proceedings of the 33rd Conference on Decision and Control,* Lake Buena Vista, FL, Dec., 1994.

[13] M.Egerstedt, "Linguistic control of Mobile Robots" *Preceedings of IEEE/RSJ International Conference on Intelligent Robotic System.* Maui, Hawaii, USA, Oct. 2001.

[14] V. Manikonda, P. S. Krishnaprasad and J. Hendler, "Behaviors, Hybrid architectures and Motion Control."*In Mathematical Control Theory.* pp. 199-226, Springer-Verlag, 1998.

[15] J. E. Hopcroft. R. Motwani and J. D. Ullman, "Introduction to Automata Theory, Languages, and Computation." 2nd Edition, Addison Wesley, 2001.

[16] M. Egerstedt, "Behavior Based Robotics Using Hybrid Automata." *Lecture Notes in Computer Science: Hybrid Systems III* pp.103-116, Springer-Verlag, March 2000.

In: Control and Learning in Robotic Systems
Editor: John X. Liu, pp. 149-180

ISBN 1-59454-356-9
© 2005 Nova Science Publishers, Inc.

Chapter 6

PATTERN-BASED HEURISTICS FOR MASTER PRODUCTION SCHEDULING WITH A CONTINUOUS MANNER

Hsiao-Fan Wang

Department of Industrial Engineering and Engineering Management,
National Tsing Hua University, Hsinchu, Taiwan 300

Kuang-Yao Wu [b]

Department of Business Management, National United University,
Miaoli, Taiwan 360

Abstract

This study concerns with modeling and solution of a Master Production Scheduling problem characterized by multi-timeframe, multi-product and multi-resource. The considered problem is a kind of scheduling problems with setup carryover and very expensive setup cost, and thus, it is expected that each individual product is set up once along the scheduling periods. Therefore, the main issue for the MPS problem is how to schedule production in a continuous mode of periods. Giving the demand quantity, production capacity and operational principles, the considered problem was formulated into a mixed-quadratic-binary program (MQBP). This MQBP model can be transformed into a mixed-binary program and solved by the existent software packages. However, this optimization approach is time-consuming, especially in large scaled problems. To facilitate subsequent solution analysis, this MQBP model was decomposed into two inter-dependent sub-models of a quadratic binary constraint (QBC) and a linear program (LP). In order to satisfy both speed and accuracy requirements, a two-phased heuristic approach was proposed. In Phase 1, the search space of the QBC sub-model was reduced into a preliminary scheduling pattern and thereby a reference model from the assignment problem is formed. In Phase 2, according to the permutation property in the reference model, a stochastic global optimization procedure that incorporates a genetic algorithm with neighborhood search techniques was applied to obtain a desirable solution. Numerical evidence has shown that the proposed pattern-based heuristic approach is effectively applicable for an uninterrupted production schedule.

Keywords: Production scheduling; Mixed-binary programming; Permutation optimization; Hybrid genetic algorithm; Neighborhood search

1 Introduction

Production scheduling is in essence a decision-making process, in which the resources, such as manpower, equipment and tools, are allocated in line with pending orders and forecast demands in a fixed horizon as the basis for manufacturing such that the minimum production cost can be obtained. It is considered as a detailed execution plan over scheduling periods, and has been studied as the *capacitated lot sizing problem* (see, e.g., Bahl *et al.*, 1987; Hung and Hu, 1998) if the setup times and setup costs are significant. Let us consider a kind of scheduling problems with *setup carryover* (Sox and Gao, 1999) and very expensive setup cost, from which it is expected that each individual product is set up once along the scheduling periods. That is, once the product is on a production line, its production procedure is persisted until completed. This leads to the so-called *continuous production-period scheduling* problem in which how to schedule production in a continuous mode of periods is addressed. This kind of requirement can be found in a common manufacturing environment such as a batch process, for example, in a real case of LED industry (Wu *et al.*, 2000). Therefore, this motivates our interests in study. Gue *et al.* (1997) also recognized the merit of the uninterrupted processing and emphasized its importance in practice and on research. However, from the literature, it can be noted that little has been developed for modeling and solution analysis of such scheduling problems.

Let us consider a case of Master Production Scheduling (MPS) problem involved multi-period, multi-product and multi-resource, of which multiple levels of capacity and total demand of production are taken into account. This study is meant to investigate the modeling of continuous production-period scheduling with the MPS case, and to develop the corresponding solution approaches for optimization. As a practical result of the real case, this study is developed in terms of the actual management of executing the production process, especially how the production can be evenly distributed throughout a sequence of production periods for a given amount of products. To present the features we concerned above, a Mixed-Quadratic-Binary Programming (MQBP) model based on the early work of Wang and Wu (2003) is introduced first in this study. For solving our concerned problem, two solution approaches of mixed-binary programming and pattern-based heuristic are developed. In the mixed-binary programming approach, a linear transformation for the quadratic form in the model is investigated so that the MQBP model can is addressed as a mixed-binary program. However, the considered mixed-binary problem is an NP-hard one (Vavasis, 1991). With the rapid increase of computer capability and the growing need for accuracy in scheduling, it becomes increasingly important to explore alternative ways of obtaining better schedules with additional computation costs. Thus, the pattern-based heuristic approach with a two-phased solution procedure is proposed for obtaining a compromised solution of the MQBP model. In the first phase, a heuristic algorithm for solution pattern mining is developed so that the search space in terms of combination can be reduced into a permutation problem. For solving the reduced one in the second phase, a stochastic global optimization procedure (Wang and Wu, 2004a) for the problems with permutation property can be employed, which incorporates a genetic algorithm (GA) with neighborhood search (NS) techniques.

The remainder of this study is organized as follows. In Section 2, a foundational process with MPS is introduced with literature review, from which the core issues in scheduling domain related this study and our other extension works can be depicted with an overall insight. Then a case in interest is described with respect to the related MPS models, from which the features are presented by a deterministic model in a mixed-quadratic-binary programming form, marked by Model (MQBP), in Section 3. To facilitate subsequent solution analysis, Model (MQBP) is decomposed into two inter-dependent sub-models: Model (QBC) with a quadratic binary constraint, and Model (LP) of a linear program in this section. In Section 4, a pattern-based heuristic approach is proposed for solving this MQBP problem. The approach first transforms the search space of Model (MQBP) into a preliminary solution pattern by a heuristic mining algorithm so that the search space is reduced into a permutation space that our developed methods are applied for. Eventually, the concerned MPS problem is treated as a permutation optimization problem and a hybridized GA is then employed as the main solver for the proposed model. In Section 5, Numerical results of the proposed approach to the considered problem are reported. Finally, the conclusions of this study are drawn with the possible extension.

2   Problem Statement

In this section, a case of MPS is described, which has inspired our interests in this research. Therefore, the role of MPS and the related literature is reviewed first. Moreover, the studied case will be compared with two representative MPS models. In the next section, a mathematical programming approach to optimizing the concerned MPS problem is reviewed.

2.1   The Role of MPS

In the last century, there has been a remarkable evolution in production control. From early Material Requirements Planning (MRP), sequential Manufacturing Resource Planning (MRP-II) to Enterprise Resource Planning (ERP), their fundamental relations in a production planning and scheduling system are constructed in a hierarchical framework. The Master Production Scheduling process is usually cited as the primary level where the market and production functions can make appropriate trade-offs among customer service, manufacturing efficiency and inventory investments. A typical flowchart of an MRP-II process is sketched in Fig. 1 for depicting the role of MPS in manufacturing environments. It is fair to say that the addition of MPS was a key ingredient in the evolution of MRP to MRP-II to ERP. A practical guide with basic concept to master scheduling can be referred to Gessner (1986) and Proud (1999).

Let us depict the role of MPS in MRP-II, a common planning tool used in industry today, which allows manufacturers to optimize customer service, inventory investment, manufacturing efficient, etc., and to provide financial and planning reports. As shown in Fig. 1, MRP-II is a hierarchical planning tool. Let us follow the process from Production Planning, providing a tactical function that integrates information across forecasting, order, manufacturing and inventory. The result of planning is a decision on production quanta that fulfill those operational requirements. MPS plays the role of linking tactical and operational planning stages. Therefore, the basic task of MPS is to determine the production quantities and production timing of the individual end items or major product options so that demand requirements are satisfied by considering available resources over a scheduling horizon. Moreover, it is an engine that drives MRP.

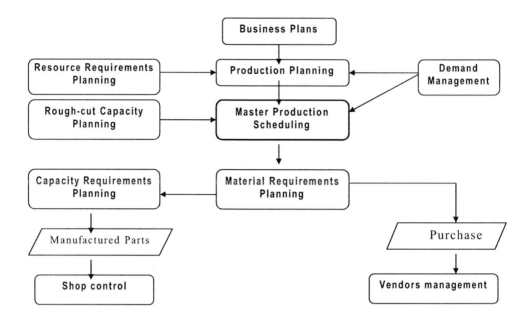

Fig. 1. A typical MRP-II process flowchart (Süer *et al.*, 1998)

Typically, the master schedule is developed using rough-cut capacity technique such as capacity bills and load profits. Because of the inadequacy of the Rough-cut Capacity Planning techniques and the unrealistic nature of the infinite capacity assumption, the inconsistent result of MPS is detected after Capacity Requirements Planning (CRP) (Süer *et al.*, 1998). Actions are often needed at the MRP level to modify the MPS schedule; and at the CRP level to revise available capacity by considering overtime, shift work, temporary workforce, etc. These changes may require MRP models to be run again due to chain effect resultant from the product structure. Therefore, several improvements over standard MRP-II approach were proposed. The advanced models and methods for MRP-II can be referred to Drexl and Kimms (1998). In the following, the literature related to our development of MPS is briefly reviewed.

2.2   Literature Review of MPS

In this section, a review of the recent development on MPS is done according to the following two aspects.

(1) The development of integrating MPS with other activities

As shown in Fig. 1, because MPS is closely related to other planning activities, the quality of a master schedule can significantly influence the total cost, schedule instability and service level of a production inventory system. Thus, several works were focused on the development of structures and models of the integration. In the upstream integration with production planning, an essential issue is to disaggregate product families or groups (e.g. Weinstein and Chung's (1999) and Zäpfel's (1998)). Inventory balance is inherently considered in their models. Moreover, the capacity constraints for MPS are imposed by Resource Requirements Planning because such integrations are based on Aggregate Production Planning. On the other hand, several studies focused on the downstream integration based MPS. Olhager and Wikner (1998) investigated a framework for integrated material and capacity based MPS, which depicted master scheduling typically is dominated by capacity issues in make-to-stock environments and dominated by material issues in make-to-order situations. Concerning capacity issues, Süer *et al.* (1998) proposed a knowledge-based framework for MPS, in which finite capacity scheduling details are used for replacing rough-cut capacity planning so that a realistic MPS can be generated and therefore, the entire MRP-II process can be operated smoothly. In addition to material issues, Chu (1995) proposed a linear program for finding an optimal MPS schedule subject to the restrictions on the supply of material, the total demand and the aggregated capacity resource. Kimms (1998) gave a linear programming model for their main concern on investment level of inventory.

According to APICS Dictionary, 9th Edition (1998), "*The master schedule must take into account the forecast, the production plan, and other important considerations such as backlog, availability of material, availability of capacity, management policy and goals, etc.*" Therefore, issues concerning MPS are complicate. However, in terms of the actual management, little has been studied on the execution of the production process, especially how the amount of production can be evenly distributed throughout a sequence of production periods for the required demand such that the set-up costs can be reduced has rarely been discussed. This is one issue to be addressed in this study.

(2) The development of analyzing schedule stability

Only having a master schedule does not guarantee the success in production and inventory control. As with all processes and tools, the master schedule must be maintained. If the schedule is improperly maintained, many of the benefits from the sales and operations planning process will be lost. Under rolling planning horizons, frequent adjustments to the master schedule are unavoidable. These changes can induce major changes in detailed MRP schedules, and thereby leading to increases in production and inventory costs and deterioration in customer service level. This phenomenon is called *schedule instability* or MRP *nervousness*. Several methods have been suggested to reduce schedule instability in MRP systems (Blackburn *et al.*, 1986). One common used method is to freeze the master schedule (Sridharan *et al.*, 1987), of which a more recent development can be referred to Xie *et al.* (2003).

From these results, it can be remarked that the factors of demand, capacity and cost structure have significant impacts on schedule stability and the way for freezing the master schedule. These results also remind us of the fact that the parameters used for MPS are not always exactly known and stable. Therefore, we should treat the MPS problems by considering the allowed tolerances of parameters. In the last section, our extension works on this issue will be discussed with a linkage of this study.

2.3   A Case of MPS

As Gessner (1986) observed in real applications, "given 20 different companies, 20 different varieties of MPS will exist, each company using their own terms to define what they do and each performing different functions according to their needs." In this study, the case we faced is a local LED manufacturer that adopts make-to-stock strategy for production and sales management. This means that goods are manufactured based on forecasts and placed in a finished goods inventory, of which the goods are delivered. Thus, this notion of scheduling requires that the shipment of the finished goods should be efficient to minimize the inventory cost. In a make-to-stock environment, the capacity investment is typically quite substantial and given a prime focus when designing the process. From the typical instances of the case, it was observed that the sales volume was high that indicated that capacity levels were tight. This leads to a cost focused capacity resource utilization. That is, it would be beneficial to consider some types of over-capacity such as the additional capacity by overtime and outsourcing.

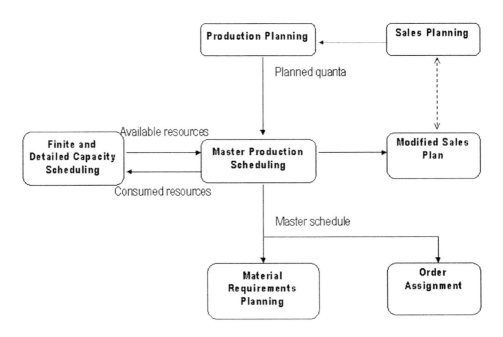

Fig. 2. A process flowchart of the considered MPS

A process flowchart related to the studied case of MPS is depicted in Fig. 2. Based on the result of production planning, the planned quantum of each product is disaggregated among several periods in the MPS. Instead of Rough-cut Capacity Scheduling, Finite and Detailed Capacity Scheduling (Süer et al., 1998) is incorporated into the MPS model, in which, because of limited resource levels such as regular manpower, overtime and out-sourcing, disaggregating products' quanta into each period needs to perform in a complementary manner. The task of MPS is then to conciliate the conflict between the targeted production requirements and the limited capacity resources, and to indicate the shortage for modifying the sale's plan if it occurs. Based on the scheduling results, the scheduling of both order assignment and material requirements can be subsequently executed for production preparation. The decision of due date for delivery is made at the Order Assignment level, and therefore inventory factors are not considered at the MPS level. Therefore, a master schedule not only provides the basis for agile promising of customer order and efficient utilization of capacity resource, but also yields the foundation for reliable purchasing of material resource.

Practically, to facilitate subsequent operations corresponding to the master schedule, the manufacturer implemented three operational principles as follows:

Principle I. For any product, the production quantities in each production period should be above a reasonable level set by production controllers. This principle is used to prevent the schedule result from the unreasonable appearance that a considerable small value for the quantity is yielded, e.g. a value of 1.2 is scheduled in a certain period verse total demand of 1230 for a certain product.

Principle II. During one period, the maximal number of the product's types that can be scheduled is limited and is determined by shop floor supervisors. The master schedule can be considered as a plan of executable work order for short-term jobs at workstations. In order to lessen the burden of site management at the shop floor in advance, this principle imposes a certain limitation in term of product types.

Principle III. For each product, the production is arranged continuously along the production periods. This means that the production periods for one product are arranged in sequence and each product completes its production procedure without interruption in a scheduling horizon. Such production sequence mainly facilitates the sales department to promise and assign orders using display bar chart.

Based on these principles, the schedule result is expected to be organizable, reasonable and traceable so that the master schedule can be smoothly implemented along the subsequent operations. A comparison of our model with two existing models is listed in Table 1. In the following section, a mathematical programming approach is employed to study the concerned MPS problem.

Table 1. A comparison between the considered MPS model and two representatives

MPS model:	This study	Chu's (1995)	Kimma's (1998)
Planed quantum:	Total demand	Total demand	Individual demand in each period
Availability of capacity:	Limited multi- resource with three levels of regular time, over- time and outsource	Limited resource in terms of aggregation	Limited multi- resource with two levels of regular time and overtime
Availability of material for each part of BOM:	Unlimited	Limited supply in each period	Unlimited
Backlog:	Discarded	Discarded	Inventory balance and cost
Operational principles involved:	Yes	No	No
Objective to be optimized:	Minimize total cost	Maximize total profit	Minimize the sum of holding and overtime costs

3   Modeling and Analysis of the Considered MPS Problem

In this section, the production environment of interest is described first, and a mathematical programming model is developed for the concerned MPS problem. Furthermore, the proposed model is decomposed into two sub-models for solution analysis. Then a heuristic approach is proposed in the next section.

3.1   Problem Features

The production system is described in general term as follows: P products are produced at W workstations, and each product has its individual workstation route and assigned working time. Each workstation represents a unit where operators and parallel multi-machines are integrated to complete a specific processing job. The problem is to schedule the production within T periods that are segregated from a planning period. This manufacturing structure is shown in Fig. 3. However, because of limited resource levels such as regular manpower, overtime and out-sourcing at workstations, disaggregating products' quanta into each period needs to perform in a complementary manner. Therefore, the key issue of MPS is to identify the amount of every product that should be produced within each period so that the overall cost can be minimized. The result of such production sequence is used for the planning of material requirements, orders assignment and shift arrangement.

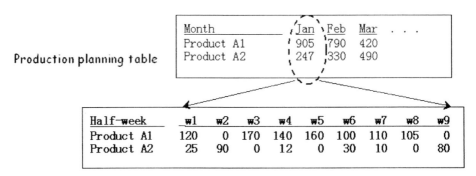

Production planning table

Month	Jan	Feb	Mar	. . .
Product A1	905	790	420	
Product A2	247	330	490	

Half-week	w1	w2	w3	w4	w5	w6	w7	w8	w9
Product A1	120	0	170	140	160	100	110	105	0
Product A2	25	90	0	12	0	30	10	0	80

Master Production scheduling table

Fig. 3. An example of disaggregating plan

Essentially, the concerned MPS problem is referred to multi-period, multi-product and multi-resource production scheduling problem with the following assumptions.

- Multi-period: the scheduling time horizon is equally divided into several periods. Under the shelter of sufficient time span, the influences of window delay, setup, transfer and cycle times are ignored.
- Multi-product: the number of product types to be scheduled is more than one. All products are referred to the ones of serial structure (Billington *et al.*, 1983). The products scheduled at a workstation within a time period can be manufactured synchronously.

- Multi-resource: Each product to be scheduled is limited to the available resources of multiple individual workstations. The unit of resources is time-based. The capacity resources are classified into three levels: regular time, overtime and outsourcing time. The production of a product consumes only the resources available in the periods. A product can be out-sourced from any workstation or the workstations associated with its route.

- Miscellanies: The WIP and stock are null in the initial time. The production amount, manufacturing cost and processing time satisfy the assumptions of linear programming.

Mathematical programming is commonly used to study the scheduling problem. The advantage of using a mathematical programming to tackle such problem lies in a freer domain in which many other constraints can be introduced if required. Based on the early work of Wang and Wu (2003), a MPS model will first be shown in the following and will be discussed later.

3.2   Formulation of the Proposed Model (MQBP)

The considered MPS problem is described by a mixed-quadratic-binary programming model as follows.

Model (MQBP). A mixed-quadratic-binary program for the considered MPS problem

(1) The sets of indices:

$PI=\{1, 2, ..., P\}$: the set of the indices of product type;

$WI=\{1, 2, ..., W\}$: the set of the indices of workstation type;

$TI=\{1, 2, ..., T\}$: the set of the indices of scheduling period.

(2) Decision variables:

For all $i \in PI, j \in WI$ and $k \in TI$,

$x_{ik}^{n} \in R$: the amount of Product i to be produced at inside workstations within Period k;

$x_{i}^{o} \in R$: the amount of Product i to be out-sourced for its whole production route;

$x_{i}^{s} \in R$: the shortage of Product i to be released for sales management;

$x_{ijk}^{v} \in R$: the amount of Product i to be produced by overtime at Workstation j within Period k;

$x_{ijk}^{u} \in R$: the amount of Product i to be out-sourced from Workstation j within Period k;

$\dot{y}_{ik} \in B$: the variable of monitoring x_{ik}^{n} (i.e. record 0 for suspending Product i from Period k, and vice versa for 1).

Let $\dot{\mathbf{x}} = (x_{ik}^{\mathbf{n}}, \forall i \in PI, \forall k \in TI,\ x_i^{\mathbf{o}}, x_i^{\mathbf{s}}, \forall i \in PI,\ x_{ijk}^{\mathbf{v}}, x_{ijk}^{\mathbf{u}}, \forall i \in PI, \forall j \in WI, \forall k \in TI) \in \mathbb{R}^{2PWT+PT+2P}$

denote the vector of the real decision variables, and $\dot{\mathbf{y}} = (\dot{y}_{ik}, \forall i \in PI, \forall k \in TI) \in \mathbb{B}^{PT}$ the vector of the binary decision variables.

(3) Coefficients:

For all $i \in PI, j \in WI$ and $k \in TI$,

$\dot{a}_{ij} \in \mathbb{R}$: the processing time for Product i to be completed at Workstation j;

$b_i^{\mathbf{d}} \in \mathbb{R}$: the output for the total production units of Product i to satisfy the demand within a production plan;

$b_i^{\mathbf{e}} \in \mathbb{R}$: the lower bound of Product i to be manufactured in production periods;

$b_{jk}^{\mathbf{r}} \in \mathbb{R}$: the available resource-hours of regular shift provided by Workstation j within Period k;

$b_{jk}^{\mathbf{v}} \in \mathbb{R}$: the available resource-hours of overtime provided by Workstation j within Period k;

$b_{jk}^{\mathbf{u}} \in \mathbb{R}$: the available resource-hours of out-sourcing that can be provided for Workstation j within Period k;

$b_i^{\mathbf{o}} \in \mathbb{R}$: the upper bound of Product i to be out-sourced during the scheduling horizon;

$c_i^{\mathbf{r}} \in \mathbb{R}$: the total cost of regular time for Product i to be produced at whole route;

$c_i^{\mathbf{s}} \in \mathbb{R}$: the penalty or cost associated with the shortage of Product i;

$c_i^{\mathbf{o}} \in \mathbb{R}$: the total cost of out-sourcing on Product i;

$c_{ij}^{\mathbf{u}} \in \mathbb{R}$: the extra cost of out-sourcing for Product i at Workstation j;

$c_{ij}^{\mathbf{v}} \in \mathbb{R}$: the extra cost of overtime for Product i at Workstation j;

$T^{\mathbf{U}}$ and $T^{\mathbf{L}}$: the upper and lower bounds in term of production periods in a sequence for all products;

$P_k^{\mathbf{U}}$ and $P_k^{\mathbf{L}}$: the upper and lower bounds in term of product types for the allowable production in Period k.

(4) Objective function: to minimize $\Phi_{\mathrm{MQBP}}(\dot{\mathbf{x}}, \dot{\mathbf{y}})=$

$$\sum_{i \in PI}\left(c_i^{\mathbf{o}} x_i^{\mathbf{o}} + c_i^{\mathbf{s}} x_i^{\mathbf{s}} + c_i^{\mathbf{r}} \sum_{k \in TI} x_{ik}^{\mathbf{n}} + \sum_{j \in WI}\left(c_{ij}^{\mathbf{u}} \sum_{k \in TI} x_{ijk}^{\mathbf{u}} + c_{ij}^{\mathbf{v}} \sum_{k \in TI} x_{ijk}^{\mathbf{v}} \right) \right). \qquad (1)$$

(5) Constraints:

(i) Planned demand:

$$x_i^o + x_i^s + \sum_{k \in TI} x_{ik}^n = b_i^d, \ \forall i \in PI; \tag{2}$$

(ii) Production capacity:

$$\sum_{i \in PI} \dot{a}_{ij} x_{ik}^n - \dot{a}_{ij} x_{ijk}^u - \dot{a}_{ij} x_{ijk}^v \le b_{jk}^r, \ \forall j \in WI, \forall k \in TI, \tag{3}$$

$$\sum_{i \in PI} \dot{a}_{ij} x_{ijk}^u \le b_{jk}^u, \ \forall j \in WI, \forall k \in TI, \tag{4}$$

$$\sum_{i \in PI} \dot{a}_{ij} x_{ijk}^v \le b_{jk}^v, \ \forall j \in WI, \forall k \in TI, \tag{5}$$

$$x_i^o \le b_i^o, \ \forall i \in PI, \tag{6}$$

$$x_{ik}^n - x_{ijk}^u - x_{ijk}^v \ge 0, \ \forall i \in PI, \ \forall j \in WI, \forall k \in TI; \tag{7}$$

(iii) Operation principles:

$$x_{ik}^n \ge b_i^e \dot{y}_{ik} \ \text{and} \ x_{ik}^n \le b_i^d \dot{y}_{ik}, \ \forall i \in PI, \forall k \in TI, \tag{8}$$

$$\sum_{k \in TI} \dot{y}_{ik} \ge T^L \ \text{and} \ \sum_{k \in TI} \dot{y}_{ik} \le T^U, \ \forall i \in PI, \tag{9}$$

$$\sum_{i \in PI} \dot{y}_{ik} \ge P_k^L \ \text{and} \ \sum_{i \in PI} \dot{y}_{ik} \le P_k^U, \ \forall k \in TI; \tag{10}$$

(iv) Suspending control:

$$\left(\dot{y}_{i1} - 0\right)^2 + \sum_{k=2}^{T} (\dot{y}_{i,k} - \dot{y}_{i,k-1})^2 + \left(0 - \dot{y}_{iT}\right)^2 = 2, \ \forall i \in PI; \tag{11}$$

(v) Domain constraints:

$$x_{ik}^n \ge 0, \forall i \in PI, \forall k \in TI, \ x_i^o, x_i^s \ge 0, \forall i \in PI, \ x_{ijk}^v, x_{ijk}^u \ge 0, \forall i \in PI, \forall j \in WI, \forall k \in TI, \tag{12}$$

$$\dot{y}_{ik} \in \{0,1\}, \forall i \in PI, \forall k \in TI. \tag{13}$$

3.3   Explanation of Model (MQBP)

In general, the requirements of a production plan and the capacity of all workstations are considered in a production scheduling problem. Based on the demands obtained from a production plan, Constraint (2) restricts the total sum of the production from inside shops, out sources and shortage for each product.

Apart from the satisfaction of the demands, the amount of production is constrained by the capacity of the workstations. When the schedule cannot be completed by using the resource of regular shift, the production manager will consider one of the following alternatives for adjustments.

(1) The planned quantum for sales may be cut-down from the shortage of the current schedule. When this case occurs, the penalty or cost (c_i^s) associated with Product i is incurred in terms of loss of sales.

(2) The entire manufacturing process of a given product may be removed. When this case occurs, the cost of out-sourcing (c_i^o) on a given product i will be considered.

(3) The production deficiency of a given workstation is compensated by overtime or out-sourcing. Then, the cost of overtime (c_{ij}^v) or out-sourcing (c_{ij}^u) for each product i at a given workstation j will be considered.

These three alternatives can be streamlined for cross-boarder consideration, as shown in Constraints (3-7). Equation (3) states that the production within a plant should not exceed the processing capacity for all workstations in a time period, and that all within-production ($x_{ik}^n - x_{ijk}^u - x_{ijk}^v$) should not fall below zero. Equation (7) is the boundary condition of (3). The limitation of the occupied overtime and out-sourcing is represented by Equations (4-6). This model provides an adaptive combination of the cost factors associated with out-sourcing production, overtime production, and shortage. Thus, this model can support management decisions in practice.

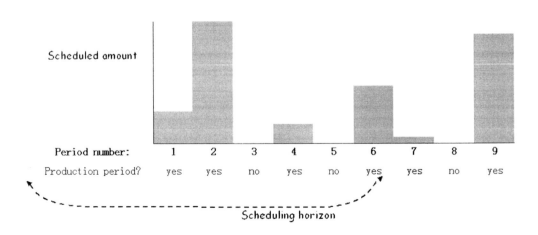

Fig. 4. An interrupted and uneven production scheduling of one product

However, the results of sequencing based on Constraints (3-7) could not avoid interruption occurred in the actual production as shown in Fig. 4. Such an ill-organized scheduling solution is no benefit to practical implementation. Especially, in a common manufacturing environment such as a batch process, it is expected that each individual product is set up once along the scheduling periods. Thus, additional operational issues are considered by three principles stated in Section 2.3.

Table 2. A sequence of production periods represented by $\dot{y}_{ik}$

	Lead	$\dot{y}_{i1}$	$\dot{y}_{i2}$	$\dot{y}_{i3}$	$\dot{y}_{i4}$	$\dot{y}_{i5}$	$\dot{y}_{i6}$	$\dot{y}_{i7}$	$\dot{y}_{i8}$	$\dot{y}_{i9}$	End
Case 1:	0	1	1	1	1	1	0	0	0	0	0
Case 2:	0	0	1	1	1	1	1	1	0	0	0
Case 3:	0	0	0	0	1	1	1	1	1	1	0

To incorporate the above constraints into the scheduling model, the 0/1 variable $\dot{y}_{ik}$ is introduced here to indicate whether a production will be operated for Product i at Period k in a continuous production mode. To elaborate such production situation visually, Table 2 is used to illustrate a possible schedule with nine periods. For product i, we encode the bit string $\hat{y}_i$ of "0 $\dot{y}_{i1}$ $\dot{y}_{i2}$ ... $\dot{y}_{iT}$ 0" where dummy value 0 is added as the leading bit and the ending bit; for $k=1,2,...,T$, $\dot{y}_{ik}=1$ for production, and vice versa for 0. To manufacture a product in a sequence of production periods, only one change of operations on each production-to-suspension and that of suspension-to-production are allowed. Thus, the total number of state changes on the adjacent periods is equal to two. That is, there are two pairs of the adjacent bits in which the absolute value of the difference between each pair equals one. A quadratic form is thus formulated in Equation (11). The aforementioned principles are also shown in Equations (8-10). Constraint (8) is the boundary condition that allows the value of x_{ik}^n to be controlled by $\dot{y}_{ik}$, and enforces the products to be scheduled as widely spread as possible in each period. In Equation (8), the lower bound b_i^e can be defined by pegging to the ratio of T over b_i^d and so forth, so that the demand of Production i can be satisfied at an aspiration level of the lower limits. In Equation (9), the number of production periods is restricted within an interval, whereas in Equation (10) is that of product types. In practice, giving of P_k^U and P_k^L depends on the number of predetermined product types, of which those products have preempted the schedule for pilot run in Period k.

In summary, Model (MQBP) provides a processing mechanism, which not only satisfies the total demand of a production period encompassing an array of sequencing demands but also levels the production output among a variety of product types. In particular, with an adaptive logical function, several management issues can be addressed. However, Model (MQBP) is in the form of a mixed-quadratic-binary program. For an efficient development of solution procedure, we shall analyze the inter-relations of both the real and the binary variables in Model (MQBP).

3.4   Decomposition of Model (MQBP)

For ease of analysis, Model (MQBP) is decomposed into two sub-models in the forms of a quadratic binary constraint program and a linear program as follows:

Model (QBC). A quadratic-binary constraint sub-model of Model (MQBP)
   Constraints: (*PT* binary variables and $3P+2T$ equations)

   (i) Operation principles: Equations (9) and (10);
   (ii) Suspending control: Equation (11);
   (iii) Domain constraints for the binary $\dot{\mathbf{y}}$: Equation (13).

Model (LP). A linear programming sub-model of Model (MQBP)

   (1) Objective function: to minimize $\Phi_{LP}(\dot{\mathbf{x}} \mid \dot{\mathbf{y}})$=Equation (1).
   (2) Constraints: ($2PTW+PT+2P$ real variables and $PTW+2PT+3TW+2P$ equations)

   (i) Planned demand: Equation (2);
   (ii) Production capacity: Equations (3), (4), (5), (6), (7);
   (iii) Operation principles: Equation (8);
   (iv) Domain constraints for the real $\dot{\mathbf{x}}$: Equation (12).

While $\dot{\mathbf{y}}$ are variables in Model (QBC), they are constants in Model (LP). Let Ω_{QBC} denote the set of the feasible solutions for Model (QBC). As a solution $\dot{\mathbf{y}} \in \Omega_{QBC}$ is obtained, it can lead to a corresponding optimum in Model (LP). Let $\Omega_{LP}(\dot{\mathbf{y}})$ denote the set of the feasible solutions for Model (LP) of the given $\dot{\mathbf{y}}$. The corresponding optimal value of $\dot{\mathbf{y}}$ is defined by

$$\Phi^{*}_{LP}(\dot{\mathbf{y}})=\Phi_{LP}(\dot{\mathbf{x}}^{*} \mid \dot{\mathbf{y}})\leq\Phi_{LP}(\dot{\mathbf{x}} \mid \dot{\mathbf{y}}), \forall \dot{\mathbf{x}} \in \Omega_{QBC}(\dot{\mathbf{y}}). \tag{14}$$

However, this optimum is just a sub-optimum (one feasible solution) for Model (MQBP). Therefore, the optimization problem of Model (MQBP) is equivalent to the problem of Model (QBC) associated with Model (LP), which is defined as follows:

$$Find\ a\ solution\ \dot{\mathbf{y}}^{*} \in \Omega_{QBC} \ni \Phi^{*}_{LP}(\dot{\mathbf{y}}^{*})\leq\Phi^{*}_{LP}(\dot{\mathbf{y}}), \forall \dot{\mathbf{y}} \in \Omega_{QBC}. \tag{15}$$

So far, an optimal solution of Model (MQBP) can be obtained via Problem (15) by enumerating all 0-1's combinations of $\dot{\mathbf{y}}$ for which $\dot{\mathbf{y}} \in \Omega_{QBC}$ as constant when Model (LP) is optimized. Alternatively, a transformation for Model (QBC) into a linear form is derived first so that Model (MQBP) can turn out to be a mixed-binary program and thereby being solvable with several optimization-based software packages, and the obtained optimal solution can be conferred for solution analysis and numerical comparison.

3.5   A Linear Transformation of Model (QBC)

Recall the quadratic equation (11) in which both the leading and the ending terms have already been in a linear form, i.e. $(\dot{y}_{i1}-0)^2=\dot{y}_{i1}$ and $(0-\dot{y}_{iT})^2=\dot{y}_{iT}$ due to $\dot{y}_{i1},\dot{y}_{iT}\in\{0,1\}$. Also, we have the middle term $(\dot{y}_{ik}-\dot{y}_{i,k-1})^2=\dot{y}_{ik}+\dot{y}_{i,k-1}$ if $(\dot{y}_{i,k-1},\dot{y}_{ik})=(0,0)$ or $(0,1)$ or $(1,0)$. Thus, the case of $(\dot{y}_{i,k-1},\dot{y}_{ik})=(1,1)$ remains to be discussed. Now, artificial binary variables $\dot{y}_{ik}^{+}$ and $\dot{y}_{ik}^{-}$ are introduced to represent the difference of $\dot{y}_{i,k-1}$ and $\dot{y}_{ik}$. For the instance of $(\dot{y}_{i,k-1},\dot{y}_{ik})=(1,1)$, $(\dot{y}_{ik}^{+},\dot{y}_{ik}^{-})=(0,0)$ can be derived by setting $\dot{y}_{ik}-\dot{y}_{i,k-1}=\dot{y}_{ik}^{+}-\dot{y}_{ik}^{-}$ and $\dot{y}_{ik}^{+}+\dot{y}_{ik}^{-}\leq1$. Hence, we have derived an equivalent problem of Model (QBC) as follows:

Model (BC). A binary constraint equivalent model of Model (QBC)
   Constraints: ($3PT-2P$ binary variables and $2PT+P+2T$ equations)

(i) Operation principles: Equations (9) and (10);
(ii) Suspending control:

$$\dot{y}_{i1}+\sum_{k=2}^{T}\left(\dot{y}_{ik}^{+}+\dot{y}_{ik}^{-}\right)+\dot{y}_{iT}=2\,,\;\forall i\in PI, \tag{16}$$

$$\dot{y}_{ik}-\dot{y}_{i,k-1}=\dot{y}_{ik}^{+}-\dot{y}_{ik}^{-}\,,\;\forall i\in PI,\forall k\in\{2,3,...,T\}, \tag{17}$$

$$\dot{y}_{ik}^{+}+\dot{y}_{ik}^{-}\leq1,\;\forall i\in PI,\forall k\in\{2,3,...,T\}, \tag{18}$$

(iii) Domain constraints:

$$\dot{y}_{ik}\in\{0,1\},\forall i\in PI,\forall k\in TI,\;\dot{y}_{ik}^{+},\dot{y}_{ik}^{-}\in\{0,1\},\forall i\in PI,\forall k\in\{2,3,...,T\}. \tag{19}$$

By replacing Equations (11) and (13) with Equations (16-19), Model (MQCP) turns out to be a MBP problem which can be solved by the mixed-integer programming solvers included in many mathematical programming software packages such as Cplex, OSL, MINTO, MIPO and so on. It is noted that although setup issues are out of consideration in this study, Equation (9) can be regarded as a constraint for explicating the problems with setup carryover (Sox and Gao, 1999) and one-setup allowance. Based on this viewpoint, a model comparable to Model (BC) can be developed.

According to the survey of Balas and Perregaard (2002), the MIPO version with lift-and-project cuts is considerably faster than the others on 29 hard MBP problems from MIPLIB. However, Vavasis (1991) has shown that the mixed-integer programming problem with 0-1 variables belongs to the class of NP-complete problems. In other words, Model (MQBP) cannot be solved within a polynomial time, and hence, it is almost impossible to be applied to practical large-scale problems. Therefore, the development of an efficient (reasonable time) and effective (near-optimum) algorithmic approach is necessary. If a pattern of solution properties, especially of $\dot{\mathbf{y}}^*$, is recognized, the efficiency of search algorithms could be improved. Such idea can be realized in the next section.

4   Pattern-Based Heuristic Approach for Model (MQBP)

According to the characteristics of the problem, a heuristic approach is proposed in this section. This approach utilizes the heuristic explicitly to derive certain acceptable scheduling results, but they are not necessarily the optimal one. Based on this fact, a pattern of solution properties is studied a priori so that the search space of Model (QBC) can be reduced as a permutation space, and afterward the reduced problem is solved by a hybrid GA (Wang and Wu, 2004a). In the next section, a numerical experiment is represented with an illustration.

4.1   Preliminary Pattern in Model (QBC)

The solution pattern can be referred to the suspending distribution of the variables in $\dot{\mathbf{y}}$. Intuitively, a $P \times T$ spreadsheet can facilitate the presentation of the pattern and enumerate the feasible combinations of $\dot{\mathbf{y}}$. The spreadsheet is formatted as illustrated in Table 3, where the identical order of the periods is marked in the horizontal axis and any permutation of the products, named $\pi=(\pi_1, \pi_2, \ldots, \pi_P)$, is allocated in the vertical axis. In Cell (j,k), the value 1 represents Product π_j to be arranged for manufacturing in Period k, and vice versa for 0. According to Equations (9) and (10), the sum of the contents in each row and that in each column must be within the intervals of $[T^L, T^U]$ and $[P_k^L, P_k^U]$, respectively. According to Equation (11), content "1" in each row will be filled continuously. In this manner, Constraints (9-11) will automatically be satisfied and a pattern with $P!$ feasible solutions can be obtained. Then a filling procedure is developed below to provide a competitive pattern.

Table 3. The pattern of the optimal solution of a 9×4 problem[*]

$\dot{y}_{ik}$	k=1	k=2	k=3	k=4
i=π_1	1	0	0	0
i=π_2	1	0	0	0
i=π_3	1	1	0	0
i=π_4	1	1	0	0
i=π_5	0	1	1	0
i=π_6	0	1	1	0
i=π_7	0	0	1	1
i=π_8	0	0	1	1
i=π_9	0	0	0	1

[*]Here, $(T^L, T^U)=(1,2)$ and $(P_k^L, P_k^U)=(1,4)$ for all k

Symbolically, we give the following definition for a pattern.

Definition 1. For all $j \in PI$ and $k \in TI$, let $\varphi(j,k)$ denote the content of Cell (j,k) in the spreadsheet, and φ be the *pattern* with the distribution of $\varphi(j,k)$. Pattern φ is feasible if Constraints (9-11) are satisfied in terms of $\dot{y}_{\pi_j,k}$ with any permutation π.

Moreover, when associating a permutation π with φ, we can obtain a corresponding solution of $\dot{\mathbf{y}}$ to be evaluated by the linear programming of Model (LP). Now let Ω_{PAT} denote the set of feasible patterns, and $\Pi = \{(\pi_1, \pi_2, \ldots, \pi_P): $ all permutations of $\pi_j, \pi_j \in PI, \forall j \in PI\}$ denote the set of permutations on PI. For a pair of Permutation $\pi \in \Pi$ and Pattern $\varphi \in \Omega_{PAT}$, let $F_{PAT}:(\pi, \varphi) \to \dot{\mathbf{y}}$ be the mapping function for which $\dot{y}_{\pi_j,k} = \varphi(j,k)$, $\forall j \in PI$, $\forall k \in TI$. By associating Model (LP) with both permutation and pattern spaces, the problem of Model (MQBP) is equivalent to

$$Find\ a\ solution\ (\pi^*, \varphi^*) \ni \Phi_{LP}^*(F_{PAT}(\pi^*, \varphi^*)) \leq \Phi_{LP}^*(F_{PAT}(\pi, \varphi)), \forall \pi \in \Pi, \forall \varphi \in \Omega_{PAT}, \quad (20)$$

where $\Phi_{LP}^*(\bullet)$ is the corresponding optimal value of Model (LP), defined in (14).

Let us consider a given pattern φ^{**} and denote $\dot{\mathbf{x}}^{**}$ as the optimal solution with the optimal value Φ_{LP}^{**} subject to Pattern φ^{**}. If a particular $x_{ijk}^{v^{**}}$ or $x_{ijk}^{u^{**}} \neq 0$, it means that the capacity of the regular time at Workstation j is not enough in Period k. Now if the capacity of the regular time at Workstation j is slack in Period $(k'-1)$ or $(k''+1)$ where $\dot{y}_{i\tau} = 1$ for Product i and $k' \leq \tau \leq k''$. If $k''-k' < T^u$ and $\sum_{i=1}^{P} \dot{y}_{i\tau} < P_\tau^u$ for $\tau = k'-1$ and $k''+1$, this slack capacity can be shared for Product i so that the producing mode is still continuous and the total cost is reduced. That is, the solution $\dot{\mathbf{x}}^{**}$ can be improved by changing the value of $\dot{y}_{i,k'-1}$ or $\dot{y}_{i,k''+1}$ from 0 to 1. On the other hand, if the capacity of the regular time at all workstations is slack in all periods (i.e. all of $x_{ijk}^{v^{**}}$ and $x_{ijk}^{u^{**}}$ are equal to 0), any change of the spread does not improve $\dot{\mathbf{x}}^{**}$ but still keeps the optimal value Φ_{LP}^{**}. Thus, by assigning the number of product types $\sum_{i=1}^{P} \dot{y}_{ik}$ and the number of continuous manufacturing periods $\sum_{k=1}^{T} \dot{y}_{ik}$ up to the upper bounds P_k^u and T^u respectively, the spread pattern is observed to be more promising on obtaining optimal production distribution.

In this manner, as the manufactured numbers are spread out over the production periods, the capacities of all machines can be leveled out and then the objective function could effectively be promoted. Based on the principle above, to fill in the blanks of the spreadsheet, one would start freezing the northwest cell with $\varphi(1,1)=1$. Then, we shift the target toward the southeast, and cram the other cells in the first column with Constraint (10) where the largest value is considered a priori. Heading to the next column, the filling process is constrained by Equations (9-10). At each cell, the value 1 is assigned a priori unless Constraints (9-10) cannot be held. Continuing this manner until the southeast cell is filled, a preliminary pattern with $P!$ feasible solutions can be obtained. Moreover, with the assistance of a northwestern diagonal line, it is robust to complete this filling procedure by visual display. Hence, the heuristic algorithm to realize such visual display of discovering a good pattern is a pattern mining procedure, which is developed in Procedure 1 below. It is also interesting to note that transforming the original search space into the optimum-wise pattern in Procedure 1 is similar to branching into a promising tree as a branch-and-bound algorithm does.

Procedure 1. Pattern Mining Algorithm

Step 1. *Initialization.*

1.1. Set the iterative index $j=0$ and the cell status $\varphi(i,k)=0$, $\forall i \in PI$, $\forall k \in TI$.

1.2. Let $\rho'_k = \left\lfloor \dfrac{(k-1)P}{T} \right\rfloor$ and $\rho''_k = \left\lfloor \dfrac{kP}{T} \right\rfloor$, $\forall k \in TI$, where $\lfloor \bullet \rfloor$ is a mod.

(This is to identify the range of the cells passed by the northwestern diagonal line for each column.)

1.3. For all $k \in TI$, let $\varphi(i,k)=1$ if $\rho'_k \geq i \geq \rho''_k$.

(This is to label the northwestern diagonal cells.)

Step 2. *Termination.* If $j \geq T$, stop; otherwise, go to next step.

Step 3. *Iteration.*

3.1. Let $\rho''_j = \rho''_j + 1$. If $P^U_j \geq \rho''_j - \rho'_j + 1 \geq P^L_j$ (i.e. Constraint (2.10) holds) and $\rho''_j \leq P$, then set $\varphi(\rho''_j, j) = 1$ and go to next step; otherwise, go to Step 3.3.

3.2. If $\varphi(\rho''_j, 1) + \sum_{k=2}^{T} \left(\varphi(\rho''_j, k) - \varphi(\rho''_j, k-1) \right)^2 + \varphi(\rho''_j, T) \neq 2$ or $\sum_{k=1}^{T} \varphi(\rho''_j, k) > T^U$

(i.e. Constraint (2.11) or (2.9) does not hold), then reset $\varphi(\rho''_j, j) = 0$ and go to next step; otherwise, return to Step 3.1.

3.3. Let $j = j+1$, then return to Step 2.

The implementation of Procedure 1 for the case shown in Table 3 is illustrated in Fig. 5. Note that the preliminary pattern obtained from Procedure 1 cannot be spread because any change of $\varphi(i,k)$ from 0 to 1 will violate Constraints (9) and (10). Thus, for any problem, the best solution based on the preliminary pattern cannot be improved by relaxing $\dot{y}_{ik}$.

Therefore, to compensate each for the optimization in permutation problems, GA and NS improvement are incorporated as the hybridization by Wang and Wu (2004a). For efficient applications, they developed the hybrid GA using position-oriented design, i.e. considering the problems with the major sensitivity of absolute position in permutation. In the following, the solution procedure is described for the application of our concerned problem.

4.3   Application of a Hybrid GA in Problem (PP)

GA is a parallel, randomized-search optimization method. Its heuristics are inspired by the biological paradigm of natural evolution. A potential solution to a problem is inferred as an individual, which can be represented as the genetic structure of a chromosome. Throughout the genetic evolution by the mechanisms of reproduction, crossover and mutation, the offsprings of good quality are born from a population of chromosomes. Generation by generation, the stronger individuals (i.e., better solutions to the problem) are likely to be the winner in a competing environment.

Many studies (Chu and Beasley, 1997, Houck *et al.*, 1996, Renders and Flasse, 1996) have shown that GAs often perform well for global searching because it is capable of quickly finding and exploiting promising regions of the search space, but they take a relatively long time to converge to a local optimum. On the other hand, local improvement procedures such as pairwise interchange for combinatorial problems and gradient descent for unconstrained nonlinear programming, quickly find the local optimum of a small region of the search space, but are typically poor at global searches. Because GAs are not suitable for fine-tuning of solutions around optima, therefore local search can be incorporated into a GA in order to improve its performance through what could be termed "learning".

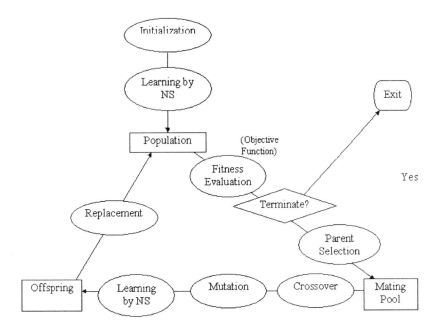

Fig. 6. Structure of a Hybrid GA (Wang and Wu, 2004a)

Wang and Wu (2004a) have proposed a hybrid GA to solve the type of permutation problems successfully. This hybridization employs (1) a GA, which performs global exploration (evolution) among the population, and (2) a NS procedure, which performs local exploitation (learning) around chromosomes. Fig. 6 illustrates the structure of Wang and Wu's hybrid GA. Basically, the NS in this context is analogous to a kind of learning that occurs during the lifetime of an individual string. After learning, the improved individual is placed back into the population and is allowed to compete for reproductive opportunities. The selection is then based on the fitness at the end of the individual life. That is, the fitness of an individual is evaluated by the learning effect of neighborhood search.

The detail design for this problem is described as follows.

(1) Initialization

The *Initialization* procedure creates a starting GA population filled with randomly generated integer-number strings, and ensures a chromosome corresponding to a permutation in problem. In our problem, a sequence of products is assigned as a string to represent a chromosome.

(2) Learning

A *neighborhood search* procedure is used as learning mechanism to reinforce all individuals in the population. This search utilizes the pairwise scanning to find the first better neighbor of the current solution and replace the current solution with this elite (dominating) neighbor. When designing a hybrid GA, attention should be paid to the equilibrium between the NS and the GA. If the current solution is replaced with its elite neighbor only once (named *once-improvement*), the burden of the computing resource on NS can be lightened to a large extend and the promise of the good pattern explored by GA is enhanced. Because, by improving once, the improvement of the individuals is not complete, the progress of the population could be suspended. Thus, the *once-improvement* stopping criterion can be adopted at the beginning so that the best solution could be exploited early. And then, when the best fitness of the previous generation cannot be improved under the *once-improvement* criterion, the *until-no-improvement* one takes over so that the population can progress.

(3) Selection

The parents are randomly selected from the current population to produce children by genetic operations. Although the roulette-wheel-selection (Mitchell, 1996) is one commonly used technique, it is not used for the proposed hybrid GA because the selected individuals in the population have been evaluated with good fitness throughout the learning procedure. Therefore, each individual is chosen as a parent by using a discrete uniform sampling scheme.

(4) Recombination

The *crossover* procedure recombines gene-codes of two parents and produce offspring. The parents here are selected from the improved individuals which could possess several positions or subsequences or orders of gene-codes consistent with the global optimal one. In viewpoint of position-oriented design, the positions of gene-codes shall be treated a prior. Such a crossover scheme is proposed as shown in Fig. 7. It works as follows:

Step 1. Any gene-code which occupies the same position of both parents has the top priority of being assigned to that position of the offspring. This step attempts to let the offspring inherit actual positions of the promising gene-codes from parents.

Step 2. The remaining sites in the offspring are assigned by the order of all gene-codes in Parent 1 within the subsequence bounded by two randomly selected positions. This step attempts to let the offspring inherit partial positions of the promising gene-codes from a Parent 1.

Step 3. Any unassigned sites are placed in the order of appearance in Parent 2. This step attempts to let the offspring inherit the promised relative order as well as maintain the permutation property of gene-codes.

Steps 2 and 3 operate as LOX does.

Parent 1	2	9	1	5	4	8	3	7	6
Parent 2	8	9	6	3	4	1	2	7	5

Random crossover sites   *   *

Copy actual position in common locations		9			4			7	

Copy the subsequence from Parent 1		9	1	5	4	8		7	

Copy the order from Parent 2	6	9	1	5	4	8	3	7	2

Fig. 7. Illustration of crossover with the same-site-copy-first principle

(6) Mutation

Mutation takes place on some newly formed children in order to prevent all solutions from converging to their particular local optima. As the survey of Murata *et al.* (1996), the mutation operations commonly used for permutation problems are adjacent two-change, arbitrary two-change, arbitrary three-change and shift change. Because our hybridization uses a 2-exchanging NS to improve offspring, shift-change has a great ability to force a new solution for escaping from the path of the NS, and thus is adopted as mutation operator. This technique is to take the form of selecting two sites at random, replace one selected site with the other and shift all remaining sites within the subsequence. It is depicted in Fig. 8.

Fig. 8. Illustration of the shift-changed perturbation scheme

(7) Replacement

The population replacement occurs for individuals showing high fitness. We adopt the elitist strategy that copies the best individual of each generation into the succeeding generation. To maintain enough diversity, the improved (learned) offspring that are duplicates of current individuals in the population are discarded rather than inserted into the population.

From the above presentation, it can be seen that the functions of crossover, mutation and learning are designed subject to position implication in permutation. In summary, the solution framework of Model (MQBP) is proposed in Fig. 9.

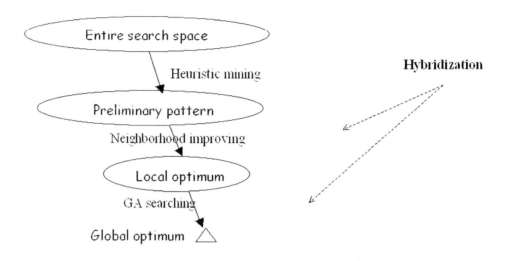

Fig. 9. The solution framework for Model (MQBP)

Now let us follow a 9×4 problem with the preliminary pattern as shown in Table 3. A numerical example of the proposed hybrid GA is illustrated in Fig. 10, which shows the implementation of the first iteration. Continuing this execution until 1000 permutation solutions had been evaluated, we have the best solution "351942867" with the optimal value 5784.6 of the objective function. This means that according to the first row of Table 3, Product 3 is to be produced only at Period 1 (i.e. $\dot{y}_{31}=1$ and $\dot{y}_{32}=\dot{y}_{33}=\dot{y}_{34}=0$), and in the similar way for the other products. By substituting these values of $\dot{\mathbf{y}}$ into Model (LP) for deciding the amount of each products to be produced, out-sourced and delayed, it yields the corresponding optimal solution with the minimal production cost of $\Phi_{LP}^{*}(\dot{\mathbf{y}})=5784.6$. In the following section, we depict the performance of our proposed pattern-based approach to this 9×4 problem as well as other four problems with different scales.

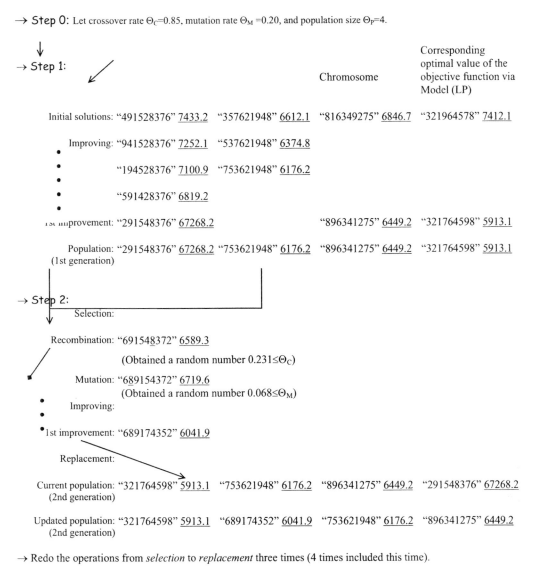

→ **Step 0**: Let crossover rate Θ_C=0.85, mutation rate Θ_M =0.20, and population size Θ_P=4.

→ **Step 1**:

Chromosome | Corresponding optimal value of the objective function via Model (LP)

Initial solutions: "491528376" 7433.2 "357621948" 6612.1 "816349275" 6846.7 "321964578" 7412.1

Improving: "941528376" 7252.1 "537621948" 6374.8

"194528376" 7100.9 "753621948" 6176.2

"591428376" 6819.2

1st improvement: "291548376" 67268.2   "896341275" 6449.2 "321764598" 5913.1

Population: "291548376" 67268.2 "753621948" 6176.2 "896341275" 6449.2 "321764598" 5913.1
(1st generation)

→ **Step 2**:
Selection:

Recombination: "691548372" 6589.3
(Obtained a random number 0.231≤Θ_C)

Mutation: "689154372" 6719.6
(Obtained a random number 0.068≤Θ_M)

Improving:

1st improvement: "689174352" 6041.9

Replacement:

Current population: "321764598" 5913.1 "753621948" 6176.2 "896341275" 6449.2 "291548376" 67268.2
(2nd generation)

Updated population: "321764598" 5913.1 "689174352" 6041.9 "753621948" 6176.2 "896341275" 6449.2
(2nd generation)

→ Redo the operations from *selection* to *replacement* three times (4 times included this time).

Fig. 10. A numerical example of the proposed hybrid GA

5   Numerical Results

To test the performance of the proposed solution approach, a case study has been conducted on an LED manufacturing plant (Wang and Wu, 2003). Five problems of different scales (W is fixed to 11) were studies. This comparative study meant to evaluate the effectiveness of the proposed method when the level of required efficiency is fixed. Because the global optima were unknown for the testing cases except that of 9×4, by considering the LP computing time and the computer file storing time, termination criterion were set at "1000 different solutions have been found". In the following, we shall first depict the convergence behavior of the applied hybrid GA for a sufficient report on the comparison of mix-binary programming and pattern-based heuristic approaches.

The performance of the applied hybrid GA with the best parameters obtained from a preliminary experiment, denoted A_{HGA}, was evaluated by comparing the three methods below in five cases:

- Random Sampling method, A_{RS}: 1000 different solutions were randomly generated.
- Neighborhood Search, A_{NS}: The tactics of "pairwise", "the first improvement", "terminated when no further improvement exists", and "re-start when terminated" were adopted.
- Pure Genetic Algorithm, A_{PGA}: It worked like A_{HGA} but excluded the learning procedure.

All of the methods were implemented 30 times with different starting-seeds. During this test, A_{HGA}, A_{PGA} and A_{NS} started with the same initial solutions of 30 groups. For a given number of evaluated solutions, ϑ, each method obtained 30 corresponding best objective values $z_i(\vartheta)$, $\forall i=1,2,...,30$. By averaging these values, $\bar{z}(\vartheta)=(z_1(\vartheta)+z_2(\vartheta)+...+z_{30}(\vartheta))/30$, and comparing them with the optimal evaluation value of the best solution z^+, as well as the worst solution z^-, the performance indicator $\zeta(\vartheta)$ for all the methods is formulated as:

$$\zeta(\vartheta) = \frac{z^- - \bar{z}(\vartheta)}{z^- - z^+}. \tag{21}$$

The best and the worst solutions were found after gathering all the evaluated solutions in the whole test. Since the indicator $\zeta(\vartheta)$ has already been normalized between 0 and 1, it can be applied for the evaluation and comparison of the implementation effectiveness of the different cases. Fig. 11 presents the results of the five cases. It can be noted that with the problems becoming larger, A_{HGA} performed better than the other methods. In conclusion, the results indicate that if HGA is

- without the local improvement of NS (as pure GA), the performance is not effective in general and even more inadequate than RS with middle and large-scale problems;
- without the global search of GA (as in pure NS), the level of effectiveness in performance decreases rapidly in large-scale problems.

Besides, it is interesting to see that RS outperforms the other methods excluding HGA in the largest case. Thus, the concept of initial, transient and steady states was proposed, in which an implemented algorithm will be followed. Furthermore, for the detailed implementation process of the algorithms, an estimation of the efficiency according to a given degree of accuracy was established so that users can trade off between the limitation of computation time and the degree of improvement on computation result. For more detail discussions, one can refer in Wang and Wu (2004a). In summary, these figures show the adaptability of the applied hybrid GA for the permutation problem that is extracted from Model (MQCP) based on preliminary pattern. Furthermore, the pattern-based heuristic approach is to be compared with the mix-binary programming one below.

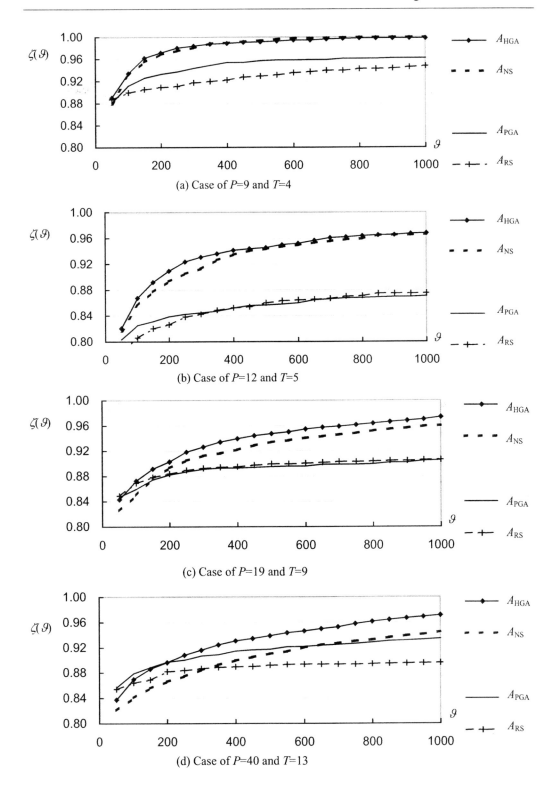

(a) Case of $P=9$ and $T=4$

(b) Case of $P=12$ and $T=5$

(c) Case of $P=19$ and $T=9$

(d) Case of $P=40$ and $T=13$

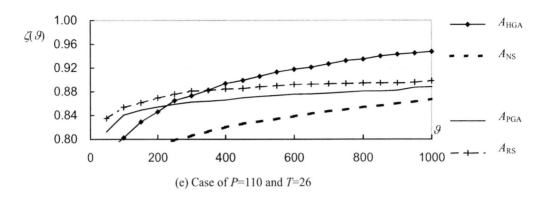

(e) Case of $P=110$ and $T=26$

Fig. 11. Performance of four algorithms for Model (PP) in different cases (Wang and Wu, 2004a)

In the mixed-binary programming approach, the Cplex MIP solver (ILOG, 1997) of a branch and bound algorithm with default settings was applied for solving the above five cases in the form of Model (QB), and this implementation was done on a Pentium-III 600MHz / 512MB RAM machine. The termination criterion of the mixed-binary programming were set at the time comparably when 1000 different solutions have been found by the pattern-based heuristic. That is, we would fix the level of the resource consumption efficiency in order to evaluate the reached effectiveness. Based on the comparison of the achieved objective values in terms of the performance indicator as shown in Equation (21), the simulation result is listed in Table 5. Note that for all of the five cases, the best solutions were identified by the hybrid GA, and therefore in Table 5, content '1.000' in the row marked 'in the best' is filled.

From Table 5, it can be noted that the pattern-based heuristic approach outperformed the mixed-binary programming one even when the former performs in the worst run. In conclusion, the outperformance of our proposed heuristic approach is referred to the following two phases: (1) through the pattern mining algorithm described in Procedure 1, the search spaces of the problems are efficiently reduced as shown in Table 4; (2) through the hybrid genetic algorithm proposed by Wang and Wu (2004a), the problem with the reduced permutation space is competently solved as shown in Fig. 11.

Table 5. Performance of two solution approaches for Model (MQBP) in different cases

(i) Characteristics $P \times T$	9×4	12×5	19×9	40×13	110×26
(ii) Computation time (second)	180	230	310	440	2870
(iii) Achieved performance indicator: • The pattern-based heuristic via Model (PP) and the hybrid GA					
(1) 30 runs in average	0.998	0.968	0.974	0.971	0.947
(2) In the best run	1.000	1.000	1.000	1.000	1.000
(3) In the worst run	0.968	0.938	0.941	0.936	0.880
• The mixed-binary programming via Model (BC) and the Cplex MIP solver	0.930	0.902	0.883	0.875	0.816

6 Summary and Extension Remarks

This paper described a case of master production scheduling problem involved multi-period, multi-product and multi-resource, of which the properties of permutation and linear optimization are characterized. In order to achieve smooth production in a sequence of production periods, and to allow such sequencing system be automatically adjusted to the demand requirement and production capability, Model (MQBP) of a mixed-quadratic-binary program was proposed. To solve this model, this study adopted a systematic planning concept to develop an entire solution-seeking process. In order to revamp the searching direction of an optimal solution, a promising initial pattern is identified to serve this purpose, and thus allows for a more expeditious response capability in business administration. In the solution analysis, a preliminary pattern by a heuristic mining algorithm was developed to reveal the permutation property from Model (QBC). To solve the permutation optimization problem, a hybrid GA is employed. By incorporating the GA of its efficiency to explore the entire search space with the NS of its convergent ability, the accuracy and computation efficiency of this model have been shown in this study. Facing large-scaled and complicated mathematical programming problems in practice, by applying the proposed pattern-based heuristic, lengthy analysis procedures can be reduced. This has been illustrated by a case of the production scheduling.

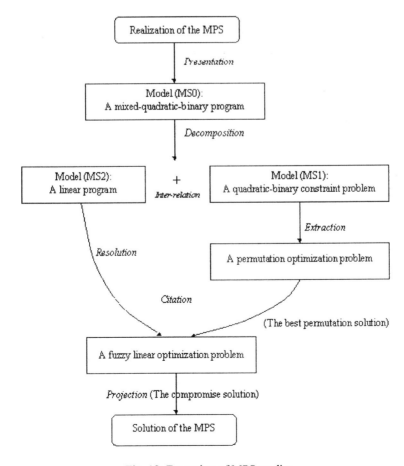

Fig. 12. Extension of MPS studies

Using the efficient heuristic approach, it is expected to maintain the interactive effectiveness of scheduling stability with respect to the way of sequence scheduling. This logical modeling has already been utilized to a local LED manufacturer, and it has been shown to facilitate the decision making process in actual production scheduling. However, in our practical implementation, the parameters of demand, capacity and cost used for the scheduling model were not exactly known and stable. This leads to the requirement of considering the allowed tolerances of the parameters. By extension, such uncertain phenomena can be incorporated into Model (LP) with fuzzy coefficients. The approach of fuzzy linear programming (FLP) based on hyperplane targets has then been proposed by Wang and Wu (2004b), when one's optimistic or pessimistic attitude is incorporated. The framework of this extension is summarized in Fig. 12. In many fields of industry and technology, solving complex optimization problems is the key to increase productivity and product quality. Permutation optimization problems are complex in computation complexity, whereas fuzzy linear optimization problems are complex in non-linear models. This extension presents a combinatorial problem in common production systems with permutation property and a linear program problem with fuzzy parameters. Both results are required as core techniques for promoting a decision-making support system in the practical case of MPS.

References

Aarts, E. and Lenstra, J.K., 1997. Local Search in Combinatorial Optimization, John Wiley & Sons, Chichester.

Bahl, H.C., Ritzman, L.P. and Gupta, J.N.D., 1987. Determining lot sizes and resource requirements: a review, *Operations Research* **35** 329-345.

Balas, E. and Perregaard, M., 2002. Lift-and-project for mixed 0-1 programming: recent progress, *Discrete Applied Mathematics* **123** 129-154.

Billington, P.J., McClain, J.O. and Thomas, L.J., 1983. Mathematical programming approaches to capacity-constrained MRP systems: review, formulation and problem reduction, *Management Science* **29** 1126-1140.

Blackburn, J.D., Kropp, D.H. and Millen, R.A., 1986. A comparison of strategies to dampen nervousness in MRP systems, *Management Science* **32** 413-329.

Chu, S.C.K., 1995. A mathematical programming approach towards optimized master production scheduling, *International Journal of Production Economics* **38** 269-279.

Chu, P.C. and Beasley, J.E., 1997. A genetic algorithm for the generalized assignment problem, *Computers & Operations Research* **24** 17-23.

Djerid, L. and Portmann, M.-C., 2000. How to keep good schemata using cross-over operators for permutation problems, *International Transactions in Operational Research* **7** 637-651.

Drexl, A. and Kimms, A., 1998. Beyond Manufacturing Resource Planning (MRP II): Advanced Models and Methods for Production Planning, Springer-Verlag, Berlin.

Gessner, R.A., 1986. Master Production Schedule Planning, John Wiley & Sons, New York.

Glover, F., 1993. A user's guide to tabu search, *Annals of Operations Research* **41** 3-28.

Gue, K.R., Nemhauser, G.L. and Padron, M., 1997. Production scheduling in almost continuous time, *IIE Transactions* **29** 391-398.

Holland, J.H., 1975. Adaptation in Natural and Artificial Systems, University of Michigan Press. (Second edition, 1992, MIT Press, Cambridge.)

Houck, C.R., Joines, J.A. and Kay, M.G., 1996. Comparison of genetic algorithms, random restart, and two-opt switching for solving large location-allocation problems, *Computers & Operations Research* **23** 587-596.

Hung, Y.F. and Hu, Y.C., 1998. Solving mixed integer programming production planning problems with steups by shadow price information, *Computers and Operations Research* **25** 1027-1042.

ILOG, 1997. *Using the CPLEX Callable Library*, ILOG CPLEX Division, Incline Village.

Iyer, S.K. and Saxena, B., 2004, Improved genetic algorithm for the permutation flowshop scheduling problem, *Computers & Operations Research* **31** 593-606.

Kimms, A., 1998. Stability measures for rolling schedules with applications to capacity expansion planning, master production scheduling, and lot sizing, *Omega* **26** 355-366.

Kirkpatrick, S., Gelatt, C.D. and Vecchi, M.P., 1982. Optimization by Simulated Annealing, *IBM Research Report RC 9355*, Yorktown Heights, NY.

Mitchell, M., 1996. An Introduction to Genetic Algorithm, MIT Press, Cambridge.

Murata, T., Ishibuchi, H. and Tanaka, H., 1996. Genetic algorithms for flowshop scheduling problems, *Computers & Industrial Engineering* **30** 1061-1071.

Olhager, J. and Wikner, J., 1998. A framework for integrated material and capacity based master scheduling. In: Drexl, A. and Kimms, A. (Eds.), *Beyond Manufacturing Resource Planning (MRP II): Advanced Models and Methods for Production Planning*, Springer-Verlag, Berlin, pp. 3-20.

Proud, J.F., 1999. *Master Scheduling: A Practical Guide to Competitive Manufacturing*, John Wiley & Sons, New York.

Ramachandran, B. and Pekny, J.F., 1998. Lower bounds for nonlinear assignment problems using many body interactions, *European Journal of Operational Research* **105** 202-215.

Rardin, R.L., 1998. *Optimization in Operations Research*, Prentice-Hall, New Jersey.

Reeves, C.R., 1993. Modern Heuristic Techniques for Combinatorial Problems, Oxford, Blackwell.

Renders, J.-M. and Flasse, S.P., 1996. Hybrid methods using genetic algorithms for global optimization, *IEEE Transactions on Systems, Man and Cybernetics* **B26** 243-258.

Sox, C.R. and Gao, Y., 1999. The capacitated lot sizing problem with setup carry over, *IIE Transactions* **31** 173-181.

Sridharan, S.V., Berry, W.L. and Udayabhanu, V., 1987. Freezing the master production schedule under rolling planning horizons, *Management Science* **33** 1137-1149.

Süer, G.A., Saiz, M. and Rosado-Varela, O., 1998. Knowledge-based system for master production scheduling. In: Drexl, A. and Kimms, A. (Eds.), *Beyond Manufacturing Resource Planning (MRP II): Advanced Models and Methods for Production Planning*, Springer-Verlag, Berlin, pp. 21-44.

Tate, D.M. and Smith, A.E., 1995. A genetic approach to the quadratic assignment problem, *Computers & Operations Research* **22** 73-83.

Tian, P., Ma, J. and Zhang, D.-M., 1999. Application of the simulated annealing algorithm to the combinatorial optimization problem with permutation property: An investigation of generation mechanism, *European Journal of Operational Research* **118** 81-94.

Vavasis, S., 1991. Nonlinear Optimization: Complexity Issues, Oxford University Press, New York.

Wang, H.F. and Wu, K.Y., 2003. Modeling and analysis for multi-period, multi-product and multi-resource production scheduling, *Journal of Intelligent Manufacturing* **14** 297-309.

Wang, H.F. and Wu, K.Y., 2004a. Hybrid genetic algorithm for optimization problems with permutation property, *Computers and Operations Research* **31** 2453-2471.

Wang, H.F. and Wu, K.Y., 2004b. Preference approach to fuzzy linear inequalities and optimizations, *Fuzzy Optimization and Decision Making*, accepted on July 2004.

Weinstein, L. and Chung, C.H., 1999. Integrating maintenance and production decisions in a hierarchical production planning environment, *Computers & Operations Research* **26** 1059-1074.

Whitley, D., 1997. Permutations. In: Bäck, T., Fogel, D. and Michalewicz, M. (Eds.), Handbook of Evolutionary Computation, IOP Publishing and Oxford University Press, New York, pp. C1.4:1-C1.4:8.

Wu, K.Y., Wang, H.F., Huang, C. and Chi, C., 2000. Multi-period, multi-product production scheduling – a case of LED manufacturing system. In: *Proceedings of the 5th Annual International Conference on Industrial Engineering – Theories, Applications and Practice*, Dec 13-15, 2000, Hsinchu, Taiwan.

Xie, J., Zhao, X. and Lee, T.S., 2003. Freezing the master production schedule under single resource constraint and demand uncertainty, *International Journal of Production Economics* **83** 65-84.

Yagiura, M. and Ibaraki, T., 1996. The use of dynamic programming in genetic algorithms for permutation problems, *European Journal of Operational Research* **92** 387-401.

Zäpfel, G., 1998. Customer-order-driven production: an economical concept for responding to demand uncertainty?, *International Journal of Production Economics* **56** 699-709.

In: Control and Learning in Robotic Systems
Editor: John X. Liu, pp. 181-203

ISBN 1-59454-356-9
© 2005 Nova Science Publishers, Inc.

Chapter 7

COMBINING LEARNING AND EVOLUTION FOR ADAPTIVE AUTONOMOUS SYSTEMS

Genci Capi[1]*, Daisuke Hironaka[1], Masao Yokota[1], and Kenji Doya[2]

[1] Faculty of Information Engineering, Fukuoka Institute of Technology
3-30-1 Wajiro-Higashi, Higashi-ku, Fukuoka, 811-0195, Japan
[2] CREST, Japan Science and Technology Agency (JST)
ATR, Computational Neuroscience Laboratories
"Keihanna Science City", Kyoto, 619-0288, Japan

Abstract

Reinforcement learning is a fundamental process by which organisms learn to achieve their goals from their interactions with the environment. The crucial issue in reinforcement learning applications is how to set meta-parameters, such as the learning rate, "temperature" for exploration, in order to match the demands of the task and reduce the learning time. In this chapter, we propose a new method, based on evolutionary computation. The basic idea is to encode the metaparameters of the reinforcement learning algorithm as the agent's genes, and to take the metaparameters of best-performing agents in the next generation. First, we investigated the influence of metaparameters on the agent learned policy by considering a battery capturing task. Then, by utilizing the capturing behavior, we considered a more complex task where the Cyber Rodent robot has to survive and increase its energy level. Our main focus was to see the effect of metaparameters and initial weight connections on learning time. The results show that appropriate settings of metaparameters found by evolution have a great effect on the learning time and are strongly dependent on each other. Furthermore, we verified in the real hardware of Cyber Rodent robot that metaparameters evolved in simulation are helpful for learning in real hardware.

1   Introduction

Reinforcement learning (RL) (Barto, 1995; Doya, 2000; Doya, Kimura, & Kawato, 2001; Sutton & Barto, 1998) provides a sound framework for autonomous agents to acquire

---

* E-mail address: capi@fit.ac.jp

adaptive behaviors based on reward feedback. In RL, learning is contingent upon a scalar reinforcement signal, which provides evaluative information about how good an action is in a certain situation, without providing an instructive supervising cue as to which would be the preferred behavior in the situation. Behavioral research indicates that RL is a fundamental means by which experience changes behavior in both vertebrates and invertebrates, as most natural learning processes are conducted in the absence of an explicit supervisory stimulus (Donahoe, 1997). The theory of RL has been successfully applied to a variety of dynamic optimization problems, such as game programs (Tesauro, 1994), robotic control (Morimoto & Doya, 2001), and resource allocation (Singh & Bertsekas, 1997).

In RL, the learning capabilities are strongly dependant on a number of parameters, such as learning rate, the degree of exploration, and the time scale of evaluation. These parameters are often called metaparameters because they regulate the way detailed parameters of an adaptive system change with learning. The appropriate settings of metaparameters depend on the environmental dynamics, the goal of the task, and the time allowed for learning. The permissible ranges of such metaparameters are dependant on particular tasks and environments, making it necessary for a human expert to tune them usually by trial and error. But tuning multiple metaparameters is quite difficult due to their mutual dependency, e.g., if one changes the noise size, one should also change the learning rate. In addition hand tuning of metaparameters is in a marked contrast with learning in animals, which can adjust themselves to unpredicted environments without any help from a supervisor.

In statistical learning theory, the need for setting the right metaparameters, such as the degree of freedom of statistical models and the prior distribution of parameters, is widely recognized. Theories of metaparameter setting have been developed from the viewpoints of risk-minimization (Vapnik, 2000) and Bayesian estimation (Neal, 1996). However, many applications of reinforcement learning have depended on heuristic search for setting the right metaparameters by human experts. The need for the tuning of metaparameters is one of the major reasons why sophisticated learning algorithms, which perform successfully in the laboratory, cannot be practically applied in highly variable environments at home or on the street.

In this chapter, we explore methods to determine metaparameters by combining learning and evolution. Learning and evolution are complementary mechanisms for acquiring adaptive behaviors, within the lifetime of an agent and over generations of agents. Although there are several works on relation between evolution and learning, such as Baldwin effect (Baldwin, 1896; French & Messinger, 1994), optimizing learning parameters (Unemi, 1996), combining of artificial neural network and genetic algorithm (Belew, McInerney & Schraudolph, 1992), there are a number of basic design issues in combining learning and evolution in artificial agents (Nolfi & Floreano, 1999), such as:

- What to be learned and what to be evolved: the policy, the initial parameters of policy, the metaparameters, or even the learning algorithm.
- What type of evolutionary scheme to be used: Lamarcian or Darwinian.
- How to evaluate the fitness: average lifetime reward (Unemi 1994), final performance, and/or learning speed.

- Centralized, synchronous evolution or distributed, asynchronous evolution: the former is standard in simulation, but latter may be more advantageous in hardware implementation (Pollack et al., 2000).
- How to combine simulation and hardware experiments.

In our method, the basic idea is to encode the metaparameters of the RL algorithm as the agent's genes, and to take the metaparameters of best-performing agents in the next generation. In difference from previous approaches, the specific questions we ask in this study are: 1) whether GA can successfully find appropriate metaparameters subject to mutual dependency, 2) how the metaparameters and initial weight connections effect the learning time and 3) whether the metaparameters optimized by GA in simulation can indeed be helpful in hardware implementation of RL.

In order to answer these questions, first we considered a battery capturing task for the Cyber Rodent (CR) robot (Capi, Uchibe & Doya 2002), which is a two wheeled robot with a wide-angle camera. Evolution considered two metaparameters: the learning rate and cooling factor of a Sarsa (λ) RL algorithm. Then, the learned capturing behavior is utilized in a survival behavior, where the robot must recharge its own battery by capturing active battery packs distributed in the environment. In this environment set-up, our main focus is on how metaparameters influence the learning time. In addition of learning rate and cooling factor, the discount factor, and initial weight connections are encoded in the genome.

Simulation and experimental results show that appropriate settings of meta-parameters can be found by evolution. In addition, evolved metaparameters show a mutual dependency and they have a great influence on the learning time. The evolved metaparameters in simulation were also applied to learn the task in real hardware of CR robot. Although there are some differences between final policies learned in simulation and real hardware, the learning processes follow the same stable pattern in both two cases. This shows that even if the environments and the physical conditions of the agents do not allow the same optimal policy, the metaparameters optimized in simulation gives the right stability to complete the learning process.

This chapter is organized as follows. In the following section, we present the CR robot. In Section 3, the task and environments are discussed. Section 4 describes the RL algorithms. The method of evolving metaparameters is discussed in Section 5. Simulation and experimental results are given in Section 6. Finally, the chapter concludes in Section 7.

2   CR Robot

The CR robot is a two-wheel-driven mobile robot as shown in Figure 1. The CR is 250 mm long and weights 1.7 kg. The CR is equipped with:

- Omni-directional C-MOS camera.
- IR range sensor.
- Seven IR proximity sensors.
- 3-axis acceleration sensor.
- 2-axis gyro sensor.

- Red, green and blue LED for visual signaling.
- Audio speaker and two microphones for acoustic communication.
- Infrared port to communicate with a nearby agent.
- Wireless LAN card and USB port to communicate with the host computer.

Five proximity sensors are positioned on the front of robot, one behind and one under the robot pointing downwards. The proximity sensor under the robot is used when the robot moves wheelie. The CR contains a Hitachi SH-4 CPU with 32 MB memory. The FPGA graphic processor is used for video capture and image processing at 30 Hz.

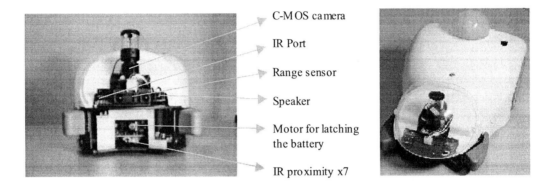

Figure 1. Cyber Rodent robot used in the experiments.

3   Tasks and Environments

3.1   Battery Capturing Behavior

In order to investigate the effect of metaparameters on the agent learned behavior and to know the interaction between them, we considered a capturing task where the CR robot has to capture the battery packs distributed in the environment. This task requires exploration in the environment and learning from delayed reward during limited lifetime. Thus the settings of metaparameters, such as the learning rate, randomness in action selection, critically affect the agent's performance.

In the first experiment, the agent is placed in a square environment of size 3.0 m x 3.0 m containing a single battery pack. At each time step the agent receives the angle to the nearest battery pack as state input. The vision range is assumed to be between $-\pi/2$ and $\pi/2$. The agent starts learning in short distance relative to the battery pack, oriented so that the battery pack is always visible. In the late stage of leaning, the position and orientation of the agent relative to the battery pack are randomly selected.

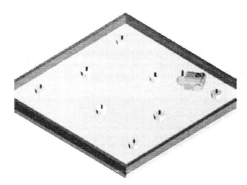

Figure 2. Environment of surviving behavior.

In addition, we considered the case when there are 15 battery packs, which decrease linearly from 15 to 1 during the agent's lifetime. The agent searches without getting repositioned. An obstacle avoidance behavior is generated when the agent gets at a 250 mm distance from walls. This allows us to study learning and metaparameters settings under more complex environmental conditions.

3.2 Surviving Behavior

In the second environment, the CR robot has to survive and increase its energy level. The environment has 8 battery packs, as shown in Figure 2. The positions of battery packs are considered fixed in the environment and the CR robot is initially placed in a random position and orientation.

The agent can recharge its own battery by capturing active battery packs, which are indicated by red LED color. After the robot captures the battery pack, it can recharge its own battery for a determined period of time (charging time), then the battery pack becomes inactive and its LED color changes to green. The battery becomes active again after the reactivation time. Therefore, in this environment the following parameters can vary:

- The number of battery packs;
- The reactivation time;
- The energy received by capturing the battery pack (by changing the charging time);
- The energy consumed by the agent for 1m motion.

Based on the energy level and the distance to the nearest active battery pack, the agent can select among three actions: 1) capture the active battery pack; 2) search for a battery pack or 3) wait until a battery pack becomes active. In the simulated environment, the batteries have a long reactivation time. In addition, the energy consumed for 1m motion is low. Therefore, the best policy is to capture any visible battery pack (the nearest when there are more than one). When there is no visible active battery pack, the agent must search in the environment.

4   Reinforcement Learning

4.1   Sarsa (λ)

The agent learns the battery capturing behavior based on *gradient decent Sarsa(λ)* RL algorithm (Sutton & Barto, 1998). At each time step the agent receives the angle to the closest battery as state input which varies from $-\pi/2$ to $\pi/2$. Thus the state s of the agent is the angle of the nearest visible battery pack. The state space is explored using a softmax method, employing the Boltzmann distribution.

To represent the action value function, we use a radial basis function network. The Gaussian functions are used as basis functions given as:

$$\phi(i,s) = e^{-\frac{\|s-c_i\|^2}{2\sigma_i^2}},$$ (1)

where c_i is the center and σ_i is the width of the ith basis function. It follows that the action value function is given by:

$$Q(s,a) = \sum_{i=1}^{n} \phi(i,s)\theta(i,a),$$ (2)

where $\theta(i,a)$ is the weight value for basis function i and action a. The Gaussian basis functions are uniformly distributed over one dimensional input space. Furthermore, we used two binary basis functions to represent the situation when the battery packs get out of agent's sight field, one for losing the battery pack to the left and one for losing to the right. The weight matrix that controls the agent behavior is updated according to the *gradient decent Sarsa(λ)* RL algorithm. *Gradient decent Sarsa(λ)* is based on *TD(λ)* and the incremental update equation is:

$$\vec{\theta}_{t+1} = \vec{\theta}_t + \alpha\delta_t\vec{e}_t,$$ (3)

where $\vec{\theta}_t$ are the network connection weights, α the learning rate, δ_t is the temporal difference error and $\vec{e}_t$ is the trace matrix distributing δ_t over the states recently visited. δ_t and $\vec{e}_t$, are calculated as follows:

$$\begin{aligned}\delta_t &= r_{t+1} + \gamma Q_t(s_{t+1}, a_{t+1}) - Q_t(s_t, a_t),\\ \vec{e}_t &= \gamma\lambda\vec{e}_{t-1} + \Delta_{\vec{\theta}_t} Q_t(s_t, a_t),\end{aligned}$$ (4)

where γ and λ are respectively the discount factor and trace decreasing factor. α is the step size by which the incremental update rule approaches the estimation of the state action value function. The reward is given by:

$$r = \begin{cases} 0 & |s| > 0.2\pi \\ \dfrac{0.1}{\pi}(0.2\pi - |s|) & |s| \leq 0.2\pi \\ 1 & \text{reaching battery} \end{cases} \tag{5}$$

The main reward is given when the agent captures the battery pack. Furthermore, a small supplementary reward is given when a battery is visible in front of the agent.

The agent chooses an action $a \Box 1, 2, \ldots, A$, where $A = 7$ is the number of possible actions. The wheel velocities are given by:

$$\omega_{lefWheel}(a) = \omega_{Const} * \frac{a-1}{A-1},$$

$$\omega_{RightWheel}(a) = \omega_{Const} * \frac{A-a}{A-1}, \tag{6}$$

where ω_{Const} is the constant angular velocity. In this way, the action spans from turning left to turning right. The probability of choosing an action a at time t is given as:

$$P(a \mid s_t) = \frac{e^{Q_t(a,S_t)/\tau}}{\sum_{i=1}^{A}\left(e^{Q_t(a_i,S_t)/\tau}\right)}, \tag{7}$$

where τ is the *temperature* of the algorithm. When the temperature is high ($\lim_{\tau \to \infty}$), the probability of choosing any action is equal. As the temperature decreases, the probability of choosing an action at time t is proportional to the action's Q-value. When the temperature approaches zero, the action with the highest Q-value is selected. After each action taken by the agent, the temperature is decreased exponentially:

$$\tau_{t+1} = \tau_{df}\,\tau_t, \tag{8}$$

where τ_{df} is the temperature decreasing factor. The RL performance is strongly related to the values of $\vec{\theta}_{initial}$, $\tau_{initial}$, τ_{df}, α, γ, λ and the total number of steps. In our work, we considered the influence of the metaparameters α and τ_{df}. The values of other parameters are considered as follows:

$$\begin{aligned} \vec{\theta}_{initial} &\in [0, 0.05], \\ \tau_{initial} &= 10, \\ \gamma &= 1, \\ \lambda &= 0.1. \end{aligned} \tag{9}$$

4.2   Actor-Critic RL

In the actor-critic, used in the surviving behavior, a computational called the actor continually produces actions. While it does so, a second computational unit called the critic continually criticizes the action taken. The actor adapts its action choices using the critic's information. The critic also adapts in the light of the changing actor. The critic's role is as a go-between, between the actions on one hand, and the reward information on the other.

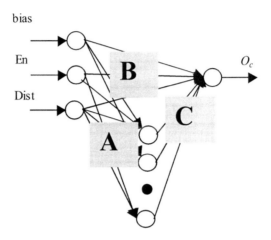

Figure 3. Critic network.

Our implementation of the actor-critic has three parts: 1) an input layer of agent state; 2) a critic network that learns appropriate weights from the state to enable it to output information about the value of particular state; 3) an actor network that learns the appropriate weights from the state, which enable it to represent the action the agent should make in a particular state. The agent can selects among three actions: 1) Capture the active battery pack; 2) Search for an active battery pack; 3) Wait for a determined period of time. The wait behavior is interrupted if a battery becomes active or after a pre-determined period of time. Both networks receive as input a constant bias input, the CR battery level and the distance to the nearest active battery pack (both normalized between 0 and 1).

4.2.1   The Critic

The architecture of critic network is shown in Figure 3. The critic has a single output cell, whose firing rate is calculated as follows:

$$O_c = \sum_{i=1}^{3} b_i x_i + \sum_{i=1}^{m} c_i y_i \tag{10}$$

where m is the number of hidden neurons, $y_i = g(\sum_{j=1}^{3} a_{ij} x_j)$ and $g(s) = \dfrac{1}{1 + e^{-s}}$.

The standard approach is for the critic to attempt to learn the value function, $V(x)$, which is really an evaluation of the actions currently specified by the actor. If x_t is the state at time t, we may define the value as:

$$V(x_t)=<r_t+ \gamma r_{t+1}+ \gamma^2 r_{t+2}+...>, \tag{11}$$

where γ is a constant discounting factor, set such that $0<\gamma<1$ and $<\cdot>$ denotes the mean over all trials.

However, the value function is not given; the critic must learn it using TD learning, i.e., the weights must be adapted so that $O_c(x)=V(x)$. Specifically, the following relationship holds between successively occurring values, $V(x_t)$ and $V(x_{t+1})$:

$$V(x_t)=<r_t>+ \gamma V(x_{t+1}). \tag{12}$$

If it were true that $O_c(p)=V(p)$, then a similar relationship should hold between successively occurring critic outputs $O_c(x_t)$ and $O_c(x_{t+1})$:

$$O_c(x_t)=<r_t>+ \gamma O_c(x_{t+1}). \tag{13}$$

TD uses the actual difference between the two sides of eq. 5 as a prediction error, which drives learning:

$$\delta_t=r_t+\gamma O_c(x_{t+1})-O_c(x_t). \tag{14}$$

The TD error is calculated as follows:

$$\hat{r}[t+1] = \begin{cases} 0 \text{ if the start state} \\ r[t+1]+\gamma^k O_c(x_t) \text{ otherwise} \end{cases}, \tag{15}$$

using the reward $r_{t+1} = (En\_level_{t+1} - En\_level_t)/50$.

TD reduces the error by changing the weights, as follows:

$$b_i[t+1] = b_i[t]+ \rho_1\hat{r}[t+1]x_i[t]$$
$$c_i[t+1] = c_i[t]+ \rho_1\hat{r}[t+1]y_i[t] \tag{16}$$

$$a_{ij}[t+1] = a_{ij}[t]+ \rho_2\hat{r}[t+1]y_i[t](1-y_i[t])\text{sgn}(c_i[t])x_j[t],$$

where ρ_1, ρ_2 are the learning rates. Under various conditions on the learning rates, this rule is bound to make $O_c(x)$ converge to the value function $V(x)$ as required.

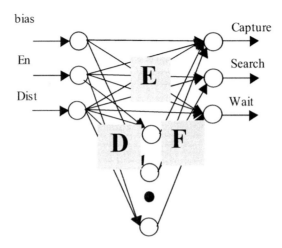

Figure 4. Action network.

4.2.2   The Actor

The actor architecture is shown in Figure 4. It can select one of three actions and so the actor make use of three action cells, p_j, $j=1,2,3$. The captured behavior is considered pre-learned, as explained in the previous section. When the search behavior is activated, the agent rotates 10 degrees clockwise. The agent do not move when the wait behavior becomes active. The output of action neurons is calculated as:

$$z_i = g(\sum_{j=1}^{3} d_{ij}x_j) , \tag{17}$$

$$p_i = g(\sum_{i=1}^{3} e_{ij}x_i + \sum_{i=1}^{n} f_{ij}z_i) , \tag{18}$$

where $g(s) = \dfrac{1}{1 + \exp^{-s}}$ and n is the number of hidden neurons.

A winner-take-all rule prevents the actor from performing two actions at the same time. The action is selected based on the softmax method as follows:

$$P(a,s_t) = \frac{e^{p_i(a,s)/\tau}}{\sum_{i=1}^{3}(e^{p_i(a,s)/\tau})} , \tag{19}$$

where τ is the temperature of the algorithm.

To choose the best action, a signal is required from the critic about the change in that results from taking an action. It turns out that an appropriate signal is the same prediction error, δ_t, used in the learning of the value function. The actor weights are adapted as follows:

$$e_i[t+1] = e_i[t] + \alpha_1 \hat{r}[t+1](q[t] - p[t])x_i[t]$$

$$f_i[t+1] = f_i[t] + \alpha_1 \hat{r}[t+1](q[t] - p[t])z_i[t] \quad (20)$$

$$d_{ij}[t+1] = d_{ij}[t] + \alpha_2 \hat{r}[t+1]z_i[t](1 - z_i[t])\mathrm{sgn}(f_i[t])(q[t] - p[t])x_j[t]$$

5 Evolution of Metaparameters

In our implementation, a real-value GA was employed in conjunction with the selection, mutation and crossover operators. Many experiments, comparing real value and binary GA, show that the real value GA generates better results in terms of the solution quality and CPU time (Michalewich, 1994). To ensure a good result of the optimization problem, the best GA parameters are determined by extensive simulations that we have performed, as shown in Table 1. A more detailed explanation of genetic operators is given in Appendix.

The main steps of proposed algorithm (Figure 5) are:

- First, a population of individuals is generated randomly based on the predefined searching space. Every individual of the population has genes encoding the metaparameters.
- Each agent learns using the encoded metaparameters.
- The fitness is evaluated for every individual of the population.
- Based on the fitness value, the best individuals are selected at the end of each generation and the genetic operators are performed until the termination criterion is satisfied.

Our main goal in battery capturing task is to see the effect of metaparameters on learned policy. Therefore, the fitness is evaluated based on the number of steps used to reach the battery pack when the agent is placed in predefined positions and the reward accumulated by the agent during last stage of learning. On the other hand, in order to see the effect of metaparameters and initial weight connections on learning time the fitness function in the surviving behavior is considered as follows:

$$Fitness = \begin{cases} \dfrac{CR_{life}}{max\_learning\_time} & \text{if the agent dies before the maximum learning time} \\ En_{level} + 1 & \text{if the agent suvives} \\ En_{max} + \dfrac{max\_learning\_time}{CR_{life}} & \text{if the agent's battery is fully recharged} \end{cases} \quad (21)$$

where CR_{life} is the CR life in seconds, *max_learning_time* is the maximum learning time, En_{level} is the level of energy if the agent survives but the battery is not fully recharged and En_{max} is the maximum level of energy. The maximum learning time is 7200 s. The value of En_{max} is 1 and En_{level} varies between 0 and 1. Based on this fitness function the agents that can not survive get a better fitness if they live longer. The agents that survive get a better fitness if the energy level at the end of maximum learning time is higher. On the other hand, the

192   Genci Capi, Daisuke Hironaka, Masao Yokota, et al.

learning stops when the agent's battery is fully recharged. The agents that learned to fully recharge their battery faster get the highest fitness value.

In order to see the effect of metaparameters and initial weight connections on learning time, we considered the following cases:

1) Evolution of meta-parameters, initial weight connection of actor and critic networks, and the number of hidden neurons.
2) Evolution of meta-parameters with randomly initialized weight connections.
3) Evolution of initial weight connections with arbitrary selected meta-parameters.
4) Learning with arbitrary selected meta-parameters and randomly initialized weight connections.

Table 1. Functions and parameters of GA.

Function Name	Parameters
Arithmetic Crossover	2
Heuristic Crossover	[2 3]
Simple Crossover	2
Uniform Mutation	4
Non-Uniform Mutation	[4 GNmax 3]
Multi-Non-Uniform Mutation	[6 GNmax 3]
Boundary Mutation	4
Normalized Geometric Selection	0.08

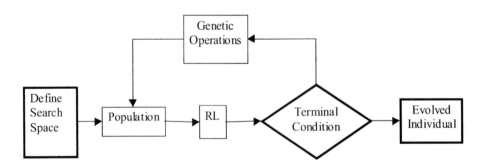

Figure 5. Schematic diagram of proposed algorithm.

6   Results

6.1   Capturing Behavior

In the simulations ω_{Const} is set to 1300 mm/s and the time step is 75 ms. Therefore, each action taken by the agent will result in a movement of 49 mm. All coming time measurements refer to simulations time. The searching space for the metaparameters is predefined, τ_{df} □ [0.9,1] and α □ [0.001,0.2].

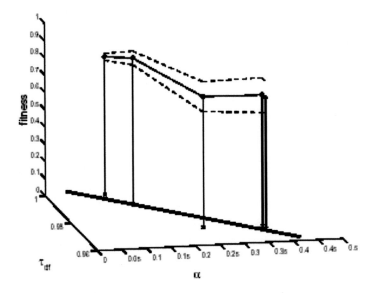

Figure 6. Relation between α and τ_{df}. The average fitness and fitness standard deviation of the last populations shows the performance of the solutions.

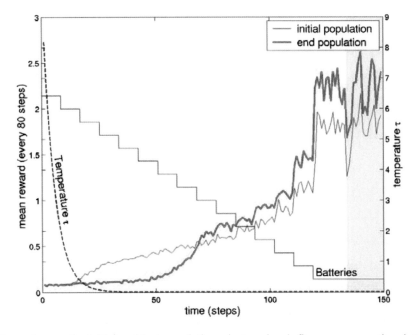

Figure 7. Reward curve for initial and last populations, battery level, fitness measure time interval and temperature curve of the best individual.

6.1.1   Single Food Capture Task

The agent starts learning in short distance relative to the battery pack (300 mm), oriented so that the battery pack is always visible. In the late stage of leaning, the position and orientation of the agent relative to the battery are randomly selected. The episode length differs in the two stages. The fitness value is calculated when the learning is completed. The derived policy is evaluated by letting the agent capturing 18 battery packs from predefined positions. The population size is 40. After 20 to 30 generations the individuals become very similar with each other and the evaluation was terminated.

Figure 6 shows the evolved values of metaparameters from five simulations. The two rightmost results in the figure have an additional term in the fitness function utilizing the time of learning. 4% of the total fitness value is based on how fast the agent learns the task and explains the lower value of τ_{df}. The additional reward is withdrawn in the figure to make a fair comparison. The figure shows the relation between α and τ_{df}. If the algorithm cools off early (small τ_{df}) the step size towards the estimated value function must be large (large α). When the algorithm is allowed to cool off slowly, the step size can be smaller and the estimation of the value function is more correct. The average fitness (connected line) and the standard deviation (dashed lines) of the last populations shows the performance of the evolved solutions. The reliability increases when τ_{df} decreases.

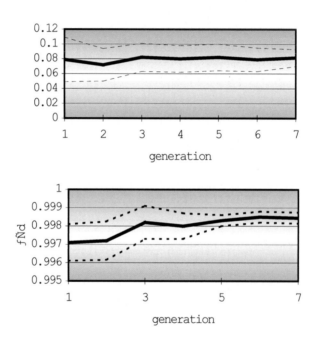

Figure 8. The course of evolution showing the evolvements of mean and standard deviation values for α and τ_{df}.

6.1.2   Multiple Food Capture Task

In the second experiment, the number of battery packs decreases linearly from 15 to 1 during the agent's lifetime. During the last 98 s of the agents lifetime, the fitness value is calculated directly from the collected reward.

Figure 7 shows the average reward curve of the initial and the last population of individuals with a lifetime of 900 s. Also, the temperature τ during the learning process of the evolved individual and the number of battery packs are presented. The result is not what we expected. We expected that the best solution should be to wait as long as possible, to cool off (τ approach zero) just before the beginning of the fitness measurement. However, the solution shows that it is better to cool off earlier in the lifetime, when there are still eight ten batteries left in the environment. This is because the agent needs to have frequent opportunities to catch batteries to successfully learn a good policy.

The course of evolution for the two parameters is presented in Figure 8. The population size is 50 individuals and the evolution converged after 7 generations. Figure 9 shows the relation between α and τ_{df} after evolution from eight experiments using different lifetimes. Again, we can clearly see the relationship of the optimal settings of α and τ_{df}.

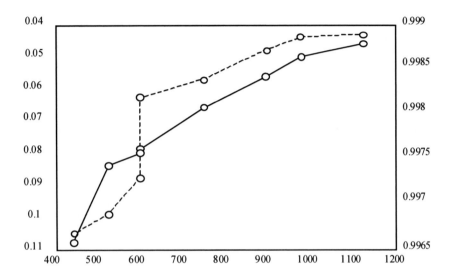

Figure 9. Evolved individuals from eight experiments using different lifetimes in the multiple food task.

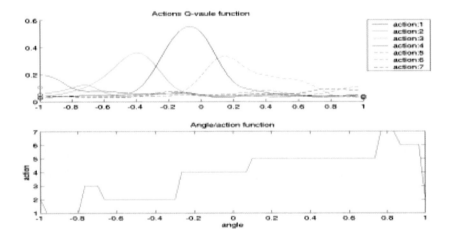

Figure 10. Action value functions and learned policy in simulation.

6.1.3   Hardware Implementation

In order to investigate how the optimized metaparameters in simulation will perform in real hardware system, we implemented the RL algorithm by using the CR robot. The environment and the learning conditions are identical to the task in section 6.1.2. Metaparameters used in the experiment are evolved by GA where each individuals lifetime is 525 s. The processing time and the procedure after capturing a battery pack are different in the simulation and hardware implementation. For smooth movement during learning, all calculations has to be carried out while performing the actions. As a compensation for calculation delay, we increased the time step to 500 ms and decreased ω_{Const} to 300 mm/s, resulting in the step length of 75 mm.

Figure 10 and Figure 11 show the learned policy in simulation and hardware implementation, respectively. Note that action number four is the straight forward action. The circles at both ends show the case when no battery packs are visible. These figures show that there are two major differences between the learned policies. First, the straight forward action is used for a wider range of angle state in the hardware solution. This is because the CR hardware has two "claws" to collect in the battery pack, even if it is not exactly straight in front of the agent. The other difference is the behavior when there is no battery pack visible. While the agent turns around to search for the battery packs in simulation, it uses the straight forward action in hardware implementation.

This is because CR hardware has shorter range of vision. While the agent in simulation can usually find a battery pack by looking around, the hardware agent has to explore the field using straight-ahead movement and wall-avoidance reflex. The course of learning is similar in both cases. The temperature decreases in the same way and time of decision occurs almost simultaneously. However, in the real robot learning, the period just before settled for a policy (cooling off) is more critical. Verifying the policy is more important due to noise components and time delays that are not present in the simulation environment.

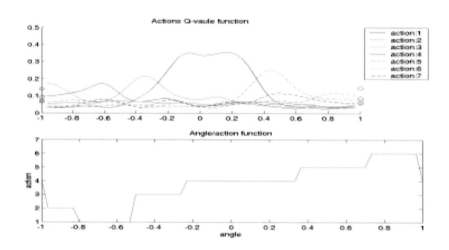

Figure 11. Action value functions and learned policy in hardware implementation.

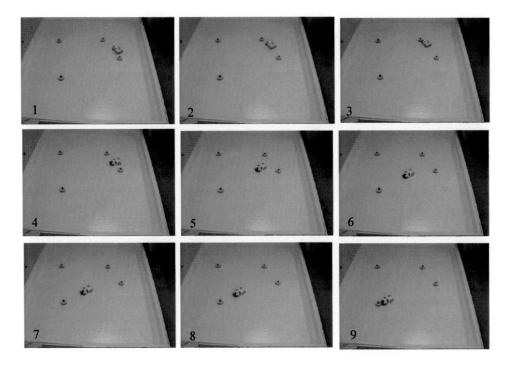

Figure 12. Video capture of last phase of learning in real hardware.

Table 2. Only learning

Parameters	Values
Initial Weights	Randomly between [-0.5 0.5]
ρ_1, ρ_2, α_1, α_2	0.2; 0.1; 0.8; 0.3
τ_0	10
γ	0.9

Table 3. Searching space and GA results.

Optimized parameters	Searching interval	Results
Initial Weights	[-1 1]	
ρ_1, ρ_2, α_1, α_2	[0 1]	0.9210; 0.3022; 0.9256; 0.6310
τ_0	[1 10]	5.3718
γ	[0 1]	0.794

The results show that final policies learned in simulation and real hardware are different. However, the learning processes follow the same stable pattern in both two cases. This shows that even if the environments and the physical conditions of the agents do not allow the same optimal policy, the metaparameters optimized in simulation gives the right stability to complete the learning process, as shown in Figure 12.

6.2   Surviving Behavior

In order to determine the energy level after each action, we measured the battery level of CR when the robot moves with a nearly constant velocity of 0.3m/s. The collected data are shown in Figure 13. The graph shows that there is a nonlinear relationship between time and energy level. Therefore, we used the virtual life time to determine the energy level after each action. Except capturing the battery pack, the search and wait actions increased the virtual life time. The virtual life time when the CR captured an active battery pack was decreased by 50 s, while the search and wait action increased the virtual time with 1 s and 5 s, respectively.

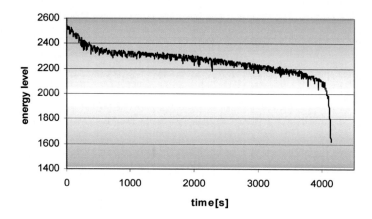

Figure 13. Battery level during CR motion.

In order to see the effect of metaparameters and initial weight connections on learning time, first learning took place with arbitrarily selected metaparameters (Table 2) and random initial weight connections. The number of hidden neurons in actor and critic networks is considered 3. The agent continued learning for nearly 5500s until the battery was fully recharged (Figure 14). Then, using the same metaparameters and neural structure, we evolved the initial weight connections. Initially, 100 individuals with different weight connections are created and the evolution terminated after 20 generations. Figure 14 shows that CR learned the surviving behavior faster compared with the case when initial weight connections are randomly selected. However, the reduced learning time is small because of the effect of selected metaparameters, especially the balance between exploration and exploitation.

On the other hand, when the metaparameters are evolved and initial weight connections are randomly selected, the learning time is significantly reduced (Figure 14). This result is important because it shows that metaparameters has larger effect on learning time compared to initial weight connections. The searching interval and GA results, when both metaparameters and initial weight connections are optimized by GA, are shown in Table 3. The performance of different individuals selected from the first generation is shown in Figure 15. In the first generation, most of the individuals could not survive during the 7200 s of learning time. Based on the fitness value, the individuals that survived longer had higher probability to be selected in the next generation. The critic and actor networks have 1 and 3 hidden neurons, respectively. The initial value of weight connections are very near with their respective values after learning. In addition, the optimized value of cooling factor is low.

Therefore, the agent starts to exploit the environment and make greedy actions soon after the learning starts, recharging the battery in a very short time.

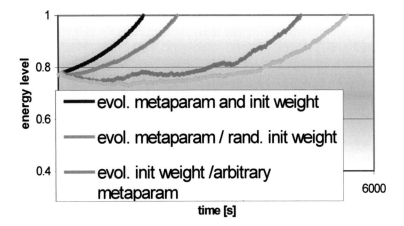

Figure 14. Energy level of the best agent for different combinations of learning and evolution.

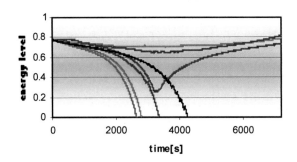

Figure 15. Performance of different agents selected from the first generation.

7   Conclusions

The interplay between learning and evolution has been previously investigated in the field of evolutionary computation. Much of this research has been directed to elucidating the relationship between evolving traits (such as synaptic weights) versus learning them (Ackey & Littman, 1991). A relatively small amount of research has been devoted to the evolution of metaparameters, interactions among metaparameters and their role on agent learned policy. In this chapter, we presented a method to optimize the metaparameters in RL based on evolutionary approach. Based on the simulations and experimental results we conclude:

- Evolutionary algorithms can be successfully applied to determine the optimal values of metaparameters.
- Meaparameters play important role on the agent learned policy.
- Metaparameters have a strong interaction with each-other.

- Optimized metaparameters can significantly reduce the learning time.
- Application of evolved metaparameters on learning in real hardware of CR robot showed good results.

In this work, the reward during learning is determined by us. However, the reward in the surviving behavior and the ratio between the main reward (capturing the battery pack) and the small reward (keeping the battery pack visible) in the capturing behavior influences the learned policy and learning time. Therefore, it will be interesting to see if evolution can also be applied to shape the reward function used during the learning process. This and some other issues will be our focus in the future.

Appendix

Genetic Operators Used in Real-Number GA

Selection Function
Selection of individuals is done based on normalized geometric ranking, which defines selection probability p_i for each individual as:

$$p_i = q'(1-q)^{r-1},$$

A(1)

where: q is the probability of selecting the best individual, r is the rank of the individual where 1 is the best, P is the population size, and $q' = \dfrac{q}{1-(1-q)^P}$.

Uniform Mutation randomly selects one variable, j, and sets it equal to an uniform random number $U(a_i, b_i)$ as follows:

$$x'_i = \begin{cases} U(a_i, b_i), \text{if } i = j \\ x_i \text{ otherwise} \end{cases}$$

(A2)

where the a_i and b_i are the lower and upper bound of searching interval.

The uniform mutation is repeated 4 times.

Boundary mutation randomly selects one variable, j, and sets it equal to either its lower or upper bound, where r is uniform random number between $(0, 1)$:

$$y'_i = \begin{cases} a_i, \text{if } i = j, r < 0.5 \\ b_i, \text{if } i = j, r \geq 0.5 \\ x_i, \text{otherwise} \end{cases}$$

(A3)

The boundary mutation is repeated 4 times.

Non-uniform mutation randomly selects one variable, j, and sets it equal to a non-uniform random number:

$$y_i^{'} = \begin{cases} x_i + (b_i - x_i)f(G), \text{if } r_1 < 0.5, \\ x_i - (x_i + a_i)f(G) \text{ if } i = j, \, r_i \geq 0.5 \\ x_i, \text{ otherwise} \end{cases} \quad , \tag{A4}$$

where $f(G) = (r_2(1 - \dfrac{GN}{GN\max}))^b$, r_1, r_2 are uniform random numbers between $(0, 1)$, GN is the current generation, GN_{max} is the maximum number of generations, and b is a shape parameter equal to 3.

The multi non-uniform mutation operator applies the non-uniform operator to all of the variables in the parent $\overline{X}$.

Real-value simple crossover is identical to the binary version.

Arithmetic crossover produces two complimentary linear combinations of the parents, where r is uniform random number between $(0, 1)$ as follows:

$$\overline{X}^{'} = r\overline{X} + (1-r)\overline{Y} \quad , \tag{A5}$$

$$\overline{Y}^{'} = (1-r)\overline{X} + r\overline{Y} \tag{A6}$$

Heuristic crossover produces a linear extrapolation of the two individuals. This operator utilizes the fitness information. A new individual, $\overline{X}^{'}$, is created using the following equation:

$$\overline{X}^{'} = \overline{X} + r(\overline{X} - \overline{Y}) \quad , \tag{A7}$$

$$\overline{Y}^{'} = \overline{X} \, , \tag{A8}$$

where r is uniform random number between $(0, 1)$ and $\overline{X}$ is better than $\overline{Y}$ in term of fitness.

If $\overline{X}^{'}$ is infeasible, i.e. feasibility equals 0 as given by equation:

$$feasibility = \begin{cases} 1, \text{if } x_i^{'} \geq a_{i,} \, x_i^{'} \leq b_i, \forall i \\ 0, \text{ otherwise} \end{cases} \, , \tag{A9}$$

then generate a new random number r and create a new solution using equation (A7), otherwise stop. To ensure halting after t failures let the children equal the parents and stop. We considered t=3.

Simple Crossover generates a random number r from a uniform distribution from 1 to m and creates two new individuals ($\overline{x}^{'}$ and $\overline{y}^{'}$) according to the following equations:

$$x_i = \begin{cases} x_i, \text{if } i < r \\ y_i, \text{otherwise.} \end{cases} \quad , \tag{A10}$$

$$y_i = \begin{cases} y_i, \text{if } i < r \\ x_i, \text{otherwise.} \end{cases} \tag{A11}$$

In Table 1, the parameter for arithmetic crossover, simple crossover, uniform mutation, boundary mutation and the first parameter of heuristic crossover, non-uniform mutation and multi-non-uniform mutation is the number of times to call the operator in every generation.

References

Ackey, D. & Littman, M. (1991). Interactions between learning and evolution. In J.D. Farmer, C. G. Langton, C. Taylor and S. Rasmussen, editors, Artifilial Life II, Addison-Wesley.

Baldwin J. M. (1986). A new factor in evolution. *American Naturalist* **30**:441451.

Barto, A. G. (1995). Reinforcement learning. In M. A. Arbib (Ed.), The handbook of brain theory and neural networks (pp. 804–809). Cambridge, MA: MIT Press.

Belew, R. K., McInerney, J. & Schraudolph, N.N. (1992). Evolving networks: using the genetic algorithm with connectionist learning. In C.G. Lanton et al eds. *Artificial Life II*, 511-547. Addison Wesley.

Donahoe, J.W. & Packard-Dorsel, V. (1997). Neural network models of cognition: Biobehavioral foundations. Elsevier Science.

Doya, K. (2000). Reinforcement learning in continuous time and space. *Neural Computation*, **12**, 215–245.

Doya, K., Kimura, H., & Kawato, M. (2001). Computational approaches to neural mechanism of learning and control. *IEEE Control Systems Magazine*, **21**(4), 42–54.

Doya, K. (2002). Metalearning and Neuromodulation. *Neural Networks*, **15**:495-506.

French, R. M., & Messinger, A. (1994). Genes, phenes and the Baldwin effect: learning and evolution in a simulated population. In *Proceedings of the fourth International workshop on the synthesis and simulation of living systems*, 277-282, MIT Press.

Ishii, S., Yoshida W. & Yoshimoto J. (2002). Control of Exploitation-Exploration Metaparameter in Reinforcement Learning. *Neural Networks*, **15**:665-687.

Michalewich, Z. (1994). Genetic Algorithms + Data Structures = Evaluation Programs. *Springer Verlag.*

Morimoto, J., & Doya, K. (2001). Acquisition of stand-up behavior by a real robot using hierarchical reinforcement learning. *Robotics and Autonomous Systems*, **36**, 37–51.

Neal, R. M. (1996). Bayesian learning for neural networks. New York: Springer.

Nolfi, S., & Floreano, D. (1999). Learning and Evolution." *Autonomous Robots*, **7**(1): 89-113.

Pollack, J. B., Lipson, H., Ficici, S., Funes, P., Hornby, G. & Watson, R. A. (2000) Evolutionary Techniques in Physical Robotics. *Evolvable Systems: from biology to hardware; proceedings of the third international conference (ICES 2000)*, Springer (Lecture Notes in Computer Science; Vol. 1801), 175-186.

Singh, S., & Bertsekas, D. (1997). Reinforcement learning for dynamic channel allocation in cellular telephone systems. In M. C. Mozer, M. I. Jordan, & T. Petsche (Eds.), (Vol. 9) (pp. 974–980). Advances in neural information processing systems, Cambridge, MA: MIT Press.

Sutton, R. S., & Barto, A. G. (1998). Reinforcement learning. Cambridge, MA: MIT Press.

Tesauro, G. (1994). TD-Gammon, a self teaching backgammon program, achieves master-level play. *Neural Computation*, **6**, 215–219.

Unemi, T., Nagayoshi, M. Hirayama, N., Nade, T., Yano K. & Masujima, Y. (1994). Evolutionary differentiation of learning abilities - a case study on optimizing parameter values in Q-learning by genetic algorithm. *Artificial Life IV*, 331-336, MIT Press.

Unemi, T. (1996). Evolution of reinforcement learning agents - toward a feasible design of evolvable robot team. In Workshop Notes of ICMAS'96 Workshop 1: Learning, Interactions and Organizations in Multiagent Environment, 1996.

Vapnik, V. N. (2000). The nature of statistical learning theory (2nd ed). New York: Springer.

In: Control and Learning in Robotic Systems
Editor: John X. Liu, pp. 205-241

ISBN 1-59454-356-9
©2005 Nova Science Publishers, Inc.

Chapter 8

VORONOI-BASED MAP LEARNING AND UNDERSTANDING IN ROBOTIC PROBLEMS

B.L. Boada[a], D. Blanco[b], C. Castejón[a] and L. Moreno[b][*]
[a] Mechanical Engineering Department
[b] System Engineering and Automation Department
Carlos III University
Avd. de la Universidad, 30. 28911. Leganés (Madrid). SPAIN

Abstract

In this chapter, a new Voronoi-based technique to model the environment is presented. The objective is to build a unique environment model, which can be used to solve different problems in robotics such as localization, path-planning and navigation. The proposed model combines geometric and topological approaches to take advantage of each method. This topo-geometric model is obtained from the Voronoi Diagram, which is built using the measurements of a laser telemeter. Apart from the topological and geometric information extracted from the Local Voronoi Diagram, a symbolic information, which represents distinctive places typical of indoor environments (HALL, CORRIDOR, etc.), is also provided to the model. The technique used to learn and recognize these places is HMM (Hidden Markov Models).

Different algorithms based on topo-geometric maps are also proposed to solve the problems of localization, navigation and path-planning. Experimental results show the effectiveness of these algorithms. However, the models based on Voronoi Diagrams present a problem when the number of objects detected by the sensor is minor to two. In this case, no Voronoi Diagram is generated. This implies that the robot has not got any information to navigate through the environment. We solve this problem by generating a Virtual Voronoi Diagram which guarantees the continuity of model.

---

[*]E-mail address: {bboada,dblanco,castejon,moreno}@ing.uc3m.es

1 Introduction

Different models of environment are required to carry out different tasks such as localization, path planning and navigation. Thus, the environment modelling is one of the main activity to develop in mobile robotics.

The most significant types of representation according to the environment information are: cell decomposition maps, geometric maps, topological maps and hybrid maps. The geometric models [1, 2, 3] are a set of geometric primitives which can be extracted directly from sensor information or from other kinds of representations like cell decomposition maps. They are suitable for localization. The main problem is that the extraction of geometric characteristics tends to be simple (segments, points, corners, edges) but in complex environments these simple features are not able to model many of the environment objects. Another problem derives from the inability of geometric feature maps to be directly used in navigation or path planning which obliges to translate these maps to other representations suitable to be used in those tasks. This conversion has an important computational cost that is usually neglected. The grid maps represent the environment as a two dimensional array of cells, each of which indicates the probability of being occupied [4, 5]. They are suitable for localization and navigation. However, it is necessary a large amount of memory in large environments. The topological representations model relationships between different types of entities [6, 7, 8, 9, 10]. The environment is described by a connected graph where the adjacency property or other relationships are modelled. The main advantage of the topological approaches is the qualitative aspects used to describe the environment instead of metric aspects, which are closer to human behavior than another type of environment modelling. The topological models are suitable for path-planning, but they are not able to maintain an accurate geometric estimate of the robot pose.

Many authors propose a representation structure with different levels of models in such way that each task uses a different model. The problem of this structure is to convert a representation in another required to develop a task. Recently, hybrid maps have been adopted to represent the environment [11, 12, 13, 14, 15, 16, 17]. They combine geometric and topological approaches, therefore hybrid methods can be used successfully for localization, path-planning and navigation.

The research line of Mobile Manipulator's Group of Carlos III University has the objective to build a unique environment model, which can be used to solve different problems in robotics such as localization [18, 19], path-planning [20] and navigation [16]. The proposed model combines geometric and topological approaches to take advantage of each method and to solve their disadvantages. This topo-geometric model is obtained from the Voronoi Diagram using the measurements of a laser telemeter. Because of sensor range and in order to reduce the error in the environment model, the robot observes a partial area of the environment. For this reason, the model extracted directly from sensor measurements is called Local Voronoi Diagram. An important consideration of the proposed model is the compactness of the environment representation. Because the local Voronoi Diagram is a retraction of the two dimensional space over a set of one-dimensional network of curves, this representation requires less data to represent the environment than those based on traditional grids. Both topological and geometrical information added to the model is extracted from this Local Voronoi Diagram. The generated model contains enough information to be used it in localization, navigation and path-planning. The Voronoi Diagram is the set of points which are equidistant to closer objects. For this reason, it is thought to be the safest way to move the robot (navigation). Moreover, the Voronoi Diagram represents the environment topology. Then, it is also useful for path-planning. Finally, the topological and geometric information stored in the map is used to correct the robot position (localization).

However, this type representation fails in opened areas or in areas where the number of observed objects is minor to one. In this case, a Local Voronoi Diagram is not generated and the robot has not any information about the environment. This prevents that the robot can move and carry out its tasks. This problem has been solved building a Virtual Voronoi Diagram which considers that there are virtual objects in the limit of view field.

Apart from the topological and geometric information, extracted from the Local Voronoi Diagram, a symbolic information is also provided to the proposed model [21]. This information is stored in the topo-geometric map to carry out a topological localization and path-planning. The method used to learn and recognize the distinctive places typical of indoor environments (corridors, halls, doors, etc.) is HMM (Hidden Markov Models). This method operates with the topo-geometric map and it does not use directly the measurements of sensors.

The main advantages of obtaining symbols from the environment are: 1) the symbolic information is very compact and 2) it is easily understood by other robots or by humans. In this manner, it can be used to send spoken instructions to the robot and to make it moves in a desired manner, i.e., *move to HALL, turn on the left in the next DOOR*, etc. Moreover, it allows to obtain topological level abstraction which it can reduce planning complexity. Another advantage of the proposed method is that it provides a stochastic solution which it allows to interpret probabilistically the scene. The probability can be used to take decisions on localization, path planning and navigation. A high probability value indicates that the information, extracted by sensors, is a good and accurate information. However, a low value implies that the robot could have an incomplete information of the environment.

2   Description of the Local Environment Model Based on Voronoi Diagram

The topo-geometric approach proposed here to model the environment tries to obtain a graph which describes the environment. This graph should include nodes, as singular or characteristic point in the environment, and edges to connect the nodes. The difference with adjacency graphs is that in our case and because the graph is the model of the environment the edges requires a geometrical description (in the proposed algorithm the space is discretized in cells and the cells crossed by the edges are stored). To obtain this kind of representation to model the environment three modelling steps are adopted. In a first step a Voronoi Diagram is obtained from the observed laser scan data at a given position, this map is referred as Local Voronoi Diagram (LVD). The Voronoi Diagram is defined by a set of points equidistant to, at least, two objects. Thus, the Voronoi Diagram provides a natural way that captures the topology of the free space. Another advantage of Voronoi Diagram is that the map is reduced to a 1-dimension shape, while the information is merged cell to cell in the grid maps. This Local Voronoi Diagram is processed in a second step to eliminate unnecessary information (edges finishing in a corner) and to add some topological additional information which we call Local Topo-geometric Map or simply Local Map. Finally, these Local Maps are then fused to obtain a global representation of the environment which is referred as Global Map. This global map can only be coherently built if the localization and data association problems are previously solved.

Many authors use the Voronoi Diagram to model the environment. The Voronoi Diagram can be generated from a grid-based map [13, 14, 22] or directly from measurements of sensor [15, 16, 17]. Other authors do not use the Voronoi Diagram to build the global hybrid map such in [12], where the topological map is built using the landmarks (corners and openings) extracted from the sensor measurements. In this method, previously to build the map it is necessary to extract features such as corners. Thus, this method can only be used in a structured environment. However, our method does not require extracting geometric characteristics such as lines or segments and consequently it can also be used in unstructured environments [23].

2.1   Definition of the Local Voronoi Diagram (LVD)

The Local Voronoi Diagram (LVD) is extracted from a laser scan at a given robot pose. The adopted technique is known as Generalized Voronoi Diagram and lets to obtain a Voronoi Diagram when input information is a set of points [24]. In two dimensional case, the Generalized Voronoi Diagram is simply the set of points equidistant to two or more objects:

$$V(x) = [(d_1 - d_2)](x) \tag{1}$$

where d_1 and d_2 are distances to two convex objects. In figure 1(a), this diagram is drawn with dashed lines.

One problem of generalized Voronoi Diagrams, when it is considered for planning or navigation, is the existence of edges finalizing at the boundary of environment, e.g. corners, or in general those edges whose distances to objects are lower than the robot radius plus the minimum admissible distance to objects. In those cases, these edges add additional evaluation costs when local navigation or global planning is done. For this reason, in the proposed algorithm these edges are not generated. In figure 1(b), the Local Voronoi Diagram is represented. Because the unnecessary edges are not generated, the nodes related to this edges are not generated. This simplifies the Local Voronoi Diagram. Moreover, the removal of this edges and nodes does not originate problems at localization time because the distance of each point of the Local Voronoi Diagram is stored together with points positions.

Let $P = \{r_1, \cdots, r_{361}\} \in \mathrm{R}^2$ be the measurements from laser telemeter in the plane. These measurements are grouped according to obstacle they belong to. Successive distances

Generalized Voronoi Diagram

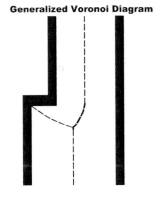

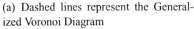

Local Voronoi Diagram

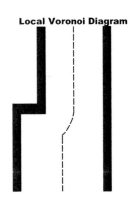

(a) Dashed lines represent the Generalized Voronoi Diagram

(b) Dashed lines represent the Local
Voronoi Diagram

Figure 1: Generalized Voronoi Diagram and Local Voronoi Diagram

measured on an object boundary increase or decrease continuously. Discontinuity, that is a
large difference between two successive distances, marks the beginning of a new object and
it will be used to distribute scan points in clusters. In our case, a discontinuity is considered
when the distance between two successive measurements is greater than 60 cm. This value
is selected to allow the robot to be able to pass through the objects.

Let $C = \{c_1, \ldots, c_n\} \subseteq P$ with $2 \leq n \leq \infty$ be a collection of n non-overlapping sets of
measurements each of them belong to the same obstacle:

$$c_i \bigcap c_j = 0, \quad i \neq j \tag{2}$$

For any point $p \in \mathrm{R}^2$, let $d_E(p, c_i)$ be the minimum Euclidean distance from p to any point
which defines c_i. A Voronoi Diagram is defined as:

$$V(c_i) = \{p | p \in \mathrm{R}^2, d_E(p, c_i) < d_E(p, c_j), i \neq j\} \tag{3}$$

And the collection:

$$V = \{V(c_1), \cdots, V(c_n)\} \tag{4}$$

is called the Generalized Voronoi generated by C.

With this definition, the LVD is built using the algorithm described in [25]. The visible
region is discretized with a suitable cell resolution, in order to reduce the computing time.
Thus, the LVD is calculated for each cell, cp:

$$V(c_i) = \{cp | cp \in \mathrm{R}^2, d_E(cp, c_i) < d_E(cp, c_j), i \neq j\} \tag{5}$$

Cells which are equidistant to two clusters are called edges, and cells which are equidistant to more two clusters are called nodes. Figure 2 shows different Local Voronoi Diagrams which represent different areas of the environment.

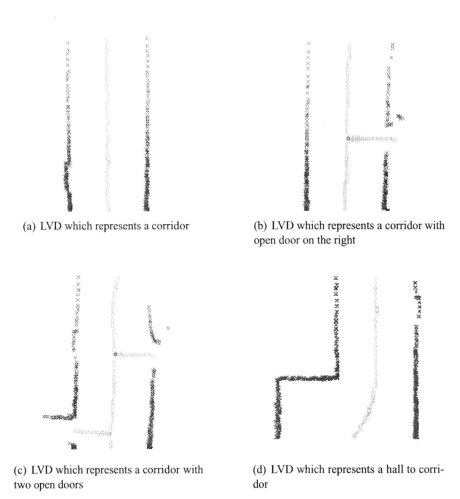

(a) LVD which represents a corridor

(b) LVD which represents a corridor with open door on the right

(c) LVD which represents a corridor with two open doors

(d) LVD which represents a hall to corridor

Figure 2: LVDs for different environments

2.2 Generation of the Local Topo-geometric Map

The Local Voronoi Diagram provides us with a set of points equidistant from objects but from planning or navigation point of view a graph representation is more convenient. For this reason, a new topological structure is extracted from the Local Voronoi Diagram. This new structure contains a topological graph structure together with the points positions and distances to objects. Because of the dual information stored in this structure, it is referred as Local Topo-geometric Map. This new representation can be used for navigation (in our case sensor-based navigation based on topological and geometrical information is used) and for path-planning (in this case the topological information contained in the map is basically used).

Let denote LM_i the local topo-geometric map obtained at time $t = i$. The local topo-geometric map is generated from the Local Voronoi Diagram. The node points from LVD become nodes in the local topo-geometric map, and the edges points from LVD are converted in edges in the local topo-geometric map:

$$LM_i = \{\sum N_i, \sum R_i\} \tag{6}$$

where N_i are the local topo-geometric map's nodes and R_i are the local topo-geometric map's edges. At least, nodes are equidistant to three objects. Two nodes are joined if at least they are equidistant to the same two clusters. The points which generate the joined edge are equidistant to these clusters.

The nodes are defined by a point and the distance of nodes to the equidistant objects, and the edges are defined by a set of points which represent the way between two nodes and the distance of edges' points to the equidistant objects:

$$\begin{aligned} N_i &= \{(x_{(i,1)}, y_{(i,1)}), d_{(i,1)}\} \\ R_i &= \{[(x_{(i,1)}, y_{(i,1)}), d_{(i,1)}], \cdots, [(x_{(i,n)}, y_{(i,n)}), d_{(i,n)}]\} \end{aligned} \tag{7}$$

where $d_{(i,j)}$ is the distance of point $(x_{(i,j)}, y_{(i,j)})$ to the closest object.

Apart from the nodes obtained directly from LVD, other nodes can be recognized by studying the edges distance variation to the equidistant objects. A change in the slope sign implies a new node or place. The new node will be situated in the point where the slope sign change occurs.

Let R_i be an edge of LVD which is formed by a set of points $R_i = \{p_i, \cdots, p_n\}$. This set of points are ordered, where p_i and p_n represent the extreme edge's points. For each point p_k, the distance to equidistant objects d_k is stored, where k represents the position of the point in the edge. A new node is generated, if it meets the following requirements (see figure 3):

$$sign(d_j - d_{j-1}) \neq sign(d_{j-1} - d_{j-2}) \tag{8}$$

$$sign(d_j - d_{j-1}) = sign(d_k - d_{j-1}), \text{ with } k > j \tag{9}$$

and

$$\| p_{j-1} - p_k \| > threshold_{\text{error measurements}} \tag{10}$$

If the previous requirements are complied with, then the new node is generated in p_{j-1}:

$$N_i = \{p_{j-1}\} \tag{11}$$

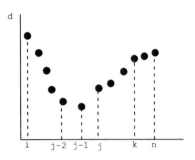

Figure 3: Distance of each edge's point to the equidistant objects

In figure 4.b the distance variation of the points which belong to the same edge to the equidistant objects is represented when a door is crossed. In figure 5.b the distance variation is represented when the edge goes through from the hall to the corridor. In figures 4 and 5 in (a) the Voronoi Diagram and (c) the topo-geometric or state map are represented for the two previous cases. The new states, door and narrowing, are shown. These new nodes provide more information or characteristics to the topo-geometric map. With this information, the robot can carry out a more accurate topological localization. On the other hand, the robot can select the suitable motion strategy. The motion sequence is executed considering the places: when the robot travels by a corridor it moves faster than when it passes by a door.

3   Non-boundary Areas. The Virtual LVD

The robot requires a knowledge of environment to navigate without colliding with any obstacle. The problem occurs in opened areas, such as halls or rooms, where the number of objects detected by the laser telemeter can be minor than two. In this case, the LVD is not built and the robot can not generate a safety path.

In opened area, although the LVD is not obtained, some information is provided about the environment. If the number of generated clusters is equal to zero, it indicates that there is not any obstacle in the surrounding area and the robot can navigate safely. On the other hand, if the number of generated clusters is equal to one, it indicates that there is one obstacle in the environment. Although the LVD is not obtained, the robot knows how far it is from obstacle. In both cases, the adopted solution to maintain the continuity in the robot navigation is to build a Virtual Voronoi Diagram which contains the previous information.

The Local Voronoi Diagram generated in opened areas is called *virtual* because it is not a real Voronoi Diagram. In this manner, the Virtual LVD is only used for navigation and not for localization.

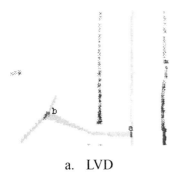

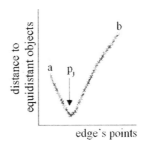

a. LVD

b. Distance of each edge's point to equidistant objects and position of the new node

c. Local topo-geometric map

Figure 4: Generation of a new node which represents a door

The algorithm to generate the Virtual LVD is divided in two main steps:

1. If the number of detected objects is less than two, then the virtual clusters are generated. The main characteristic of the virtual clusters is that they are defined by a set of artificial points which are located in the boundary of the field of view.

   The criterion to generate the virtual clusters is: "to generate so many virtual clusters in such way that a Voronoi Diagram can be obtained allowing a safely navigation". The navigation is considered as safely when the robot can pass through the clusters (real or virtual) without colliding with them.

2. Once the total number of clusters is mayor to one, the Virtual LVD is generated in the same way that in the section 2.1.

In figure 6 the Virtual LVD is represented for the case that no obstacles are detected. In this case, two virtual clusters are generated (green color) in such way they are formed by points which are located in the boundary of the field of view and the distance between them allows that the robot passes through them without colliding. The Virtual LVD is represented in blue color.

In figure 7, only an obstacle is detected and it is the lateral side of robot (red color). In this case only a virtual cluster is generated (green color) in such way it is in the opposite side of real cluster, it is formed by points which are located in the boundary of the field

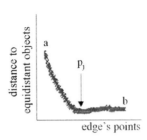

a. LVD   b. Distance of each edge's point to equidistant
                         objects and position of the new node

c. Local topo-geometric map

Figure 5: Generation of a new node which represents a door

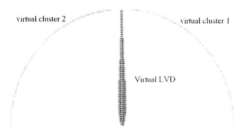

Figure 6: Generation of the Virtual LVD in the case that no obstacles are detected.

of view and the distance between the virtual and real clusters allows that the robot passes through them without colliding.

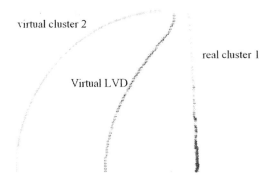

Figure 7: Generation of the Virtual LVD in the case that a obstacles is detected in a lateral side.

In case that only an obstacle is detected and it is in front of robot, such is shown in figure 8, two clusters are generated in both sides of real cluster. In this case, two edges of Virtual LVD are generated. The supervisor will select the edge used for navigation.

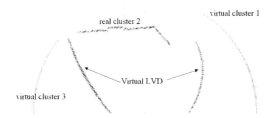

Figure 8: Generation of the Virtual LVD in the case that a obstacles is detected in front of robot.

4   Path Planning and Navigation

To navigate in complex environments, an autonomous mobile robot needs to reach a compromise between the need for reacting to unexpected events and the need for having efficient and optimized trajectories. Sensor based path planning and navigation can be used to achieve this goal. Local Voronoi Diagrams (LVD) can be used directly for safe navigation.

Classical path planning methods address the generation of collision-free trajectories in the robot workspace. The main shortcoming of these classical methods is that the roadmap is constructed off-line, and the environment information must be provided in advance. The uncertainties in workspace model and the unexpected obstacles in the environment are not considered. In a realistic approach, the robot's sensor information must to be taken into account. The perception of the environment and sensor-based path planning during robot motion, provides the robot the capability of moving in a safe way, avoiding unexpected obstacles.

There are some path planning algorithms inspired by the Voronoi Diagram concept such as *MAPRM* [26]; also, the EquiDistance Diagram (EDD) [27] is based on Voronoi roadmap idea; Nagatani, Choset and Thrun (see [28, 29]) build a Generalized Voronoi Graph (GVG), the robot explores an unknown environment and builds an incremental GVG; in [16], the Voronoi Diagram is used for path planning and navigation.

In sections 2.1 and 2.2, an efficient and fast algorithm to build up LVD using sensor data and to extract topo-geometrical map has been presented. The Voronoi Diagram, by definition, gives safe trajectories that maximizes the clearance between the robot and the obstacles. The Voronoi Diagram represents the way, or set of points which are equidistant to different objects in the environment. In this manner, the LVD is thought to be the safest way to move the robot. The Voronoi Diagram represents the topology of the free space where the robot can navigate through a set of edges and nodes, this topology is stored in the Local Topo-geometrical Map. The problem is reduced to one dimensional trajectory planning over the Topo-geometrical graph.

This map allows the robot to plan dynamically in partially known environments and to avoid collision with obstacles. This local map allows to plan a local path without employing the classical reactive behaviors.

The LVD and the topo-geometric map are built on-line, then the local path planning provides the robot the capability to react to changes in the environment and to anticipate critical situations. The robot needs not to move along Voronoi edges to compute a Local Voronoi Diagram, it constructs the LVD of a visible region around the robot and this diagram can be used for local path planning and navigation into that region. The geometric information improves the planning method, the distance from points of edges to equidistant objects, stored with LVD, allows to distinguish between a corridor and a hall, narrowing corridors, etc., and then, to plan suitable motion strategies.

5   Localization in Unknown Environments. The SLAM Problem

When a robot moves in an unknown environment, the path planning can only be developed if there is a modelled environment [30]. In order to build an environment map, a robot must know where it is relative to past locations. Due to errors in odometry, which increase in each robot displacement, it is necessary to localize the robot in relation to already known elements. Thus, an environment model is necessary to localize the mobile robot. The problem of building a suitable map of an unknown environment while the robot is travelling is referred to as Simultaneous Localization And Mapping (SLAM) problem [3].

The SLAM problem has been addressed using different techniques: probabilistic and non-probabilistic methods. Many localization techniques use probabilistic methods to estimate the robot position as a distribution of probabilities over the space of possible solutions.

Many authors uses the Extended Filter Kalman (EFK) to resolve the localization problem [1, 31, 32]. The Kalman Filter has the advantage that the representation of the probability distribution is a Gaussian distribution. However, three significant problems of EKF are 1) the EKF's inability to represent the multimodal PDF (Probabilistic Density Function). It means that the robot can not recover and detect its position from a lost one. However, Roumeliotis et al. [33] propose a extension to EKF it allows for multimodal probability distributions through multiple hypothesis tracking. 2) The computational complexity is high. The computational requirement is of order $\sim O(N^2)$, where N is the number of landmarks in the map. In large environments, the number of landmarks detected will make the computational requirement be beyond the power capacity of the computer resources. However, Guivant et al. [34] reduce the EKF computational requirements without introducing any penalties in the accuracy of the results. 3) EKF requires the robot to be localized all the time with a certain accuracy. Due to the iterative process the error in the newly observed landmarks, the error in vehicle location and the errors in other landmarks of the map are correlated. This correlation needs to be maintained throughout the mission to obtain good convergence of the algorithm in large environments. EKF has been used with grid maps [35] and geometric maps [36, 32].

Others probabilistic methods are called the particle filtering approaches. They are able to handle multimodal distributions. Many of them are based on Markov methods [37, 38]. Here, the space of possible robot positions is discretized, often in grid maps. These methods permit both precision and multimodality by using small sized grid cells. However, they require a significant processing power and memory in large areas. Recently, the Monte Carlo localization has been proposed [35, 39, 40]. This method is an efficient alternative by using a sampling-based method that can represent arbitrary distributions. The advantage of Monte Carlo localization is that the computational requirements can be scaled as needed by adjusting the number of samples used to represent the distribution. The particle filtering approaches have been used with grid maps [35], topological maps and hybrid maps [38].

The problem of probabilistic methods is that it is necessary to calculate a Probabilistic Density Function. Thus, some authors propose non-probabilistic localization algorithms [28, 16, 41]. Most of these methods have the problem that they fail in large and cyclic environments. Our localization method is a non-probabilistic method which is suitable for large and cyclic environments. In this chapter the geometrical characteristics (the nodes's position, points which form the edges, edges' orientation, and the points' distance to equidistant objects), and the topological characteristics extracted from the Voronoi Diagram are used to solve the localization problem. The geometric information is used to eliminate possible ambiguities and to improve the localization process.

Several approaches also use the characteristics of Voronoi Diagram to localize the robot. The coordinates of points from LVD (Local Voronoi Diagram) are used in [18] to estimate the robot position. A global Voronoi Diagram of the environment is given to the robot and the localization is carried out matching the LVD with the global one. The *meet points'* position, points which are equidistant to more than two objects, is used in [42] to localize the robot. Nodes' positions and edges' orientation are used in [16] according to resolve the localization problem. Victorino et al. [43] use the nodes of Voronoi Diagram to determine when the localization process takes place.

5.1   A Genetic Solution for the SLAM Problem

In figure 9 the proposed SLAM algorithm is described. The proposed SLAM algorithm consists of two main steps: a localization step (the local matching and the global matching approaches) and a mapping step (global map approach). In a new robot position, a new local topo-geometric map is generated LM_K and the odometric reading is stored ξ_K. Then, the local matching algorithm is executed using the new and the previous stored information (local topo-geometric maps and odometric readings) in order to estimate the current robot position $\widehat{X}_K$. The local matching algorithm only corrects the last robot position. If the robot revisits a place, the global matching algorithm is also executed and all the previous robot positions are re-estimated $(\widehat{X}_1, \widehat{X}_2, \cdots, \widehat{X}_K)$.

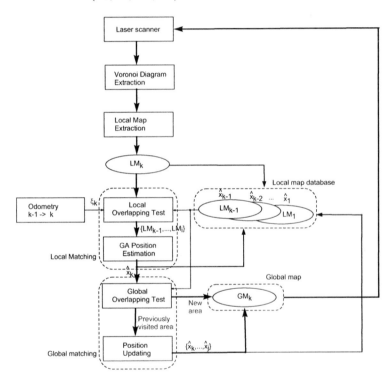

Figure 9: SLAM algorithm

Both local and global matching algorithms use a genetic-based localization approach to estimate the robot position. The robot position is estimated matching the new local map with the previous local maps. Considering small displacements, it is very probable that an important part of local topo-geometric map matches with some part of the previous local topo-geometric maps. In our experiments the robot is displaced about 2-4 meters. After the localization algorithm, the mapping procedure is executed. In the mapping algorithm, the global topo-geometric map is built (GM_K) fusing the new local map with the previous global map (GM_{K-1}). Nevertheless, if the global matching step is executed, the global map is built fusing all the local maps because all the robot positions are modified.

5.1.1 Genetic-based Localization Approach

The localization step uses a Genetic Algorithm to estimate the robot position. The Genetic Algorithm evaluates the robot's pose by means of a fitness function, which expresses the degree of topo-geometric matching between the newly obtained local topo-geometric map and the previously obtained local maps. This approach avoids to work with the probability density function explicitly, because the algorithm tests and it optimizes pose solutions instead of evaluating the posterior probability distribution given the observed data to estimate the best pose. Instead of using only geometrical data, the fitness function considers topological and geometrical information, and odometric readings to estimate the robot position.

The topo-geometric matching function between two local maps is based on the concept of distance between *equivalent nodes* and the distance between *equivalent edges*. We consider that two nodes or two edges are equivalent when the next matching hypothesis are carried out:

- Topo-geometrical hypothesis:

    - Two edges are considered equivalent when they have the similar orientation and their points' distances to the equidistant objects are similar.

    - Two nodes are equivalent when their points' distances to the equidistant objects are similar.

- Distance hypothesis: Two nodes or two edges are considered equivalents when they are closed in the space. This hypothesis is important to eliminate false solutions.

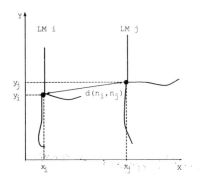

Figure 10: Distance between two equivalent nodes

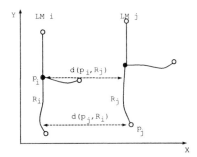

Figure 11: Distance between two equivalent edges

Taking into consideration the previous hypothesis, the distance between two equivalent nodes is calculated using the following equation (see figure 10):

$$d_{\text{eq. nodes}} = \sqrt{(x_i - x_j)^2 + (y_i - y_j)^2} \tag{12}$$

and the distance between two equivalent edges is calculated using the equation (see figure 11):

$$d_{\text{eq. edges}} = factor_\theta(R_i, R_j) \cdot \frac{\sqrt{d(p_i, R_j)^2 + d(p_j, R_i)^2}}{2} \tag{13}$$

where $factor_\theta(R_i, R_j)$ is a coefficient which evaluates the different orientations between two edges (see figure 12):

$$factor_\theta(R_i, R_j) = \frac{|\theta(R_i) - \theta(R_j)|}{\theta(R_i) + \theta(R_j)} \tag{14}$$

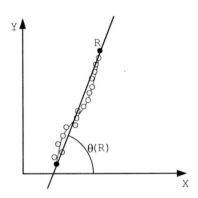

Figure 12: Edge's orientation

The selected fitness function is then defined as:

$$f = \frac{1}{\frac{\sum^n d_{\text{equivalent nodes}} + \sum^m d_{\text{equivalent edges}}}{n+m} + 1} \tag{15}$$

where $d_{\text{equivalent nodes}}$ is the distance between two equivalent nodes and $d_{\text{equivalent edges}}$ is the distance between two equivalent edges defined previously.

The GA will search the solution in the search space. The space of all feasible solutions is called search space. Each solution in the search space represents one feasible solution and is marked by its fitness for the problem. When a robot moves and due to slippage, wheel differences and other small errors, the position reached by the robot is not exactly the position indicated by the odometric values. After a number of such movements, the robot location with respect to the world frame becomes so uncertain that the robot is unlikely to succeed in actions based purely on its current information. This uncertainty area represents the search space where robot is probably located. The GA is able to search in all the space state but this is infeasible due to the high computational cost required to search the whole state space. For this reason the genetic algorithm search area is limited to the most probable area. This area is estimated using the kinematic model of the robot described in [44]. The covariance matrix obtained with this model, plus a safety coefficient determines the area where the genetic algorithm is going to search.

5.1.2 Localization Step

The localization step of the algorithm operates at two levels: when a local topo-geometric map is obtained at a given position, the map is compared with previous local maps. The genetic matching between them provides us with the best topo-geometrical pose estimate. This estimate will be sufficient if the environment is small, but considering large and cyclic environments, the small residual errors are accumulated leading us to an improper global model. To deal with this problem (data association problem) a secondary matching level is required, this is referred as global matching step. The objective of the global matching step is to evaluate if a local map corresponds to a previously visited area not evaluated in the local matching. Obviously not all previous local maps can be matched because of the high computational cost. To avoid that, only those previous local maps that overlap a search area, defined by the covariance matrix of a conventional probabilistic robot motion model, are considered. The local matching step is executed at each iteration, but the global matching step is only executed when the robot considers it has revisited a place.

Local matching   The local matching algorithm is formed by three main steps:

1. The first step is a local overlapping test which checks if the field of view for the last robot location overlaps with the field of view of any previous robot location. In spite of the re-localization reduces considerably the robot's pose error, the remaining residual error has to be considered in the overlapping test. To introduce this error, the field of view area is incremented in size taking into consideration the error introduced at motion. The adopted robot motion model is based on a Gaussian noise [44]. This area is also used to search the best robot's position estimate by the genetic algorithm, in the following step.

2. If the maps overlap, the genetic-based localization algorithm is executed. Due to the convenience of working with local information, the genetic-based localization algorithm estimates the current robot location referred to robot's estimated pose for previous overlapping map. If the current local map overlaps with several previous local maps, the genetic-based localization approach is executed for each overlapping. All those possible relative estimates are stored.

A first estimation about the current robot position, $X_K = (x_i, y_i, \theta_i)^T$, is given by odometric readings ξ_K:

$$X_K = T(\theta_{K-1}) \cdot \xi_K + X_{K-1} \tag{16}$$

where $T(\theta)$ is the transform matrix:

$$T(\theta) = \begin{pmatrix} \cos(\theta) & \sin(\theta) & 0 \\ -\sin(\theta) & \cos(\theta) & 0 \\ 0 & 0 & 1 \end{pmatrix} \tag{17}$$

If the robot positions are known in each time, X_j with $j = 1, \cdots, K-1$, the current robot position X_K could be calculated with respect to the previous ones:

$$X_L^{(j,K)} = T(\theta_j) \cdot (X_K - X_j), \text{ with } j = 1, \cdots, K-1 \tag{18}$$

$X_L^{(j,K)}$ is the robot position at time $t = i$ referred to position X_j. To indicate that this position is a local position due it is not referred to the global frame world, the subscript L is added. This value will belong to the initial population of GA.

A best solution is calculated from each GA, matching the current local map with each overlapped previous local map. The solutions obtained from GA ($\widehat{X}_L^{(j,i)}$ with $j = 1, \cdots, i-1$), are referring to the previous robot positions, and are called *local solutions*. All these local solutions are stored because they will use when the global matching step will be executed. The updated estimate of robot position is calculated from the local solution obtained from overlapped local map before. It is necessary for this local solution to be referred to the world frame:

$$\widehat{X}_i^j = T(\widehat{\theta}_j) \cdot \widehat{X}_L^{(j,i)} + \widehat{X}_j, \text{ with } j < i \tag{19}$$

where $\widehat{X}_j$ is the estimated robot position at time $t = j$ referring to the world frame, $\widehat{X}_L^{(j,i)}$ is the solution obtained from GA, and $\widehat{X}_i^j$ is the estimated robot position at time $t = i$ referring to the world frame, obtained by matching the local map g_i with the local map g_j.

3. The best solution, $\widehat{X}_i$, selected from the solutions set ($\widehat{X}_i^j$ with $j = 1, \cdots, i-1$), is which has the highest fitness value.

Global matching step   Although the local matching step corrects the robot location, there could be residual errors which cause that the local maps do not match when a cycle is closed. In order to avoid this situation, the local matching step is extended to all local maps which overlap with the current local map as is shown previously. Once a loop is detected, at local matching level, the global matching step is activated.

This step is divided in two main steps.

1. The first step is a loop detection test, which determines if the robot is revisiting a place or not. This test is done for all overlapping maps in order to detect latent loops in the environment.

   Loop detection hypothesis: It is supposed that the local map LM_i matches with local maps $MS_i = \{LM_n, \cdots, LM_k\}$. This set of local maps is obtained in the local matching step. Then, a cycle is closed if $MS_i \nsubseteq (MS_{i-1} + LM_{i-1})$.

   Let us suppose that the robot is at time K in the position shown in figure 13. At that position, the observed local map is LM_K. After the local overlapping test is done, it is observed that the set of overlapping maps contains $MS_K = \{LM_{K-1}, LM_1\}$. This is originated because the local map LM_K overlaps with LM_{K-1} (the previous map according to the robot movement) and with LM_1 (this map is overlapped due to the presence of a cycle in the environment). The set of overlapping maps at time $K - 1$ was $MS_{K-1} = \{LM_{K-2}\}$. Then, the loop detection hypothesis shows that map LM_1 at map set MS_K is not contained in the set $\{MS_{K-1}, LM_{K-1}\}$, which indicates the presence of a cycle.

2. If a cycle is closed, then all robot locations are corrected using the local estimated positions obtained in the local matching step. The best local n robot's pose solutions, relative to the previous overlapping local map obtained from GA at local matching step, are stored together with the local maps. In case of a cycle detection, a bigger number of best candidate local robot's poses are stored (m). This is required because the accumulated possible errors between this overlapping maps is higher. In our experimental tests good results have been obtained with n=2 and m=20.

   The optimization of map operates as follows: combining one local solution from the set of best possible robot's local poses stored with each local map and evaluating again the fitness function between the local maps which close the cycle. Thus, we obtain the matching degree between the initial and final local maps of the cycle. This process is iterated for all combinations of local solutions stored in the local maps. Finally, the combination of local robot's poses, which obtains the best fitness value matching the initial and final local maps at the detected cycle, is used to generate the final global map. From these local solutions, the global robot positions at time $t = i$ are calculated using equation 19. All the estimated robot positions referring to the global frame world are designated as $\widehat{X}_{i,m}$. Then, all the previous robot positions can be calculated again:

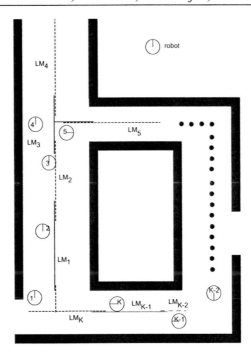

Figure 13: The robot explores the unknown environment building in each new location the local topo-geometric map

$$\widehat{X}_{i-1} = -T\left(\widehat{\theta}_i\right) \cdot T\left(-\widehat{\theta}_{L,n}^{(i-1,i)}\right) \cdot \widehat{X}_{L,n}^{(i-1,i)} + \widehat{X}_{i,m} \text{ with } \begin{array}{l} n = 1,2,3 \\ m = 1,\cdots,10 \end{array} \tag{20}$$

Equation 20 is obtained for $i = K, \cdots, 2$. If the next requirements are complied with then, 1) $\widehat{X}_{i-1}$ belongs to the uncertain space and 2) there is overlapping between the current local map and the revisited local map, then $\widehat{X}_{i-1}$ is the best obtained global solution according to their fitness value. A cycle is closed if the global matching step is executed until $i = 2$ because the first robot position is considered as $(0,0,0)$ and without error.

The global matching step is executed each time a local map corresponds to a revisited area. However, if next movements of the robot continues through a previously visited area the current local map verifies the cycle condition and executes the global matching optimization step to correct all local maps poses. To avoid an unnecessary computational cost, when the solutions found in two consecutive local map's optimization process do not change substantially, the global matching step considers that the global map is consistent and is not executed in the following visits to this environment cycle. If a new environment cycle is observed, the global matching step is activated until global map consistency is reached.

5.1.3 Mapping Step

The global topo-geometric map is generated by matching local topo-geometric maps. If the local matching step is only executed, the last robot's position is only corrected. Then, the new global topo-geometric map is generated by matching the current local map with the global map obtained in the previous time. If the global matching step is also executed, the new global topo-geometric map is generated matching all the previous local maps.

Initially, neither the environment nor the position is known by the robot. It is considered that the initial robot's position is $(x = 0m, y = 0m, \theta = 0^0)$. If the robot has no stored map of the environment, the first generated global map is equal to the generated local map.

In order to generate the global map, the points from the current local map are transferred to the global coordinate frame. Let (x^i_j, y^i_j) be a point of local topo-geometric map g_i. This point is referred to the global coordinate frame using the equation:

$$\begin{pmatrix} x_j^{(i,1)} \\ y_j^{(i,1)} \end{pmatrix} = \begin{pmatrix} \cos(\widehat{\theta}_i) & \sin(\widehat{\theta}_i) \\ -\sin(\widehat{\theta}_i) & \cos(\widehat{\theta}_i) \end{pmatrix} \cdot \begin{pmatrix} x_j^i \\ y_j^i \end{pmatrix} + \begin{pmatrix} \widehat{x}_i \\ \widehat{y}_i \end{pmatrix} \tag{21}$$

where $\widehat{X}_i = (\widehat{x}_i, \widehat{y}_i, \widehat{\theta}_i)$ is the estimated robot's position at time $t = i$ in the localization step, and $(x_j^{(i,1)}, y_j^{(i,1)})$ is a point of the local map transferred to the initial robot's position.

If only the local matching algorithm is executed, then the global map is built matching the current local map with the previous generated global map. If the global matching algorithm is also executed, then all the robot's positions have been reestimated. The global map is built matching all the local maps.

The merging process to generate the global nodes and edges is:

Node generation   The information of equivalent nodes are merged in a new node which has as coordenates the average position of them. Let $N_i = \{x_i, y_i\}$ of local map g_i and $N_j = \{x_j, y_j\}$ of local map g_j, be two equivalent nodes referred to the global coordinate frame, then the new generated global node has as coordinates:

$$N_k = \{x_k, y_k\} = \{\frac{x_i + x_j}{2}, \frac{y_i + y_j}{2}\} \tag{22}$$

The non-equivalent nodes are added to the global topo-geometric map.

Edge generation   The information of equivalent edges are merged in a new edge. Let $R_i = \{R_i^{non-eq.} \cup R_i^{eq.}\}$ of local map g_i and $R_j = \{R_j^{non-eq.} \cup R_j^{eq.}\}$ of local map g_j, be two edges with equivalent points referred to global coordinate frame, then the new generated global edge is formed by the points:

$$R_k = \{R_i^{non-eq.} \cup R_j^{non-eq.} \cup R_j^{eq.}\}, \text{ with } j > i \tag{23}$$

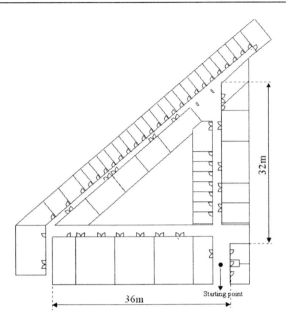

Figure 14: Environment map

The non-equivalent edges are added to the global topo-geometric map.

In order to prove the effectiveness of the proposed SLAM algorithm, an unknown and cyclic environment is explored by the robot (see figure 14). The dimensions of the environment are about 32×36 meters. The robot starts to move at "starting point". It turns left hand side and moves along the corridor. Then, it turns right hand side and travels to the next hall. In this hall, the robot moves along the corridor to arrive at the initial hall.

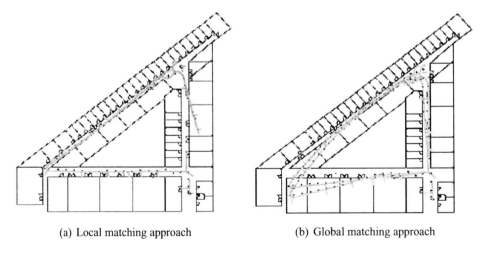

(a) Local matching approach   (b) Global matching approach

Figure 15: Solutions from the localization algorithm obtained in 59th robot displacement

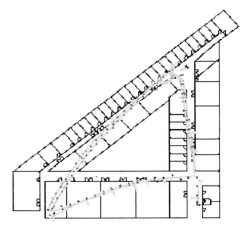

Figure 16: Best solution obtained from the localization algorithm in 60th robot displacement

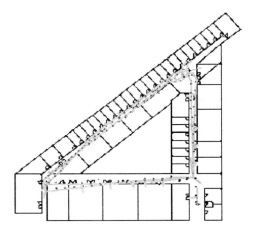

Figure 17: Best solution obtained from the localization algorithm in 61th robot displacement

In 59th robot position, the robot observes the beginning of corridor again. Thus, the global matching algorithm is executed after the local matching algorithm. The solutions obtained from local matching and global matching algorithms in 59th robot position are shown in figure 15. We can observe that if the local matching algorithm is only executed, the global map is not closed (see figure 15(a)). It is necessary to revise all previous positions to get closed the map (see figure 15(a)). When the global matching algorithm is executed different possible solutions can be obtained. The global map is generated considering the solution which has the highest fitness values and by fusing the information of all local maps.

Figures 16-18 represent the solutions obtained in consecutive robot displacements when the localization algorithm is executed. In these cases, the robot travels by a visited area. For this reason, the global matching algorithm is executed after the local matching algorithm.

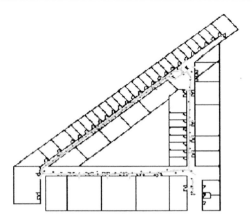

Figure 18: Best solution obtained from the localization algorithm in 62th robot displacement

Once the global map is generated, the problem is reduced to a localization problem.

6   Adding Symbolic Information to the Map

Apart from the topological and geometric information stored in the model, a symbolic infor-mation is also added to the map. The symbolic information allows to communicate different robots which use different programming languages or to give them orders in a language eas-ily understood by humans. Moreover, the symbolic information can be used in a first level to carry out a topological path-planning and navigation.

The symbols added to the map represent distinctive places typical of indoor environ-ments such as doors, corridors, narrowings, intersections, etc. The different places are recognized using both topological and geometric information stored in the map such as the presence of nodes, the number of edges adjacent to one node and the distance of points which form the edges to the equidistant obstacles.

The distinct places, which are learnt and recognized using Hidden Markov Models (HMM), are shown in figure 19. The advantage of using HMM is that they give a sto-chastic and symbolic solution about the recognition process. This information can be used to take decisions on localization, path-planning and navigation, and to be used for other robots and for humans.

Before recognizing characteristics, a learning phase is executed. In the learning phase a model for each characterized place is learnt. The proposed approach is used to recognize distinctive places typical of indoor environment and not to recognize any situation.

6.1   HMM

Hidden Markov Model (HMM) [45] [46] modeling is a stochastic technique for the study of the complete-incomplete data problems associated with time series. A process in the HMM

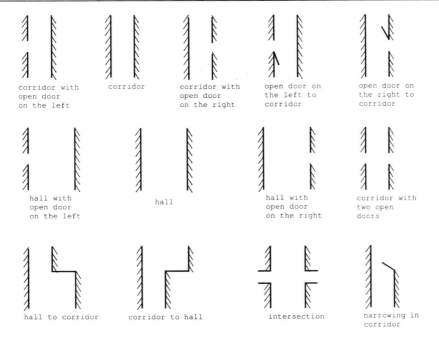

Figure 19: Distinctive places

class can be described as having a finite set of states with an output process which produces finite symbols. The objective is to estimate the chain state, given the observations.

HMMs have been used for modeling observed patterns, especially in speech recognition [47]. In this paper HMMs have been used for recognizing distinctive places in indoor environments.

6.1.1   Elements of an HMM

An HMM is characterized by:

- **N**: Number of states in the model.

- **M**: Number of observation symbols.

- **The state transition probability distribution** $A = \{a_{ij}\}$.

- **The observation symbol probability distribution** $B = \{b_j(k)\}$.

- **The initial distribution** π_i.

- **Observation sequence** $O = \{O_1 O_2 \cdots O_T\}$.

6.1.2 The Model

Each HMM is used to represent a characteristic place (see figure 19). The chosen model has 3 states and 5 transitions (see figure 20). A state is reachable from another state when a characteristic symbol is recognized. The transition between one state to another is only dependent on the previous state. This topology has the advantage of being simple and the obtained results are very successfully.

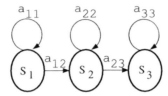

Figure 20: HMM model

6.1.3 Observation Symbols

The observation symbols are generated considering the information stored in the local topo-geometric map:

- Number of edges adjacent to one node.

- Slope of edges. The robot's orientation is selected as reference to obtain the slope of each edge. According to reduce the computing time, the slope is calculated considering only the extreme points of each edge, p_i and p_j:

$$\theta_{edge} = \frac{y_n - y_i}{x_n - x_i} \qquad (24)$$

where p_i is the closest point to the robot. In figure 21 the slope segmentation is represented.

- Distance from points of edges to equidistant objects. In table 1 the distance segmentation is represented.

An important advantage of the proposed method is that not only the topological information is used in the recognition approach (number of edges adjacent to one node) but also the geometric information (slope of edges and the distance from edges' points to equidistant objects). The geometric information improves the recognition algorithm because possible ambiguities are eliminated, for example, the distance from points of edges to equidistant objects allows to distinguish between a corridor and a hall.

Other important advantage is that only the information stored in the map is used to recognize the places. It is not necessary to operate directly with the measurements of laser telemeter and to extract primitives such as lines or arcs.

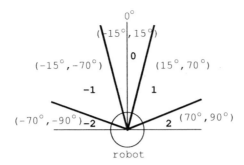

Figure 21: Slope segmentation

Table 1: Distance segmentation

Distance between edge and object (cm)	Distance symbol	
$150<D<190, \Delta D <0$	0	hall to corridor
$D\leq150$	1	corridor
$D\geq190$	2	hall
$150<D<190, \Delta D >0$	3	corridor to hall

The total of observation symbols are shown in table 2. The number of observation symbols is equal to 22. Most of places typical of indoor environments can be characterized with these symbols.

6.1.4   Observation Sequence

Given the observation symbols and the state model, it is necessary to generate the observation sequence. The observation sequence is obtained by segmenting the topological map. The two criteria to generate the sequence of observations are: presence of nodes and length of edges.

The observation sequence of graph of figure 22 is $O = \{11,0,17\}$, which represents a transition from corridor to hall. The observation sequence of graph of figure 23 is $O = \{11,1,19,11\}$, which represents a corridor with an open door on the right. In the case of corridor (see figure 24), there is only one observation symbol. The sequence of observations is $O = \{11,11,11\}$ because the minimum length of observation sequence has to be three. If there are several sequences in a graph, the graph is segmented in several simple observation sequences as in figure 25. For this example, the observation sequences are: $O_1 = \{11,1,3,11\}$ (corridor with open door on the left), $O_2 = \{11,11,11,11\}$ (corridor), $O_3 = \{11,1,19,11\}$ (corridor with open door on the right) and $O_4 = \{11,0,15\}$ (narrowing).

Table 2: Observation Symbols

Slope Symbols	Distance Symbols	Observation Symbols
—	—	0 (node with 2 edges)
—	—	1 (node with more 2 edges)
-2	0	2
-2	1	3
-2	2	4
-2	3	5
-1	0	6
-1	1	7
-1	2	8
-1	3	9
0	0	10
0	1	11
0	2	12
0	3	13
1	0	14
1	1	15
1	2	16
1	3	17
2	0	18
2	1	19
2	2	20
2	3	21

Figure 22: Observation sequence corresponding to "corridor to hall"

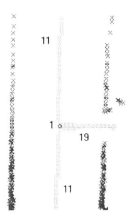

Figure 23: Observation sequence corresponding to corridor with open door on the right

Figure 24: Observation sequence corresponding to corridor

6.2   Learning Phase and Reestimation Procedure. Baum-Welch Algorithm

Two probabilities are defined in order to obtain the model parameters $\lambda = (A, B, \pi)$ from an observation sequence, ξ_t and γ_t [45]:

$$\xi_t(i,j) \quad = Prob(q_t = S_i, q_{t+1} = S_j | (O_1, \cdots, O_t), \lambda)$$

$$= \frac{\alpha_t(i) \cdot a_{ij} \cdot b_j(O_{t+1} \cdot \beta_{t+1}(j))}{Prob(O|\lambda)} \tag{25}$$

$$\gamma_t(i) \quad = \sum_{j=1}^{N} \xi_t(i,j) \tag{26}$$

where α_t is the forward variable and β_{t+1} is the backward variable [45].

The most difficult problem in HMM is adjusting the model parameters $\lambda = (A, B, \pi)$ to maximize the probability of observation sequence given the model. The iterative algorithm used to re-estimate the parameters is the Baum-Welch algorithm [48].

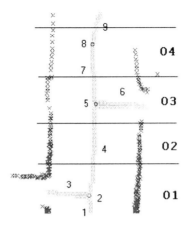

Figure 25: Segmentation corresponding to topological map

A set of reasonable estimation formulas to $\pi, A,$ and, B are [45]:

$$\overline{\pi_i} = \text{expected frequency in state } S_i, \text{at time t=1} \tag{27}$$

$$\overline{a_{ij}} = \frac{\sum_{t=1}^{T-1} \xi_t(i,j)}{\sum_{t=1}^{T-1} \gamma_t(i)} \tag{28}$$

$$= \frac{\text{expected } n^o \text{ of transitions from } S_i \text{ to } S_j}{\text{expected } n^o \text{ of transitions from } S_i}$$

$$\overline{b_j(k)} = \frac{\sum_{t-1;t \in O_t = v_k}^{T} \gamma_t(j)}{\sum_{t=1}^{T} \gamma_t(j)} \tag{29}$$

$$= \frac{\text{expected } n^o \text{ of times in } S_j \text{ and observing } v_k}{\text{expected } n^o \text{ of times in } S_j}$$

The Baum-Welch iterative algorithm is an expectation-maximization (EM) algorithm [48]. An EM algorithm is guaranteed to provide monotonically increasing convergence of $Pr(S|\lambda)$. The EM is generally a first order or linearly convergent algorithm [49]. Then, like the Baum-Welch algorithm, the learning phase is guaranteed to converge.

This procedure is executed for all the places which we want to learn. For each typical place a model is generated and its parameters has to be learnt $\lambda_j = (A_j, B_j, \pi_j)$. As is shown in figure 19, the number of models which has to be learnt is equal to 13.

6.3 Recognition Phase. Viterbi Algorithm

A formal technique to find the optimal state sequence $Q = \{q_1, q_2, \cdots, q_t\}$ associated with the given observation sequence $O = \{O_1, O_2, \cdots, O_t\}$ is the Viterbi algorithm [45]. It is necessary to define the variable:

$$\delta_t(i) = [\max_{q1,\cdots,qt-1} P[q_1,\cdots,q_t = i, O_1,\cdots,O_t|\lambda]] \tag{30}$$

By induction the $\delta_t(i)$ value is calculated:

$$\delta_{t+1}(j) = [\max_i \delta_t(i)a_{ij}]b_j(O_{t+1}) \quad 1 \le t \le T-1 \\ 1 \le i,j \le N \tag{31}$$

for t=1,

$$\delta_1(i) = \pi_i \cdot b_i(O_1) \tag{32}$$

Our algorithm works as is shown in figure 26.

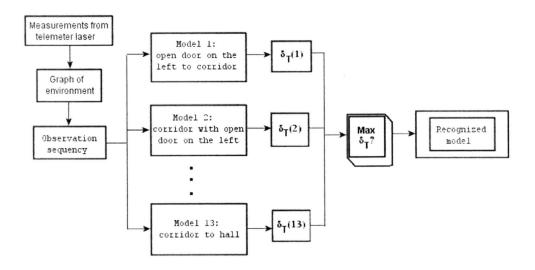

Figure 26: Recognition algorithm

When a new observation sequence is recognized, its model parameters are calculated and the recognized model parameters are estimated again using the equations 27, 28 and 29.

Figure 27 shows the map generated when the robot travels by a corridor. The dimensions explored during this experiment are about 49x4 meters. In each robot position a local map is generated and the symbolic is extracted using HMM and it is added to the map. The local map with the complete information is fused to the global map.

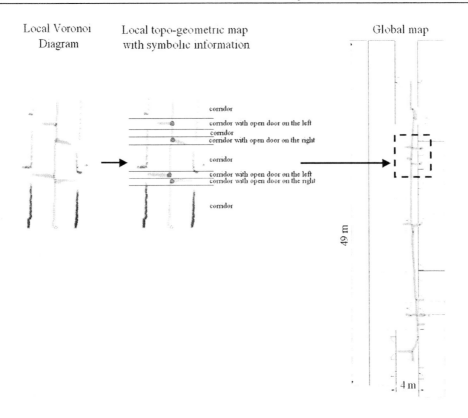

Figure 27: Generated global map for a typical indoor environment: this map is a graph which contains nodes and edges. These nodes and edges contains geometric information relative to their position and the distance to the equidistant objects. Apart from this geometric information, the nodes and edges also contains symbolic information. The local map is fused with the previous map to generate the new global map.

7   Conclusion

A new methodology to model the environment, based on Voronoi Diagram, is implemented. The proposed solution models the environment by means of a topo-geometric map, obtained from Local Voronoi Diagram, which can be directly used for localization, planning and navigation tasks in the mobile robotics. This avoids the conversion required by other modelling methods, which are effective in map modelling but requires to be concerted to other representations in order to navigate or plan robot's paths.

Moreover, this kind of representation requires a limited quantity of information because it compacts the representation of the environment by means of a network of curves. The amount of information required to model big areas grows much slower than grid map representations, because of the one dimensional nature of Voronoi Diagrams.

A symbolic information is also added to the topo-geometric model. The symbols represents distinctive places typical of indoor environments such as halls, corridors, doors, etc. The advantage of this information is to allow a topological level abstraction which it can reduce planning complexity. The technique used to learn and to recognize the different symbols is HMM (Hidden Markov Models). HMM allow to obtain a stochastic solution which is important to take decisions and, moreover, the computing time to recognize distinctive places from topo-geometric maps is very short (about 5-6 ms). Thus, this algorithm can be executed while the robot is moving.

Apart from the advantages of this model, however, it is not always possible to model the environment by means of Voronoi Diagram such in opened areas. This involves the no generation of a environment model which can provoke that the robot is not able to navigate or plan. The adopted solution is to generate a Virtual Voronoi Diagram to maintain the continuation of model.

The proposed methods for modelling, localization, navigation and path-planning have been successfully implemented and tested under real conditions in a RWI-B21 mobile robot using the measurements of a laser telemeter, although the ideas of these algorithms can be implemented using other sensors such as ultrasonics, computer vision, etc.

References

[1] J.A. Castellanos, J.M. Martínez, J. Neira, and J.D. Tardós, "Simultaneous map building and localization for mobile robots: A multisensor fusion approach", in *Proceedings of the 1998 IEEE International Conference on Robotics and Automation*, Leuven, Belgium, May 1998, pp. 1244–1249.

[2] L. Zhang and B. K. Ghosh, "Geometric feature based 2 1/2d map building and planning with laser, sonar and tactile sensors", in *Proceedings of the 2000 IEEE/RSJ International Conference on Intelligent Robots and Systems*, 2000.

[3] John J. Leonard and Hugh F. Durrant White, "Simultaneous map building and localization for an autonomous mobile robot", in *IEEE/RSJ International Workshop on Intelligent Robots and Systems IROS'91*, Osaka, Japan, Nov. 1991, pp. 1442–1447.

[4] Alan C. Schultz, William Adams, Brian Yamauchi, and Mike Jones, "Unifying exploration, localization, navigation and planning though a common representation", in *Proceedings of the 1999 IEEE Conference on Robotics and Automation*, 1999, pp. 2651–2658.

[5] Emmanuel Piat and Simon Lacroix, "Hierarchical path planning on probabilistically labeled polygons", in *Proceedings of the 1998 IEEE International Conference on Robotics and Automation*, Leuven, Belgium, May 1998, pp. 3147–3152.

[6] Sebastian Thrun, "Learning metric-topological maps for indoor mobile robot navigation", *Artificial Intelligence*, vol. 99, no. 1, pp. 21–71, 1998.

[7] David Kortenkamp and Terry Weymouth, "Topological mapping for mobile robots
     using a combination of sonar and vision sensing", in *Proceedings of the Twelfth
     National Conference on Artificial Intelligence.*, 1994, pp. 979–984.

[8] Clayton Kunz, Thomas Willeke, and Illah R. Nourbakshs, "Automatic mapping of
     dynamic office environments".

[9] Ioannis M. Rekleitis, Vida Dujmovic, and Gregory Dudek, "Efficient topological ex-
     ploration", in *Proceedings of the 1999 IEEE Conference on Robotics and Automation*,
     Detroit, Michigan, May 1999, pp. 676–681.

[10] Karen Zita Haigh and Manuela M. Veloso, "Learning situation-dependent costs: Im-
     proving planning from probabilistic robot execution", *Robotics and Autonomous Sys-
     tems*, vol. 29, pp. 145–174, 1999.

[11] A. Okabe, B. Boots, and K. Sugihara, *Spatial Tessellations. Concepts and Applica-
     tions of Voronoi Diagrams.*, John Wiley and Sons, 1992.

[12] N. Tomatis, I. Nourbakhsh, and R. Siegwart, "hybrid simultaneous localization and
     map building: a natural integration of topological and metric", *Robotics and Au-
     tonomous Systems*, vol. 44, pp. 3–14, 2003.

[13] M. Piaggio and R. Zaccaria, "Using roadmaps to classify regions of space for au-
     tonomous robot navigation", *Robotics and Autonomous Systems*, vol. 25, pp. 209–217,
     1998.

[14] S. Thrun and A. Bücken, "Integrating grid-based and topological maps for mobile
     robot navigation", in *Proceedings of the Thirteenth National Conference on Artificial
     Intelligence AAAI'96*, 1996.

[15] H. Choset and J. Burdick, "Sensor based planning, part ii: Incremental construction
     od the generalized voronoi graph", in *IEEE Conference on Robotics and Automation*,
     1995.

[16] D. Van Zwynsvoorde, T. Simeon, and R. Alami, "Building topological models for
     navigation in large scale environments", in *Proceedings of the 2001 IEEE Interna-
     tional Conference on Robotics and Automation*, 2001.

[17] T. Kanbara, A. Hayashi, S. Li, J. Miura, and Y. Shirai, "A method of planning move-
     ment and observation for a mobile robot considering uncertainties of movement, vi-
     sual sensing, and a map", in *Proceedings of the Ninth International Conference on
     Advanced Robotics'99 (ICAR)*, 1999, pp. 375–381.

[18] D. Blanco, B.L. Boada, and L. Moreno, "Localization by voronoi diagrams correla-
     tion", in *International Conference on Robotics and Automation (ICRA'01)*, 2001.

[19] B.L. Boada, D. Blanco, C. Castejón, and L.E. Moreno, "A genetic solution for the slam problem", in *11th International Conference on Advanced Robotics (ICAR'03)*, 2003.

[20] D. Blanco, B.L. Boada, C. Castejón, C. Balaguer, and L.E. Moreno, "Sensor-based path planning for a mobile manipulator guided by the humans", in *11th International Conference on Advanced Robotics (ICAR'03)*, 2003.

[21] B.L. Boada, D. Blanco, and L. Moreno, "Symbolic place recognition in voronoi-based maps by using hidden markov models", *Journal of Intelligent and Robotic Systems*, vol. 39, pp. 173–197, 2004.

[22] B.L. Boada, D. Blanco, and L. Moreno, "Localization and modelling approach using topo-geometric maps", in *IEEE/RSJ International Conference on Intelligent Robots and Systems (IROS'02)*, 2002.

[23] C. Castejón, B.L. Boada, D. Blanco, and L.E. Moreno, "Traversability analysis technics in outdoor environments: a comparative study", in *11th International Conference on Advanced Robotics (ICAR'03)*, 2003.

[24] R. Mahkovic and T. Slivnik, "Generalized local voronoi diagram of visible region", in *Proceedings of the 1998 IEEE International Conference on Robotics and Automation*, Leuven, Belgium, May 1998, pp. 349–355.

[25] D. Blanco, B.L. Boada, L. Moreno, and M.A. Salichs, "Local mapping from on-line laser voronoi extraction", in *IEEE/RSJ International Conference on Intelligent Robots and Systems (IROS'00)*, 2000.

[26] Steven A. Wilmarth, Nancy M. Amato, and Peter F. Stiller, "MAPRM: A probabilistic roadmap Planner with sampling on the medial axis of the free space", in *Proceedings of the 1999 IEEE International Conference on Robotics and Automation*, 1999, pp. 1024–1031.

[27] S.S. Keerthi, C.J. Ong, E. Huang, and E.G. Gilbert, "Equidistance diagram- a new roadmap method for path planning", in *Proceedings of the 1999 IEEE Conference on Robotics and Automation*, 1999, pp. 682–687.

[28] Keiji Nagatani, Howie Choset, and Sebastian Thrun, "Towards exact localization without explicit localization with the generalized voronoi graph", in *Proceedings of the 1998 IEEE International Conference on Robotics and Automation*, Leuven, Belgium, May 1998, pp. 342–348.

[29] Howie Choset, *Sensor Based Motion Planning: The Hierarchical Generalized Voronoi Graph*, PhD thesis, California Institute of Technology, Pasadena, California, March 1996.

[30] M.A. Salichs and L. Moreno, "Navigation of mobile robots: Open questions", *Robotica*, vol. 18, pp. 227–234, 2000.

[31] Kai O. Arras, Nicolas Tomatis, Bjrn T. Jensen, and R. Siegwart, "Multisensor on-the-fly localization: Precision and reliability for applications", *Robotics and Autonomous Systems*, vol. 34, pp. 131–143, 2001.

[32] Andrew J. Davison and David W. Murray, "Simultaneous localization and map-building using active vision", *IEEE Transactions on Pattern Analysis and Machine Intelligence*, vol. 24, no. 7, pp. 865–880, July 2002.

[33] S.I. Roumeliotis and G.A. Bekey, "Bayesian estimation and kalman filtering: A unified framework for mobile robot localization", in *IEEE International Conference on Robotics and Automation*, Washington, D.C., 2000, pp. 2985–2992.

[34] José E. Guivant and Eduardo Mario Nebot, "Optimization of the simultaneous localization and map-building algorithm for real-time implementation", *IEEE Transactions on Robotics and Automation*, vol. 17, no. 3, pp. 242–257, 2001.

[35] George Kantor and Sanjiv Singh, "Preliminary results in range-only localization and mapping", in *IEEE International Conference on Robotics and Automation*, Washington, D.C., 2002, pp. 1818–1823.

[36] José E. Guivant, Favio R. Masson, and Eduardo M. Nebot, "Simultaneous localization and map building using natural features and absolute information", *Robotics and Autonomous Systems*, vol. 40, pp. 79–90, 2002.

[37] Dieter Fox, Wolfram Burgard, and Sebastian Thrun, "Active markov localization for mobile robots", *Robotics and Autonomous Systems*, vol. 25, pp. 195–207, 1998.

[38] N. Tomatis, I. Nourbakhsh, K. Arras, and R. Siegwart, "A hybrid approach for robust and precise mobile robot navigation with compact environment modeling", in *Proceedings of the 2001 IEEE International Conference on Robotics and Automation*, 2001, pp. 1111–1116.

[39] D. Fox, W. Burgard, F. Dellaert, and S. Thrun, "Monte carlo localization: Efficient position estimation for mobile robots", in *Proceedings of AAAI-99*, 1999.

[40] Favio Masson, Jose Guivant, and Eduardo Nebot, "Robust navigation and mapping architecture for large environments", *Journal of Robotic Systems*, vol. 20, no. 10, pp. 621–634, 2003.

[41] Stephen Se, David Lowe, and Jim Little, "Vision-based mobile robot localization and mapping using scale-invariant features", in *Proceedings of the 2001 IEEE International Conference on Robotics and Automation*, 2001, pp. 2051–2058.

[42] H. Choset, "Topological simultaneous localization and mapping (SLAM): Toward exact localization without explicit localization", *IEEE Transactions on Robotics and Automation*, vol. 17, no. 2, pp. 125–137, 2001.

[43] A. C. Victorino, P. Rives, and J.-J. Borrely, "Localization and map building using a sensor-based control strategy", in *Proceedings of the 2000 IEEE/RSJ International Conference on Intelligent Robots and Systems*, 2000.

[44] R. Smith, M. Self, and P. Cheesman, "Estimation uncertain spatial relationships in robotics", *Uncertainty in Artificial Intelligence*, vol. 2, pp. 435–461, 1988.

[45] Lawrence R. Rabiner, "A tutorial on hidden markov models and selected applications in speech recognition", *Proceedings of the IEEE*, vol. 77, no. 2, pp. 257–286, February 1989.

[46] Robert J. Elliot, Lakhdar Aggoun, and John B. Moore, *Hidden Markov Models. Estimation and Control*, Springer-Verlag, 1995.

[47] X. D. Huang, Y. Ariki, and M. A. Jack, *Hidden Markov Models for Speech Recognition*, Edinburg University Press, 1990.

[48] A. P. Dempster, N. M. Laird, and D. B. Rubin, "Maximum likelihood from incomplete data via the EM algorithm", *J. Roy. Stat. Soc.*, vol. 39, no. 1, pp. 1–38, 1977.

[49] L. Xu and M.I. Jordan, "On convergence properties of the em algorithm for gaussian mixtures", *Neural Computation*, vol. 8, no. 1, pp. 129–151, Jan 1996.

In: Control and Learning in Robotic Systems
Editor: John X. Liu, pp. 243-258

ISBN 1-59454-356-9
©2005 Nova Science Publishers, Inc.

Chapter 9

MULTI-STAGE LEANING CONTROLLER
FOR UNDERACTUATED ROBOT MANIPULATORS

Lanka Udawatta [†*]
[†] Department of Electrical Engineering, University of Moratuwa
Moratuwa 10400, Sri Lanka
Keigo Watanabe [†††]
[††] Department of Advanced Systems Control, Graduate School of Science
and Engineering Saga University, 1-Honjomachi, Saga 840-8502, Japan

Abstract

This article presents a novel multi-stage learning controller for designing a neural and fuzzy reasoning based switching controller, in order to control underactuated manipulators. In the first phase of learning, parameters of both antecedent and consequent parts of a fuzzy indexer are optimized by using evolutionary computation. Design parameters of the fuzzy indexer are encoded into chromosomes, i.e., the shapes of the Gaussian membership functions and corresponding switching laws of the consequent part are evolved to minimize the angular position errors. Such parameters are trained for different initial configurations of the manipulator and the common rule base is extracted. In the next stage of training, neural network based compensator is designed for reducing the remaining nonlinear dynamics. Then, these trained fuzzy rules with compensator parameters can be brought into the operation of underactuated manipulators. Simulation results show that the new methodology is effective in designing controllers for underactuated robot manipulators.

Keywords: Multi-stage learning, Underactuated robots, Fuzzy reasoning, Neural networks, Nonlinear dynamics.

---

[*]E-mail address: lanka@ieee.org
[†]E-mail address: watanabe@me.saga-u.ac.jp

1 Introduction

Today we have various kinds of robots ranging from simple pet robots to complex humanoid robots though the word "robot" has its root in the Slavic languages and means worker. To create such a system requires combination of computer science, systems control, electronics and mechanical engineering. Underactuated robots are a special class of available robots up to date and have received much attention in the field of robotics during the past two decades [1]–[3]. There are numerous examples of underactuated robot systems in recent literature such as brachiating robots, passive walker, gymnastic robots like Acrobots, Pendabots, snake robots and so on. In the brachiating robot controller in [4], a simplified two-link robot with one actuator at the elbow connecting two arms each having grippers was employed in order to get the motion of a gibbon. Humans commonly utilize underactuation to perform tasks more precisely or more easily. For example, the martial art star Bruce Lee can project the tip of his favorite weapon, a numbchuck, on a precise point in space to produce a tremendous impact force. The numbchuck is composed of 2 hardwood segments connected by 2 passive joints, and the overall combination of the numbchuck with Bruce Lee's arm is an underactuated system which needs multi-stage learning for better operations [5]. In the case of human tasks, we need training, individual talents and experience for better performance. On the other hand, if the underactuated robots will be employed to perform tasks, design engineer needs to construct precise control algorithms and artificial learning techniques for further advancements.

Analysis of dynamics for underactuated manipulators is more complex than that for regular robot manipulators. Such manipulators consist of both active and passive joints and the passive joints result in a lack of controllability of the total system. Moreover, there is an inertial coupling between the motions of active and passive joints, so that mapping such as Jacobian matrix, depends not only on the kinematic properties, but also on the inertia properties of the links [6]–[10]. The constraints of a dynamical system which cannot be represented in the form of $F(q,t) = 0$ are defined as nonholonomic, where q and t denote the coordinates and time respectively. Moreover, the first-order nonholonomic systems can be categorized into two:

1. Nonholonomic systems under kinematic constraints.

2. Nonholonomic systems under dynamical constraints.

Examples of nonholonomic control systems have been studied in the context of robot manipulators, mobile robots, wheeled vehicles, and space robotics [1]. The passive joints, which can rotate freely, have the advantages such as reduction of weight, fault tolerance capability, law energy consumption, and reduction of cost of manipulators [11]. There has been a considerable number of studies on underactuated manipulators during the past decade; theoretically important [12]–[16] and practically interesting [15]. However, such a class of nonlinear systems, leading to a challenging research task, often exhibits complex nonlinear dynamics, nonholonomic behavior, and lack of linearizability.

The main purpose of this chapter is to introduce a new multi-stage learning scheme for controlling underactuated robot manipulators. We introduce two-stage learning controller using neural network and fuzzy reasoning techniques. In this study, one fundamental approach of our study is to use elemental controllers in order to fulfill the ultimate control objective by fuzzy logic based switching. Using these elemental controllers, we investigate to arrange the optimum control action for a given time frame with the available set of elemental controllers with fuzzy logic based switching. In fact only fuzzy based switching controller can control the system [17, 18], we further improve controller performance, reducing remaining nonlinear dynamics, by introducing neural network compensator with enhanced training. The rest of this chapter is organized as follows: Design of elemental controllers with an example and the concept of the proposed multi-stage learning controller are described in Section 2 and Section 3, respectively. In Section 4, system training topology using evolutionary computation is presented. Then, the computer simulation results are given in Section 5. Finally, summary with concluding remarks is given in Section 6.

2   Formulation of Elemental Controllers

Consider the dynamical model of a robot manipulator given below:

$$M(q)\ddot{q} + V(q,\dot{q}) = F. \tag{1}$$

Here, the generalized coordinate vector q and input force/torque vector F are given by $q \in \Re^n$ and $F \in \Re^n$ respectively. $M(q)$ is the $n \times n$ symmetric, positive-definite inertia matrix and $V(q,\dot{q})$ represents Coriolis, centrifugal, friction, and gravitational components. Suppose that an underactuated robot system has m_u number of actuators to control and n degrees-of-freedom $(1 \leq m_u < n)$ links. The equation (1) can be rearranged as follows:

$$\ddot{q} = \frac{1}{D}\hat{M}(q)\{-V(q,\dot{q}) + F\} \tag{2}$$

where $D = \det(M)$ and $\hat{M}$ is the cofactor of M. It is assumed that M is invertible, and expanding (2) gives

$$\begin{bmatrix} \ddot{q}_1 \\ \vdots \\ \ddot{q}_{m_u} \\ \vdots \\ \ddot{q}_n \end{bmatrix} = \frac{\hat{M}}{D}\left\{ -\begin{bmatrix} v_1 \\ \vdots \\ v_{m_u} \\ \vdots \\ v_n \end{bmatrix} + \begin{bmatrix} F_1 \\ \vdots \\ F_{m_u} \\ 0 \\ \vdots \\ 0 \end{bmatrix} \right\} \tag{3}$$

where $\hat{M}$ is given by

$$\hat{M} = \begin{bmatrix} \hat{M}_{11} & \hat{M}_{12} & \cdots & \hat{M}_{1(n-1)} & \hat{M}_{1n} \\ \hat{M}_{12} & \hat{M}_{22} & \cdots & \hat{M}_{2(n-1)} & \hat{M}_{2n} \\ \vdots & \vdots & \ddots & \vdots & \vdots \\ \hat{M}_{1(n-1)} & \hat{M}_{2(n-1)} & \cdots & \hat{M}_{(n-1)(n-1)} & \hat{M}_{(n-1)n} \\ \hat{M}_{1n} & \hat{M}_{2n} & \cdots & \hat{M}_{(n-1)n} & \hat{M}_{nn} \end{bmatrix}.$$

The equation (3) can be expressed as:

$$\begin{bmatrix} \ddot{q}_1 \\ \ddot{q}_2 \\ \vdots \\ \ddot{q}_n \end{bmatrix} = \frac{1}{D} \begin{bmatrix} \hat{M}_{11} & \hat{M}_{12} & \cdots & \hat{M}_{1n} \\ \hat{M}_{12} & \hat{M}_{22} & \cdots & \hat{M}_{2n} \\ \vdots & \vdots & \ddots & \vdots \\ \hat{M}_{1n} & \hat{M}_{2n} & \cdots & \hat{M}_{nn} \end{bmatrix} \left\{ - \begin{bmatrix} v_1 \\ v_2 \\ \vdots \\ v_n \end{bmatrix} + \begin{bmatrix} F_1 \\ F_2 \\ \vdots \\ F_{m_u} \\ 0 \\ \vdots \\ 0 \end{bmatrix} \right\}$$

$$= \frac{1}{D} \begin{bmatrix} \hat{M}_{11}(F_1 - v_1) + \cdots + \hat{M}_{1n}(-v_n) \\ \vdots \\ \hat{M}_{1m_u}(F_1 - v_1) + \cdots + \hat{M}_{m_u n}(-v_n) \\ \vdots \\ \hat{M}_{1n}(F_1 - v_1) + \cdots + \hat{M}_{nn}(-v_n) \end{bmatrix}. \tag{4}$$

In this derivation, we represented all the active joints torque/forces as $[F_1, F_2, ..., F_{m_u}]^T \triangleq \mathcal{F} \in \Re^{m_u}$, and it was assumed that m_u-actuators are directly allocated to the roots of m_u-links starting from first joint. Generally, whether m_u-actuators are arranged at which links differ from each problem. Now, we have $_nC_{m_u}$ number of combinations of m_u-dimensional controllers. Moreover, the number of different combinations of n-variables and m_u-actuators at a time, without repetitions, is

$$\binom{n}{m_u} = \frac{n(n-1)\cdots(n-m_u+1)}{m_u!}. \tag{5}$$

One of the above combinations of available controllers should be selected at a given time in order to actuate the robot system (1). For simplicity, we present how to derive an elemental controller for selected m_u number of link coordinates out of n-links as shown below:

$$\ddot{z} = \frac{1}{D} \begin{bmatrix} \hat{S}_{11}(F_1 - v_1) + \hat{S}_{12}(F_2 - v_2) + \ldots + \hat{S}_{1m_u}(F_{m_u} - v_{m_u}) + \ldots + \hat{S}_{1n}(-v_n) \\ \vdots \\ \hat{S}_{m_u 1}(F_1 - v_1) + \hat{S}_{m_u 1}(F_2 - v_2) + \ldots + \hat{S}_{m_u m_u}(F_{m_u} - v_{m_u}) + \ldots + \hat{S}_{m_u n}(-v_n) \end{bmatrix}$$

$$= -\frac{1}{D} \begin{bmatrix} \hat{S}_{11} & \cdots & \hat{S}_{1m_u} & \cdots & \hat{S}_{1n} \\ \vdots & \vdots & \vdots & \vdots & \vdots \\ \hat{S}_{m_u 1} & \cdots & \hat{S}_{m_u m_u} & \cdots & \hat{M}_{m_u n} \end{bmatrix} V(q, \dot{q}) + \frac{1}{D} \begin{bmatrix} \hat{S}_{11} & \cdots & \hat{S}_{1m_u} \\ \vdots & \ddots & \vdots \\ \hat{S}_{m_u 1} & \cdots & \hat{S}_{m_u m_u} \end{bmatrix} \mathcal{F}$$

where $z \stackrel{\triangle}{=} [z_1, z_2, ..., z_{m_u}]^T$ denotes one possible combination of m_u-link coordinates out of $\{q_1, q_2, ..., q_n\}$ and S_{ij} ($i = 1, ..., m_u; j = 1, ..., n$ or m_u) denotes their associated cofactors consisting of inertia element. Define $\mathcal{N}$ and $\mathcal{M}$ matrices as

$$\mathcal{N} = \frac{1}{D} \begin{bmatrix} \hat{S}_{11} & \cdots & \hat{S}_{1m_u} & \cdots & \hat{S}_{1n} \\ \vdots & \vdots & \vdots & \vdots & \vdots \\ \hat{S}_{m_u 1} & \cdots & \hat{S}_{m_u m_u} & \cdots & \hat{S}_{m_u n} \end{bmatrix} \tag{6}$$

$$\mathcal{M} = \frac{1}{D} \begin{bmatrix} \hat{S}_{11} & \cdots & \hat{S}_{1m_u} \\ \vdots & \ddots & \vdots \\ \hat{S}_{m_u 1} & \cdots & \hat{S}_{m_u m_u} \end{bmatrix}. \tag{7}$$

Since the inertia matrix M is positive definite, $\mathcal{M}$ is also $m_u \times m_u$ full rank matrix for any control law. Now, we have a set of control inputs for any control law as below:

$$\mathcal{F} = \mathcal{M}^{-1} \left\{ \ddot{z}^* + \mathcal{N} \, v(q, \dot{q}) \right\} \tag{8}$$

where $\ddot{z}^*$ is the modified acceleration, which can be constructed by using a simple PD servo such as:

$$\ddot{z}^* = \ddot{z}_d + K_v (\dot{z}_d - \dot{z}) + K_p (z_d - z) \tag{9}$$

where z_d, $\dot{z}_d$ and $\ddot{z}_d$ are the desired position, velocity and acceleration vectors of selected link coordinates to be controlled, and $K_v > 0$ and $K_p > 0$ are the derivative and position gain matrices. We can synthesize all the $_nC_{m_u}$ elemental controllers using the above procedure.

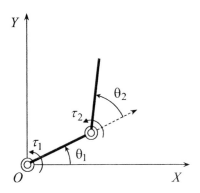

Figure 1: Two-DOF robot manipulator

To demonstrate how to construct the elemental controllers, 2-DOF RR (i.e., revolutionary-revolutionary joints) underactuated manipulator given in **Fig. 1** is taken into consideration. When defining $q^T = [q_1 \; q_2] = [\theta_1 \; \theta_2]$, the values $M(q)$, $V(q, \dot{q})$ and F of the equation (1) result in

$$M(q) = \begin{bmatrix} l_1^2 m_2 + 2l_1 l_2 m_2 C_2 + l_1^2(m_1 + m_2) & l_2^2 m_2 + l_1 l_2 m_2 C_2 \\ l_2^2 m_2 + l_1 l_2 m_2 C_2 & l_2^2 m_2 \end{bmatrix}$$

$$V(q, \dot{q}) = \begin{bmatrix} -l_1 l_2 m_2 S_2 \dot{\theta}_2 - 2l_1 l_2 m_2 S_2 \dot{\theta}_1 \dot{\theta}_2 + f_1 \dot{\theta}_1 \\ l_1 l_2 m_2 S_2 \dot{\theta}_1^2 + f_2 \dot{\theta}_2 \end{bmatrix}, \quad F = [\tau_1 \quad 0]^T$$

where $C_2 = \cos\theta_2$ and $S_2 = \sin\theta_2$. Note here that all the symbols have their usual meanings. Equation (2) gives the desired second-order derivatives ($\ddot{q}$):

$$\ddot{q}_1 = -\frac{M_{22}}{D}v_1 + \frac{M_{12}}{D}v_2 + \frac{M_{22}}{D}\tau_1 \tag{10}$$

$$\ddot{q}_2 = \frac{M_{12}}{D}v_1 - \frac{M_{11}}{D}v_2 - \frac{M_{12}}{D}\tau_1. \tag{11}$$

Here, D is given by

$$D = M_{11}M_{22} - M_{12}^2 \tag{12}$$

where M_{ij} denotes the (i,j)-th element of $M(q)$.

In this example, $n = 2$ and $m_u = 1$. Then, we have controller combinations $_nC_{m_u} = {_2C_1}$, i.e., depending on which link is controlled, two computed torque controllers can be designed by one actuator. In Section 3, we will introduce training of the neural network compensator f_c. Note here that when the elemental controllers are operating independently the f_c value is set to zero. Therefore, in general, we can represent these two control laws with the proposed compensator $f_c = [\tau_{c1} \ \tau_{c2}]^T$ as follows:

- Control law 1 for link 1

$$\tau_1 = \frac{D}{M_{22}}\left(\ddot{q}_1^* + \frac{M_{22}}{D}v_1 - \frac{M_{12}}{D}v_2\right) + \tau_{c1}$$

$$\ddot{q}_1^* = \ddot{q}_{d1} + K_{v1}(\dot{q}_{d1} - \dot{q}_1) + K_{p1}(q_{d1} - q_1) \tag{13}$$

- Control law 2 for link 2

$$\tau_1 = -\frac{D}{M_{12}}\left(\ddot{q}_2^* - \frac{M_{12}}{D}v_1 + \frac{M_{11}}{D}v_2\right) + \tau_{c2}$$

$$\ddot{q}_2^* = \ddot{q}_{d2} + K_{v2}(\dot{q}_{d2} - \dot{q}_2) + K_{p2}(q_{d2} - q_2). \tag{14}$$

Here, the desired values of the vector $[\ddot{q}_1 \ \dot{q}_1 \ q_1]$ are $[\ddot{q}_{d1} \ \dot{q}_{d1} \ q_{d1}]$. Similarly, the desired values of the vector $[\ddot{q}_2 \ \dot{q}_2 \ q_2]$ will be $[\ddot{q}_{d2} \ \dot{q}_{d2} \ q_{d2}]$. Servo controller PD gains for $\{q_1, \dot{q}_1\}$ and $\{q_2, \dot{q}_2\}$ are given by $\{K_{p1}, K_{v1}\}$ and $\{K_{p2}, K_{v2}\}$ respectively.

Applying the control laws 1 and 2 to the systems (10) and (11), it follows that

- Control law 1

$$\ddot{q}_1 = \ddot{q}_1^* + \frac{M_{22}}{D}\tau_{c1} \tag{15}$$

$$\ddot{q}_2 = -\frac{1}{M_{22}}v_2 - \frac{M_{12}}{M_{22}}\ddot{q}_1^* - \frac{M_{12}}{D}\tau_{c1} \tag{16}$$

- Control law 2

$$\ddot{q}_1 = -\frac{1}{M_{12}}v_2 - \frac{M_{22}}{M_{12}}\ddot{q}_2^* + \frac{M_{22}}{D}\tau_{c2} \tag{17}$$

$$\ddot{q}_2 = \ddot{q}_2^* - \frac{M_{12}}{D}\tau_{c2}. \tag{18}$$

The control objective here is to move the robot, towards the desired reference, using the available control laws.

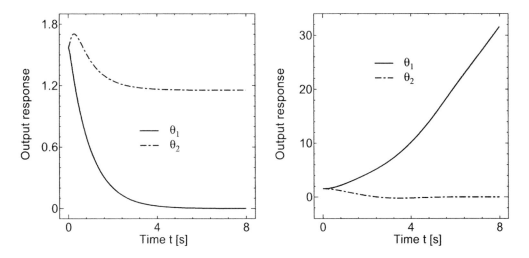

Figure 2: Time responses of control law 1 (left) and control law 2 (right)

Two control laws derived in the above will be applied to the 2-DOF RR underactuated robot independently and the desired value is set to $x_d = [0 \; 0 \; 0 \; 0]^T$, starting from $x(0) = [\pi/2 \; \pi/2 \; 0 \; 0]^T$. Note here that we do not employ the compensator at this stage; rather we use it in the training phase. If we analyze both cases separately, it can be seen from **Fig. 2** that the variable θ_2 can not be controlled under the control law 1. On the other hand, θ_1 cannot be controlled under the control law 2. Here, the manipulator parameters have been used as follows:

$$m_1 = 1.0 \; [\text{kg}], \quad m_2 = 1.5 \; [\text{kg}]$$
$$l_1 = 0.3 \; [\text{m}], \quad l_2 = 0.2 \; [\text{m}]$$
$$f_1 = 0.3 \; [\text{Nm·s/rad}], \quad f_2 = 0.3 \; [\text{Nm·s/rad}]$$
$$K_{v1} = 1.0, \quad K_{v2} = 1.0$$
$$K_{p1} = 1.0, \quad K_{p2} = 1.0.$$

Controllers in these two cases are considered to be the partly stable controllers (PSCs) such that they can be used for controlling each link whenever it is necessary to fulfill a control objective in the sense of controlling the total robot system.

3   Multi-stage Leaning Controller

The proposed architecture in this section is called as multi-stage learning controller that provides suitable solution space for controlling underactuated robots. The idea behind this concept is to train a system at several stages and capture the remaining nonlinear dynamics further while increasing the number of training processes. In this example, even though the fuzzy based switching controller can control the system up to some extend, we further improve controller performance, reducing the effect of remaining nonlinear dynamics, by introducing neural network compensator. This system has two learning stages and the resulting system can be implemented as shown in **Fig. 3**.

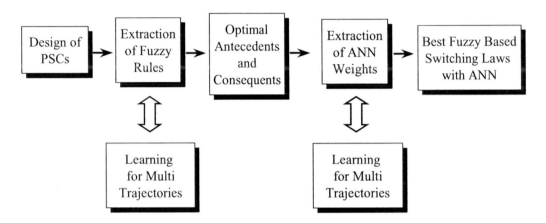

Figure 3: Multi-stage leaning controller

In the first phase of training, fuzzy reasoning based learning controller is proposed. Assume that all the elemental controllers are designed in advance; an example for deriving the elemental controllers, is given in Section 2. Let us consider a simplified fuzzy reasoning that generates an index number for the selection of a suitable control law. That is, the proposed fuzzy indexer has a set of input-output data from an underactuated system with n position error variables, $\{e_i \,|\, i = 1, 2, \cdots, n\}$, and a scalar index $s \in s = \{1, 2, \cdots, N\}$, where $N = {}_nC_{m_u}$ denotes the number of control laws. The input space is created with a fuzzy rule base such that the position error is taken and converted into a grade of membership $\{\mu_{ij}(e_i) \overset{\triangle}{=} \mu_{A_{ij}}(e_i) \,|\, i = 1, 2, \ldots, n\}$ employing a fuzzy set A_{ij} at the jth rule. Here, antecedent linguistic values, i.e., labels for fuzzy sets in each position error are assigned such as P: positive, Z: zero, N: negative, etc., totally $l = 1, 2, \ldots, M$. Then, one of the partly stable elemental controllers (i.e., control laws) will be assigned in the consequent part as below:

$$R_j : \text{ IF } e_1 \text{ is } A_{1j} \text{ AND } e_2 \text{ is } A_{2j} \text{ AND } \ldots e_n \text{ is } A_{nj}$$
$$\text{THEN   Control law is } s_j, \quad j = 1, 2, \cdots, M^n \tag{19}$$

where $s_j \in s$.

Following the conventional weighted average method, we can readily obtain the reasoning result of s, which is denoted by $\hat{s}$ such as

$$\hat{s} = \sum_{j=1}^{M^n} w_j s_j \tag{20}$$

$$w_j = \frac{h_j}{\displaystyle\sum_{k=1}^{M^n} h_k}, \quad h_j \stackrel{\triangle}{=} \prod_{i=1}^{n} \mu_{ij}(e_i). \tag{21}$$

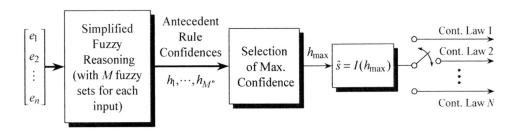

Figure 4: Fuzzy indexer for switching partly stable controllers

Note however that in this case the reasoning result generates a real number that is inconvenient for selecting an integer number in s. From this fact, we here use the following reasoning result:

$$\hat{s} = I(h_{\max}), \quad h_{\max} \stackrel{\triangle}{=} \max\{h_1, \cdots, h_{M^n}\} \tag{22}$$

where $I(h_{\max})$ means that the reasoning result is taken as the direct consequent index value which is assigned to the rule number, whose antecedent confidence has a maximum among them. **Figure 4** shows the block diagram of the proposed fuzzy indexer for switching partly stable controllers (PSCs). Training of desired parameters will be explained in Section 4.

In the second phase of training, we further reduce the effect of remaining nonlinear dynamics by introducing a neural network compensator. By introducing a neural network compensator for generating compensative torque f_c to (8) such that

$$\mathcal{F} = \mathcal{M}^{-1}\left\{\ddot{\boldsymbol{z}}^* + \mathcal{K}\, \boldsymbol{V}(\boldsymbol{q},\dot{\boldsymbol{q}})\right\} + \boldsymbol{f}_c, \tag{23}$$

the f_c value can be generated. In our past research works [5, 17], we have proved with good performance that these controllers can be employed to control the underactuated system by switching [20]. As in Section 2, we show the generation of the $\boldsymbol{f}_c = [\tau_{c1}\ \tau_{c2}]^T$ for a case of two-elemental controllers. Here, we can directly use any kind of basic neural network compensator as explained in [19].

The idea behind this system is to capture the remaining nonlinear dynamics after the fuzzy based training process, reducing the effect of remaining nonlinear dynamics. For example, **Figure 5** shows the generation of the $\boldsymbol{f}_c = [\tau_{c1}\ \tau_{c2}]^T$ for a case of two-elemental

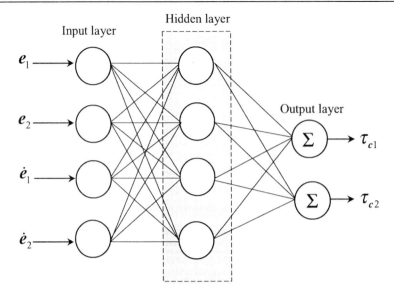

Figure 5: Neural compensator architecture

controllers, taking position and velocity errors. In Fig. 5, the layer where the input information is presented is known as the input layer. The layer where the processed information is
retrieved is called the output layer where it gives f_c. All layers between the input and output
layers are known hidden layers. For all nodes in the network, except the input layer nodes,
the total input of each node is the sum of weighted outputs of the nodes in the previous
layer.

4   Learning Process

In the training phase, for a given initial configuration, we try to extract a common fuzzy
rule base, which is optimized off-line by using genetic algorithm (GA). The membership
functions for ith input are calculated by

$$\mu_{ij} = \exp\{\ln(0.5)(e_i - w_{cij})^2 / w_{dij}^2\}$$
$$i = 1, \ldots, n, \quad j = 1, \ldots, M \tag{24}$$

where w_{cij} and w_{dij} represent the center and deviation of the jth Gaussian function of the
ith input variable, respectively. Parameters of such Gaussian membership functions and the
controller indices are encoded into chromosomes in order to obtain the optimum rule base
of (19). **Figure 6** shows the genes of a chromosome representing the Gaussian parameters
of (24) and controller indices for two PSCs. Here, 1 represents the first elemental controller
and 2 is for the second elemental controller. Fitness function evaluating criteria is based on
minimizing the total error, i.e., shapes of the Gaussian functions and the corresponding best
assigned elemental controllers are evolved.

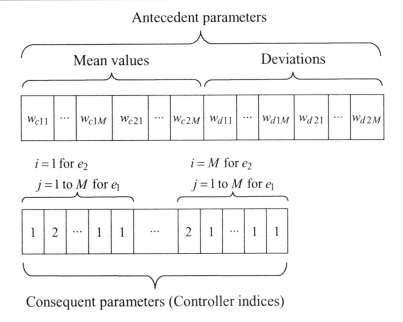

Figure 6: Coding of genes with fuzzy parameters in antecedent and consequent parts

In what follows, we introduce this rule extraction criterion for multi-trajectories. Let the trajectory $\Gamma_i (i = 1, 2, \ldots, \pi) : \Re^{2n} \mapsto \Re^{2n}$ be an optimal trajectory to (1) with initial configuration $\Phi_i (i = 1, 2, \ldots, \pi)$ starting from Π neighborhood towards the desired target x_d. The angular position errors are used as the inputs to the fuzzy membership functions in the antecedent part and the one index of the PSCs is assigned in the consequent part of the fuzzy reasoning. Therefore, it is necessary to find a reliable method to extract the optimum fuzzy set from a set of input-output training data. At this stage, fuzzy parameters are trained for different initial configurations $\Phi_i (i = 1, 2, \ldots, \pi) \in \Pi$ of the manipulator for a given desired configuration and common rule base is extracted such that it minimizes the performance index J:

$$J = \sum_{j=1}^{\pi} \sum_{k=1}^{T_N} \left\| x_d(k) - x^{\Phi_j}(k) \right\|_{W(k)}^2 \tag{25}$$

where x^{Φ_j} is the state vector starting from the initial configuration Φ_j where k is the discrete-time instant, T_N is the final discrete-time instant, $x \overset{\triangle}{=} [q_1 \ q_2 \ \cdots \ q_n \ \dot{q}_1 \ \dot{q}_2 \ \cdots \ \dot{q}_n]^T$ denotes the state variable vector, and x_d is the desired reference vector. Note that the weighting matrix $W(k) = \text{diag}\{w_1(k), w_2(k), \ldots, w_{2n}(k)\}, k = 1, 2, \ldots, T_N$ are selected so that they relax the condition at initial stage while keeping higher weights at the latter part of the time frame. Then, this extracted fuzzy rule base with compensator parameters is brought into the online operation of an arbitrary selected trajectory $\Gamma \in \Re^{2n}$ starting from $\Phi \subset \Pi$.

5   Results

As explained in Section 4, first phase of training for a given set of initial configurations extracts the optimum fuzzy rule base. The example, 2-DOF underactuated planar that we focused in the Section II is brought into to illustrate the learning methodology. We used five different initial configurations $\Phi_i (i = 1, 2, \ldots, 5)$ for optimizing (25) with a common desired value $x_d = [0 \ 0 \ 0 \ 0]^T$ as follows:

$$\Phi_1 : \quad x_0^{\Phi_1} = [\pi/3 \ -\pi/6 \ 0 \ 0]^T$$
$$\Phi_2 : \quad x_0^{\Phi_2} = [\pi/4 \ \pi/6 \ 0 \ 0]^T$$
$$\Phi_3 : \quad x_0^{\Phi_3} = [\pi/4 \ -\pi/6 \ 0 \ 0]^T$$
$$\Phi_4 : \quad x_0^{\Phi_4} = [\pi/2 \ -\pi/6 \ 0 \ 0]^T$$
$$\Phi_5 : \quad x_0^{\Phi_5} = [\pi/4 \ 0 \ 0 \ 0]^T.$$

Here, $x_0^{\Phi_i}$ represents the initial condition of the trajectory Γ_i and $t_{T_N} = 32$ [s]. After the GA optimization process, for an example, the trained Gaussian parameters were obtained for the angular error e_1 as shown in **Fig. 7** (NB=negative big, N=negative, NS=negative small, Z=zero, PS= positive small, P= positive, and PB=positive big).

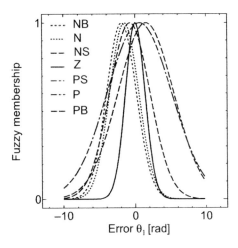

Figure 7: Trained membership functions for angular error e_1

In the second stage, we encoded the weights of the neural compensator with one hidden layer and the activation function was selected as the conventional sigmoid. After the training process, the switching laws for the corresponding linguistic variables with neural compensator parameters are obtained. According to this training, **Fig. 8** shows the time responses of angular positions θ_1 and θ_2 for five different initial configurations $\Phi_i (i = 1, 2, \ldots, 5)$, respectively.

The objective of this test was to check whether the proposed controller can comprise and operate accurately or not, even when the initial configurations are different from the

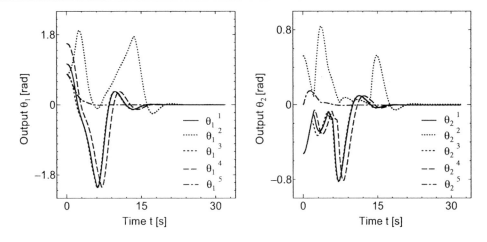

Figure 8: Time responses of trained θ_1 and θ_2

training data. In controlling the system, a testing was conducted for an arbitrarily selected point, starting from $x_0 = [2\pi/5 \;\; -\pi/5 \;\; 0 \;\; 0]^T$, towards the desired value $x_d = [0 \;\; 0 \;\; 0 \;\; 0]^T$. According to **Fig. 9**, the state variables of the manipulator have converged to the desired values. This ensures that the controlled variables are met the desired values within an acceptable level of time frame insuring the robustness.

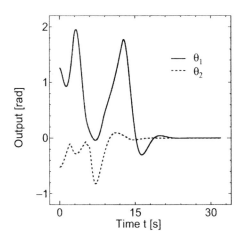

Figure 9: Time responses of θ_1 and θ_2 for untrained points

6   Conclusions

In this chapter, an attempt has been made to construct a multi-stage learning controller using fuzzy logic based switching with a neural network compensator for the operations of under-actuated manipulators. To employ elemental controllers, the rule base of the fuzzy indexer was optimized by using evolutionary computation. That is, parameters in both antecedent and consequent parts of the fuzzy rule base were acquired in the first training phase. In the next training, compensator parameters were trained for different initial configurations of the manipulator again and the common network parameters were extracted. Then, this trained rule base was able to be brought into the online operation of the underactuated manipulator. To illustrate the design procedure, we applied the concept to control an under-actuated manipulator and results showed the enhanced performance of the proposed control methodology. In addition, the following factors can be considered for future research:

- From a theoretical point of view, a better approach to the problem would be the use of "elemental controllers," with an analytically proved switching function, which guarantees the stability.

- Inclusion of velocity error as another input to the antecedent part of the fuzzy indexer for complicated systems.

- More fuzzy rules can be introduced to the system for better performance or for complicated systems.

- Fuzzy clustering method might be useful in absorbing the best rule base for the online operations.

- Compensate system uncertainties such as friction in the training process, including them as constraints.

- Instead of the off-line trained compensator back propagation for online will be useful in implementation the online fuzzy switching.

- Although acceptable results were obtained with few trajectories in the training phase more robust rule base can be successfully achieved by including more trajectories.

References

[1] I. Kolmanovsky and N.H. McClamroch, "Development in Nonholonomic Control Problems," *IEEE Control Systems Magazine*, vol. 15, no. 6, pp. 20–36, 1995.

[2] M.W. Spong, "Underactuated Mechanical Systems," in *Control Problems in Robotics and Automation*, B. Sicilano and K.P. Valavanis (Eds.), LNCIS, vol. 230, Springer Verlag, pp. 135–150, 1998.

[3] M. Reyhanoglu, A.V.D. Schaft, N.H. McClamroch, and I. Kolmanovsky, "Dynamics and Control of a Class of Underactuated Mechanical Systems," *IEEE Trans. on Automatic Control*, vol. 44, no. 9, pp. 1663-1671, 1999.

[4] J. Nakanishi, T. Fukuda, and D. E. Koditschek, "A Brachiating Robot Controller," *IEEE Trans. on Robotics and Automation,* vol. 16, pp. 109–123, 2000.

[5] L. Udawatta, K. Watanabe, K. Kiguchi, and K. Izumi, "Developments in Underactu-ated Manipulator Control Techniques and Latest Control Using AI," in *Proc. of 7th International Symposium on Artificial Life and Robotics, Oita, Japan*, vol. 2, 2002, pp. 425–428.

[6] H. Arai, K. Tanie, and N. Shiroma, "Nonholonomic Control of a Three DOF Planar Underactuated Manipulator," *IEEE Trans. on Robotics and Automation,* vol. 14, no. 5 pp. 681–695, 1998.

[7] M. Bergerman, Y. Xu, and Y. Liu, "Control of Cooperative Underactuated Manip-ulators: A Robust Comparison Study," in *Proc. of 1st Workshop on Robotics and Mechatronics, Shatin, Hong Kong*, vol. 1, 1998, pp. 279–286.

[8] F. Bullo, N.E. Leonard, and A.D. Lewis, "Controllability and Motion Algorithms for Underactuated Lagrangian Systems on Lie Group," *IEEE Trans. on Automatic Con-trol,* vol. 45, no. 8 pp. 1437–1453, 2000.

[9] A. Jain and G. Rodriguez, "An Analysis of the Kinematics and Dynamics of Underac-tuated Manipulators," *IEEE Trans. on Robotics and Automation,* vol. 9, pp. 411–422, 1993.

[10] K. Kobayashi, J. Imura, and T. Yoshikawa, "Nonholonomic Control of 3-D.O.F. Ma-nipulator with a Free Joint," *Trans. on SICE*, vol. 33, no. 8, pp. 799–804, 1997.

[11] M.D. Berkemeier and R.S. Fearing, "Tracking Fast Inverted Trajectories of the Un-deractuated Acrobot," *IEEE Trans. on Robotics and Automation,* vol. 15, no. 4 pp. 740–750, 1999.

[12] A.D. Luca, S. Iannitti, and G. Oriolo, 'Stabilization of a PR Planar Underactuated Robot," in *Proc. of IEEE Int. Conf. on Robotics and Automation (ICRA'2001)*, 2001, pp. 2090–2095.

[13] B.C.O. Maciel, M. Bergerman, and M.H. Terra, "Optimal Control of Underactuated Manipulators via Actuation Redundancy," in *Proc. of IEEE Int. Conf. on Robotics and Automation (ICRA'2001)*, 2001, pp. 2114–2119.

[14] T. Mita and T.K. Nam, "Control of Underactuated Manipulators using Variable Pe-riod Deadbeat Control," in *Proc. of IEEE Int. Conf. on Robotics and Automation (ICRA'2001)*, 2001, pp. 2735–2740.

[15] G. Muscato, "Fuzzy Control of an Underactuated Robot with a Fuzzy Microcon-troller," *Journal of Microprocessors and Microsystems,* vol. 23, pp. 385–391, 1999.

[16] Y. Nakamura, T. Suzuki, M. Koinuma, "Nonlinear Behavior and Control of a Non-holonomic Free-Joint Manipulator," *IEEE Trans. on Robotics and Automation*, vol. 13, no. 6, pp. 853–862, 1997

[17] L. Udawatta, K. Watanabe, K. Izumi, and K. Kiguchi, "Energy Optimization of Switching Torque Computed Method For Underactuated Manipulators," in *Proc. of Intelligent Autonomous Vehicle, IAV 2001, Sapporo, Japan*, 2001, pp. 394–399.

[18] L. Udawatta, K. Watanabe, K. Izumi, and K. Kiguchi, "Control of 3-DOF Underactuated Robot Using Switching Computed Torque Method," in *Knowledge-Based Intelligent Information Engineering Systems Allied Technologies, Volume 69*, edited by N. Baba, L.C. Jain and R.J. Howlett, IOS Press, 2001, pp. 1325–1329.

[19] A. Ishiguro, T. Furuhashi and S. Okuma, "A neural network compensator for uncertainties of robotics manipulators," *IEEE Trans. Industrial Electronics*, vol. 39, no. 6, pp. 565–569, 1992.

[20] K. Watanabe, L. Udawatta, K. Izumi, and K. Izumi, "Control of underactuated manipulators using fuzzy logic based switching controller," *Journal of Intelligent and Robotic Systems*, vol. 38, no. 2, pp. 155–173, 2003.

In: Control and Learning in Robotic Systems
Editor: John X. Liu, pp. 259-279

ISBN 1-59454-356-9
© 2005 Nova Science Publishers, Inc.

Chapter 10

ADAPTIVE ABILITIES IN HYBRID GENETIC ALGORITHM

YoungSu Yun[*]

School of Automotive, Industrial & Mechanical Engineering
Daegu University, Gyeongsan, Gyeongbuk, 712-714, Korea

Mitsuo Gen[**]

Graduate School of Information, Production & Systems
Waseda University, Kitakyushu 808-0135, Japan

Abstract

The aim of this paper is to compare adaptive abilities in hybrid genetic algorithm. For the hybrid genetic algorithm, a rough search technique and iterative hill climbing technique are applied to genetic algorithm. The rough search technique is used to initialize the population of the genetic algorithm, and the iterative hill climbing technique is to find a better solution within the convergence area of the genetic algorithm loop.

For constructing the adaptive abilities, we use a fuzzy logic controller and compare it with two conventional heuristics. The proposed fuzzy logic controller and two conventional heuristics can adaptively regulate the rates of the crossover and mutation operators in the genetic algorithm during genetic search process. They are tested and analyzed under the hybrid genetic algorithm proposed in this paper. Finally, a best algorithm is recommended.

Key words: Adaptive abilities, hybrid genetic algorithm, rough search technique, iterative hill climbing technique, and fuzzy logic controller.

1 Introduction

Genetic algorithms (GAs) have been known to offer significant advantages over conventional methods by using simultaneously several search principles and heuristics. Nevertheless, GAs

---

[*] E-mail address: joy629@hitel.net; Phone: +82(53)850-4431; Fax: +82(53)850-6549
[**] E-mail address: gen@waseda.jp

can suffer from excessively slow convergence and premature convergence before providing an accurate solution because of their fundamental requirements; namely, not using a priori knowledge and not exploiting local search information. GAs may also have a weakness in taking too much time to adjust fine-tuning structure of the GA parameters, i.e., crossover rate, mutation rate, and others. This kind of "blindness" may prevent them from being really of practical interest for a lot of applications.

To improve the weaknesses of not using the priori-knowledge and not exploiting the local search information in GAs, various hybrid methodologies of GAs using conventional heuristics have been developed (Gen and Cheng, 1997, 2000; Li and Jiang, 2000). In order to reduce the time for adjusting fine-tuning structure of the GA parameters, genetic operators with adaptive abilities have been also developed (Yun and Gen, 2003; Yun, Gen and Seo, 2003; Mak, Wong and Wang, 2000).

In the hybrid methodologies using conventional heuristics, several researchers have employed a method offering the priori-knowledge before initializing the population of GAs in order to improve the fundamental requirement of GAs on not using a priori-knowledge: Davis (1991) suggested a theoretical algorithm for hybridization in order to offer the priori-knowledge to GAs, Li and Jiang (2000) proposed a rough search technique using a simulated annealing for exploiting the knowledge.

To exploit the local search information and find a better solution within GAs loop, various hybrid methods using conventional local search techniques have been developed: Li and Jiang (2000) proposed a new stochastic approach called the SA-GA-CA, which is based on the proper integration of a simulated annealing (SA), a GA, and a chemotaxis algorithm (CA) for solving complex optimization problems. Yen et al. (1998) described two approaches for incorporating a simplex method into a GA loop as an additional operator. These approaches for hybridization usually use the complementary properties of GAs and conventional heuristics.

Adaptive methodologies to adjust the fine-tuning structure of the parameters used in GAs have two modes: one employs artificial intelligent techniques such as fuzzy logic controllers (FLCs), and the other employs heuristics. The genetic parameters controlled by these methods are adaptively regulated during genetic search process. Therefore, much time for the fine-tuning of the parameters can be saved, and the GA search ability can be improved in finding a global optimum.

For the first mode, Gen and Cheng (2000), in their book, surveyed various adaptive methods using several FLCs. Herrera and Lozano (2001) suggested and summarized the adaptive genetic operators based on co-evolution with fuzzy behaviors. Subbu, Sanderson and Bonissone (1998) proposed a fuzzy logic controlled genetic algorithm, and the controlled algorithm can regulate the changes of population size, crossover rate, mutation rate, and the number of generations. Lee and Takagi (1993) focused on population size, and Song et al. (1997) used two FLCs; one for the crossover rate and the other for the mutation rate. These parameters are considered as the input variables of GAs and are also taken as the output variables of the FLC. For successfully applying a FLC to a GA, the key is to produce well-formed fuzzy sets and rules. Recently, Cheong and Lai (2000) suggested an optimization mechanism for the sets and rules. The GAs controlled by these fuzzy logics have faster search speed and much better search quality than the GAs without them.

For the second mode, Mak, Wong, and Wang (2000) adaptively controlled the crossover and mutation rates according to the performance of GA operators in a manufacturing cell formulation problem. This heuristic scheme is based on the fact that it encourages well-performing operators to produce more efficient offspring, while reducing the chance for poorly performing operators to destroy the potential individuals, during genetic search process. Srinivas and Patnaik (1994), Wu, Cao and Wen (1998) also controlled the rates according to the fitness values of the population at each generation. These two works recommended the use of adaptive crossover and mutation rates to maintain diversity in the population of each generation and to sustain the convergence property of the GA.

Although most of the works mentioned in the first and second modes have proved their effectiveness in each work, any comparison or analysis among them under a same condition has not been provided in previous works. Therefore, the aim of our study is to compare the performances among them under the proposed hybrid genetic algorithm with a same condition.

By applying the priori-knowledge, local search information, and the rates of adaptive operators, described above, recently hybridized GAs are more effective and more robustness than conventional GAs or conventional heuristics. Based on this fact, a hybridized GA with the priori-knowledge and local search information is proposed in this paper, and then several adaptive schemes are included in this hybrid GA, respectively. These algorithms with the adaptive schemes use one FLC and two conventional heuristics for adaptively regulating crossover and mutation rates during genetic search process. Finally, we propose the best algorithm among them.

In Section 2, the hybrid concepts and logics underlying the proposed algorithms are proposed. Adaptive genetic operators using one FLC and two conventional heuristics are provided in Section 3. In Section 4, four algorithms (one hybrid GA and three adaptive hybrid GAs using the above concepts) are proposed. To compare the performance of the proposed algorithms, three test problems are presented and analyzed in Section 5, followed by the conclusion in Section 6.

2   Adaptive Genetic Operators (AGOs)

In this section, we suggest adaptive crossover and mutation operators using one FLC and two heuristics. The rates of the two operators are adaptively regulated according to various fitness values suggested by each method.

2.1   AGO Using FLC

To adaptively regulate genetic operators using FLC, we use the concept of Song $et\ al.$ (1997). Its main scheme is to use two FLCs: the crossover FLC and mutation FLC are implemented independently to adaptively regulate the rates of crossover and mutation operators during the course of GA.

The heuristic updating strategy for the crossover and mutation rates is to consider the changes of the average fitness of each GA population in continuous two generations. For example, in a minimization problem, we can set the change of the average fitness at generation t-1, $\Delta \overline{eval}(V;t-1)$ and the change of the average fitness at generation t, $\Delta \overline{eval}(V;t)$ as follows:

$$\overline{\Delta eval}(V;t-1) = (\overline{eval}1(V;t-1) - \overline{eval}2(V;t-1)) \times \lambda$$

$$= (\frac{\sum_{k=1}^{pop\_size} eval(v_k;t-1)}{pop\_size} - \frac{\sum_{k=pop\_size+1}^{pop\_size+off\_size} eval(v_k;t-1)}{off\_size}) \times \lambda$$

$$\overline{\Delta eval}(V;t) = (\overline{eval}1(V;t) - \overline{eval}2(V;t)) \times \lambda$$

$$= (\frac{\sum_{k=1}^{pop\_size} eval(v_k;t)}{pop\_size} - \frac{\sum_{k=pop\_size+1}^{pop\_size+off\_size} eval(v_k;t)}{off\_size}) \times \lambda$$

where $V = \{v_1, v_2, ..., v_k\}^T$, pop_size is the population size satisfying constraints, off_size is the offspring size satisfying constraints, and λ is a scaling factor for regulating the average fitness. These values can be considered to regulate the crossover rate p_C and the mutation rate p_M as follows:

Procedure: regulation of p_C and p_M using average fitness
begin
    if $\varepsilon \le \overline{\Delta eval}(V;t-1) \le \gamma$ **and** $\varepsilon \le \overline{\Delta eval}(V;t) \le \gamma$ **then**
        increase p_C and p_M for next generation ;
    if $-\gamma \le \overline{\Delta eval}(V;t-1) \le -\varepsilon$ **and** $-\gamma \le \overline{\Delta eval}(V;t) \le -\varepsilon$ **then**
        decrease p_C and p_M for next generation ;
    if $-\varepsilon < \overline{\Delta eval}(V;t-1) < \varepsilon$ **and** $-\varepsilon < \overline{\Delta eval}(V;t) < \varepsilon$ **then**
        rapidly increase p_C and p_M for next generation ;
    end
end

where ε is a given real number in the proximity of zero, γ and $-\gamma$ are respectively a given maximum and minimum values of a fuzzy membership function. The implementation strategy for the crossover FLC is as follows:

- ***Input and output of crossover FLC***
    The inputs of the crossover FLC are the $\overline{\Delta eval}(V;t-1)$ and $\overline{\Delta eval}(V;t)$ and its output is a change in the crossover rate $\Delta c(t)$.

- ***Membership functions of*** $\overline{\Delta eval}(V;t-1), \overline{\Delta eval}(V;t),$ and $\Delta c(t)$
    The membership functions of the fuzzy input and output linguistic variables are illustrated in Figures 1 and 2, respectively. The $\overline{\Delta eval}(V;t-1)$ and $\overline{\Delta eval}(V;t)$ are respectively normalized into the range [-1.0, 1.0]. The $\Delta c(t)$ is normalized into the range [-0.1, 0.1] according to their corresponding maximum values.

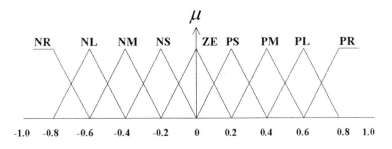

Figure 1. Membership functions for $\Delta \overline{eval}(V;t-1)$ and $\Delta \overline{eval}(V;t)$

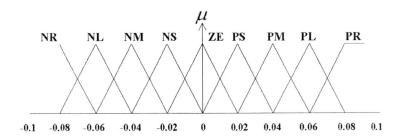

Figure 2. Membership function of $\Delta c(t)$

Where: NR – Negative larger, NL – Negative large, NM – Negative medium, NS – Negative small, ZE – Zero, PS – Positive small, PM - Positive medium, PL – Positive large, PR – Positive larger.

- **Fuzzy decision table**

Based on a number of experiments and the domain expert opinions, the fuzzy decision table was drawn as shown in Table 1.

- **Defuzzification table for control actions**

For simplicity, the defuzzification table for determining the action of the crossover FLC was setup. It is formulated as shown in Table 2.

The inputs of the mutation FLC are the same as those of the crossover FLC and the output of which is the change in the mutation rate, $\Delta m(t)$. The combination strategy used between the FLC and GA is shown in Figure 3.

Table 1. Fuzzy decision table for crossover

$\Delta c(t)$		$\Delta \overline{eval}(V;t-1)$								
		NR	NL	NM	NS	ZE	PS	PM	PL	PR
	NR	NR	NL	NL	NM	NM	NS	NS	ZE	ZE
	NL	NL	NL	NM	NM	NS	NS	ZE	ZE	PS
	NM	NL	NM	NM	NS	NS	ZE	ZE	PS	PS
	NS	NM	NM	NS	NS	ZE	ZE	PS	PS	PM
$\Delta \overline{eval}(V;t)$	ZE	NM	NS	NS	ZE	PM	PS	PS	PM	PM
	PS	NS	NS	ZE	ZE	PS	PS	PM	PM	PL
	PM	NS	ZE	ZE	PS	PS	PM	PM	PL	PL
	PL	ZE	ZE	PS	PS	PM	PM	PL	PL	PR
	PR	ZE	PS	PS	PM	PM	PL	PL	PR	PR

Table 2. Defuzzification table for control action of crossover

z		i								
		-4	-3	-2	-1	0	1	2	3	4
	-4	-4	-3	-3	-2	-2	-1	-1	0	0
	-3	-3	-3	-2	-2	-1	-1	0	0	1
	-2	-3	-2	-2	-1	-1	0	0	1	1
	-1	-2	-2	-1	-1	0	0	1	1	2
j	0	-2	-1	-1	0	2	1	1	2	2
	1	-1	-1	0	0	1	1	2	2	3
	2	-1	0	0	1	1	2	2	3	3
	3	0	0	1	1	2	2	3	3	4
	4	0	1	1	2	2	3	3	4	4

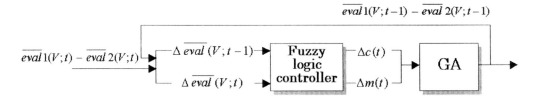

Figure 3. Coordinated strategy of the FLC and GA

The detailed procedure for its applying is as follows:

Step 1. The input variables of the FLC for regulating the GA operators are the change in the average fitness of the continuous two generations (t-1 and t) as follows:

$$\Delta \overline{eval}(V;t-1), \; \Delta \overline{eval}(V;t)$$

Step 2. After normalizing $\overline{\Delta eval}(V;t-1)$ and $\overline{\Delta eval}(V;t)$, assign these values to the indexes i and j corresponding to the control actions in the defuzzification table (see Table 2).

Step 3. Calculate the changes of the crossover rate $\Delta c(t)$ and the mutation rate $\Delta m(t)$ as follows:

$$\Delta c(t) = Z(i,j) \times \alpha, \; \Delta m(t) = Z(i,j) \times \beta$$

where the contents of $Z(i,j)$ are the corresponding values of $\overline{\Delta eval}(V;t-1)$ and $\overline{\Delta eval}(V;t)$ for defuzzification (Wang, Wang and Wu, 1997). The parameters α and β are given values to regulate the increasing and decreasing ranges for the crossover and mutation rates.

Step 4. Update the changes of the crossover and the mutation rates using the following equation:

$$p_C(t) = p_C(t-1) + \Delta c(t), \; p_M(t) = p_M(t-1) + \Delta m(t)$$

where $p_C(t)$ and $p_M(t)$ are the crossover and the mutation rates at generation t, respectively.

2.2   AGO Using Conventional Heuristics

For the AGO using heuristics, we present two concepts from the previous two studies (Mak, Wong and Wang, 2000; Srinivas and Patnaik, 1994). The inputs of which for crossover and mutation operators use one or several fitness values resulting from genetic search process in each algorithm, and the outputs of which adaptively regulate the rates of the operators. The detailed logics and procedures for applying them are as follows.

2.2.1   AGO Using the Fitness of Offspring in Each Generation: Adaptive Scheme 1

For this heuristic, we use the concept of Mak, Wong, and Wang (2000). They used the fitness value of offspring in each generation in order to construct adaptive crossover and mutation operators: this scheme increases the occurrence rates of the crossover and mutation operators, if it consistently produces a better offspring during genetic search process; however, it also reduces the occurrence rates of the operators, if it produces a poorer offspring. This scheme based on the fact that it encourages the well-performing crossover and mutation operators to produce more offspring, while also reducing the chance for the poorly performing operators to destroy the potential individuals during genetic search process. It is the main scheme for constructing the AGO. The detailed procedure, when maximization is assumed, is as follows:

Procedure: regulation of p_C and p_M using the fitness of offspring

begin

    if $(\overline{f_{pop\_size}}(t) / \overline{f_{off\_size}}(t)) - 1 \geq 0.1$ **then**

        $p_C(t) = p_C(t-1) + 0.05, \quad p_M(t) = p_M(t-1) + 0.005$;

    if $(\overline{f_{pop\_size}}(t) / \overline{f_{off\_size}}(t)) - 1 \leq 0.1$ **then**

        $p_C(t) = p_C(t-1) - 0.05, \quad p_M(t) = p_M(t-1) - 0.005$;

    if $-0.1 < (\overline{f_{pop\_size}}(t) / \overline{f_{off\_size}}(t)) - 1 < 0.1$ **then**

        $p_C(t) = p_C(t-1), \quad p_M(t) = p_M(t-1)$

    end

end

where f_{off_size} and f_{pop_size} are the fitness values in offspring and parents, respectively. In the cases of $(f_{off\_size} / f_{pop\_size}) - 1 \geq 0.1$ and $(f_{off\_size} / f_{pop\_size}) - 1 \leq -0.1$, the adjusted rates should not exceed the domain from 0.5 to 1.0 for the $p_C(t)$ and the domain from 0.00 to 0.10 for the $p_M(t)$.

The procedure mentioned above is evaluated in every generation during genetic search process and changes the occurrence rates of crossover and mutation operators.

2.2.2 AGO Using Various Fitness Values in Each Generation: Adaptive Scheme 2

For this heuristic, we use the concept of Srinivas and Patnaik (1994). This concept considers both exploitation and exploration properties in the convergence process of GA; the capacity to converge to an global optimum after locating the region containing the optimum, and the capacity to explore new regions of the solution space in search of the global optimum.

The balance between these characteristics of GA is adaptively regulated by the values of p_C and p_M in each generation; increasing values of p_C and p_M promote exploration at the expense of exploitation. By this basic scheme, the p_C and p_M are increased when the population tends to get stuck at a local optimum and are decreased when the population is scattered in the search space of GA. The detailed schemes for regulating the p_C and p_M are as follows:

$$p_C = \begin{cases} \dfrac{k_1 (f_{max} - f_{cro})}{f_{max} - f_{avg}}, & f_{cro} \geq f_{avg} \\ k_3, & f_{cro} < f_{avg} \end{cases} \quad \text{and} \quad p_M = \begin{cases} \dfrac{k_2 (f_{max} - f_{mut})}{f_{max} - f_{avg}}, & f_{mut} \geq f_{avg} \\ k_4, & f_{mut} < f_{avg} \end{cases}$$

where f_{max} and f_{avg} are the maximum and average fitness values in each generation, respectively. f_{cro} is the larger of the fitness values of the individuals to be crossed. f_{mut} is the fitness value of the ith individual to which the mutation with a rate p_M is applied. The values of k_1, k_2, k_3, k_4 are 1, 0.5, 1, and 0.5, respectively.

The adjusted rate should not exceed the domain from 0.5 to 1.0 for the p_c and the domain from 0.001 to 0.05 for the p_M .

3 Proposed Hybrid Concepts and Logics

In this section we propose the concepts and logics for constructing the searching procedures of a proposed hybrid method. First, we apply a rough search technique for initializing GA population. Secondly, the GA procedure is applied. Finally, the local search technique to find a better solution is proposed.

3.1 Rough Search Technique

Rough search technique, one of local search technique, was first mentioned by Li and Jiang (2000). It can make large jumps in search space and is able to "jump out" of the local optimal solutions, also converges rapidly. However, it is not the most effective method in searching optimal solution, when it is used alone (Davis, 1991). Therefore, it should be hybridized with other optimization techniques for finding global optimal solution. For example, if it will be used to initialize the population of GA, then this hybrid strategy can generate a better solution than using conventional GA alone or rough search technique alone.

According to this concept, we consider the rough search technique as a local search technique for the proposed hybrid algorithms. The aim of this technique is to initialize the population of GA, and the detailed implementation procedure is as follows:

Step 1: Randomly generate as many strings as the population size of GA within the allowable limit of variables.

Step 2: Check constraints. If they are satisfied, evaluate the fitness $f(x)$ of each string and store the one with the best fitness $f(x)$ as a string for the GA initial population.

Step 3: If the iteration does not run out, go to Step 1. If we find a fitness value better than the previous fitness $f(x)$ stored in Step 2, store the string as the next string for the GA initial population. This process is repeated until the pre-defined iteration number is satisfied.

Step 4: If we do not have as many strings as the population size of GA after applying Step 3, the remainders are randomly generated within the allowable limits for the variables.

3.2 Genetic Algorithm

For the representation of GA, we use a real-number representation instead of a bit-string one, since the former has several advantages of (i) being better adapted to numerical optimization for continuous problems, (ii) speeding up the search over bit-string representation, and (iii) easing the development of approaches for hybridizing with other conventional methods (Davis, 1991). The proposed GA is used as a main algorithm of the proposed hybrid algorithms and the detailed heuristic procedure for its application is as follows:

Step 1: Initial population
   We use the population resulting from the rough search technique proposed in Section 3.1.

Step 2: Genetic operators
   Selection: elitist selection strategy in enlarged sampling space (Gen and Cheng, 2000)
   Crossover: non-uniform arithmetic crossover operator (Michalewicz, 1994)
   Mutation: uniform mutation operator (Michalewicz, 1994)

Step 3: Stop condition
   If a pre-defined maximum generation number is reached or an optimal solution is located during genetic search process, then stop; otherwise, go to Step 2.

3.3   Local Search Technique

Local search techniques usually use local information about the current set of data (state) to determine a promising direction for moving some of the data set, which is in turn used to form the next set of data. The advantage of local search techniques is that they are simple and computationally efficient. However, they are easily entrapped in a local optimum. In contrast, global search techniques such as GA explore the global search space without using local information about promising search direction. Consequently, they are less likely to be trapped in local optima, but their computational cost is higher.

Many researchers have reported that combining GA and local search technique into a hybrid approach often produces certain benefits (Lee, Yun and Gen, 2002; Li and Jiang, 2000; Yen et al., 1998). The reason is that the hybrid approach can combine the merits of the GA with those of the local search technique. That is, a hybrid approach is less likely to be trapped in a local optimum than a local search technique. Due to the local search technique, a hybrid approach often converges faster than the GA does.

In this paper, we use a method to hybridize GA with local search technique. This approach seen in most of conventional hybrid GAs is to incorporate a local search technique into GA loop (Gen and Cheng, 1997; Yen et al., 1998). With this approach, the local search technique is applied to each newly generated offspring to move it to a local optimum before injecting it into the new population. For the proposed hybrid GA, we use the iterative hill climbing technique suggested by Michalewicz (1994) and improve it for its application. This technique can guarantee the desired properties of local search technique for hybridization as explained above.

The main difference between the conventional iterative hill climbing technique and the improved iterative hill climbing technique is that the latter selects an optimal string among the strings satisfying the constraints of the hybrid GA, while, the former selects a current string at random, which allows the latter to have various search abilities and good solutions unmet by the former. The detailed procedure of the improved iterative hill climbing technique, when minimization is assumed, is given as follows:

Procedure: improved iterative hill climbing technique in GA loop
begin
        select an optimum string $\mathbf{v}_c$ in each GA loop;
        randomly generate as many strings as the population size in the neighborhood of $\mathbf{v}_c$;
        select the string $\mathbf{v}_n$ with the optimal value of the objective function f among the set of new
        strings;
        if $f(\mathbf{v}_c) > f(\mathbf{v}_n)$ **then**
            $\mathbf{v}_c \leftarrow \mathbf{v}_n$
        end
end

4   Proposed Algorithms for Experimental Comparison

In this section, we propose four hybrid algorithms for experimental comparison; one hybrid GA without any adaptive ability, and three hybrid GAs with various adaptive abilities suggested in Section 3. The detailed procedures for the proposed algorithms are given as follows:

4.1   Hybrid Genetic Algorithm (HGA): Rough Search + GA with Local Search

HGA uses the rough search technique for initializing the GA population and then applies the GA with the improved iterative hill climbing technique. The heuristic implementation procedure is as follows:

Step 1: apply the rough search technique for making the initial population of GA.

Step 2: apply genetic operators (i.e., selection, crossover and mutation operators) using the population obtained by Step 1.

Step 3: select an optimal solution among the feasible solutions resulting from Step 2.

Step 4: apply the improved iterative hill climbing technique using the optimal solution obtained by Step 3.

Step 5: go to Step 2 until the pre-defined stop condition is satisfied.

This procedure is also summarized in Figure 4 (a).

4.2 Adaptive Hybrid GA 1 (AHGA1): Rough Search + GA with Local Search + Adaptive Scheme 1

The first adaptive hybrid GA applies the rough search technique, the improved iterative hill climbing technique, and the adaptive scheme 1 suggested in sub-section 2.2.1. The heuristic procedure is as follows:

The procedures from Step 1 to Step 4 are the same ones with Section 4.1.

Step 5: apply the adaptive scheme 1 for regulating the GA parameters (i.e., the crossover and the mutation rates)

Step 6: go to Step 2 until the pre-defined stop condition is satisfied.

This procedure is also summarized in Figure 4 (b).

4.3 Adaptive Hybrid GA 2 (AHGA2): Rough Search + GA with Local Search + Adaptive Scheme 2

The second adaptive hybrid GA applies the rough search technique, the improved iterative hill climbing technique, and the adaptive scheme 2 suggested in sub-section 2.2.2. The heuristic procedure is as follows:

The procedures from Step 1 to Step 4 are the same ones with Section 4.1.

Step 5: apply the adaptive scheme 2 for regulating the GA parameters (i.e., the crossover and the mutation rates)

Step 6: go to Step 2 until the pre-defined stop condition is satisfied.

This procedure is also summarized in Figure 4 (c).

4.4 Adaptive Hybrid GA 3 (AHGA3): Rough Search + GA with Local Search + Adaptive Scheme Using FLC

The last adaptive hybrid GA applies the rough search technique, the improved iterative hill climbing technique, and the adaptive scheme using FLC suggested in Section 2.1. The heuristic procedure is as follows:

The procedures from Step 1 to Step 4 are the same ones with Section 4.1.

Step 5: apply the adaptive scheme using FLC for regulating the GA parameters (i.e., the crossover and the mutation rates)

Step 6: go to Step 2 until the pre-defined stop condition is satisfied.

This procedure is also summarized in Figure 4 (d).

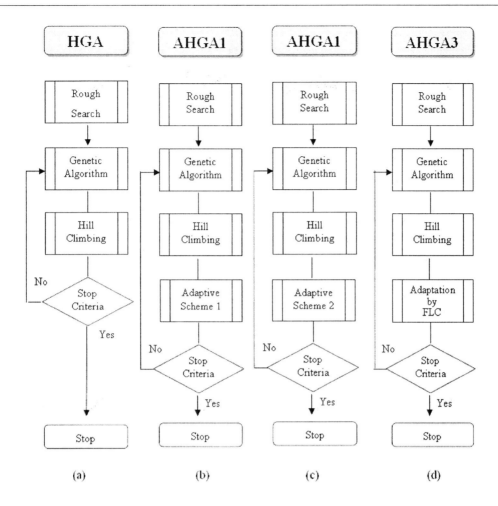

Figure 4. Various hybrid GAs

5   Numerical Examples

In this section, the four hybrid GAs proposed in Section 4 are tested in order to compare their performances. They are applied to three test problems. The results obtained are compared with each other and analyzed using various measures of performance.

5.1   Experimental Conditions

The parameters for the proposed algorithms are as follows: population size = 20, initial crossover rate = 0.5 initial mutation rate = 0.05, total generation number for stop condition = 2000, and the search range of the improved iterative hill climbing technique = 1. These parameters are applied to all the test problems as the same condition. Altogether 30 iterations were executed in order to eliminate the randomness of their searches. The procedures of all the proposed algorithms were implemented in the Visual Basic language under IBM- PC Pentium-800Mhz computer with 512M RAM.

5.2   Test Problem 1

Test problem 1 is to maximize the function of two continuous variables called "binary f6" (Davis, 1991) as follow:

$$f(x_1,x_2)=0.5-\frac{(\sin\sqrt{x_1^2+x_2^2})^2-0.5}{1.0+0.001(x_1^2+x_2^2)^2}$$

This function reaches its global maximum of 1.0 at the point ($x_1 = x_2 = 0.0$), and the search ranges of x_1, x_2 have the range from -100.0 to +100.0. The results obtained by the proposed four algorithms are shown in Table 3.

Table 3. The computational results for test problem 1

	No. of Getting Stuck	Gen. No. to Optimum	Time to Optimum
HGA	6	675.8667	2.03
AHGA1	2	501.3667	1.87
AHGA2	1	312.5333	1.70
AHGA3	0	431.5000	1.57

* No. of Getting Stuck: total number that gets stuck at a local optimum.
* Gen. No. to Optimum: average number of generations to the stop condition.
* Time to Optimum: average CPU time to the stop condition.

In terms of 'No. of Getting Stuck' of Table 3, all of the results of AHGA1, AHGA2, and AHGA3 are better than that of the HGA. Especially, the AHGA3 does not get stuck at a local optimum even on one occasion among 30 runs. This implies that if we apply an adaptive scheme to the HGA, the number of getting stuck at a local optimum can be reduced. In terms of 'Gen. No. to Optimum' and 'Time to Optimum,' the results of three algorithms with various adaptive schemes (AHGA1, AHGA2 and AHGA3) are converged to the global optimum more rapidly than that of HGA without any adaptive scheme.

In the comparison among the adaptive algorithms, the results of the AHGA2 and AHGA3 are slightly better than that of the AHGA1, which implies that the search schemes of the AHGA2 and AHGA3 for locating the global optimum outperforms that of the AHGA1. This is determined by various parameters used in GA, that is, in the three algorithms with various adaptive schemes, it is a good way that we compare the various behaviors of the rates of the crossover and mutation operators used in each algorithm in order to compare their adaptive schemes. Furthermore, by these operators, the fitness values that converge to the global optimum can be changed. We can confirm this situation by considering the behaviors of average fitness during the genetic search processes in each algorithm. Figures 5 and 6 show the various behaviors of the rates of the crossover and mutation operators in the three algorithms with various adaptive schemes.

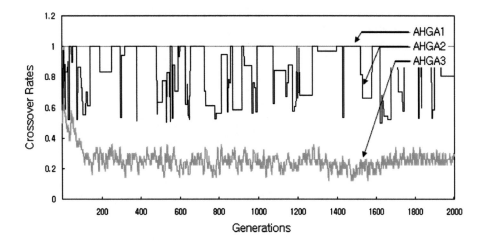

Figure 5. Evolutionary behaviors of crossover rate

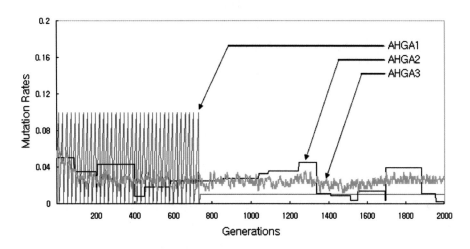

Figure 6. Evolutionary behaviors of mutation rate

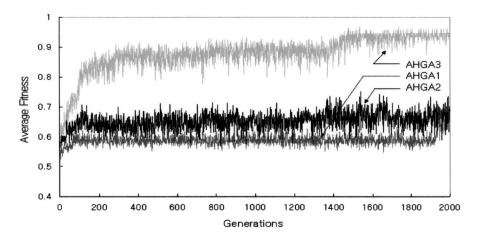

Figure 7. Evolutionary behaviors of average fitness

In Figures 5 and 6, the graph of the AHGA1, except for the initial generations of Figure 6, does not show various behaviors at all, while those of AHGA2 and AHGA3 show various behaviors, over the course of generations. Especially, in the comparison between the AHGA2 and AHGA3, the graph of the AHGA3 shows much better various behaviors and lower values than that of the AHGA2. This implies that the AHGA3 can search more variously and quicker than the AHGA2 both in the search space and search speed. These behaviors of the crossover and mutation operators affect that of the average fitness values during genetic search process. Figure 7 proves this fact, that is, the average fitness values of the AHGA3 outperforms those of the AHGA2.

5.3   Test Problem 2

Test problem 2 is to minimize a function of two variables called the Rosenbrock function as follows:

$$f(x_1, x_2) = 100(x_1^2 - x_2)^2 + (1 - x_1)^2$$

This function is De Jong's F2 (De Jong, 1975); it has a global minimum of zero at $x_1 = x_2 = 1.0$, and each of x_1 and x_2 have the continuous values within the range from -2.048 to +2.048. The results obtained using each algorithm are shown in Table 4.

Table 4. The computational results for test problem 2

	No. of Getting Stuck	Gen. No. to Optimum	Time to Optimum
HGA	6	937.1000	1.70
AHGA1	5	651.3667	1.37
AHGA2	5	798.5333	1.87
AHGA3	2	645.9667	1.33

In Table 4, the result of the AHGA3 is superior to those of the HGA, AHGA1, and AGHA2 in terms of 'No. of Getting Stuck.' In terms of 'Gen No. to optimum,' the AHGA1 and AHGA3 have almost same results, while their results are significantly better than those of the HGA and AHGA2. These rapid search schemes of the AHGA2 and AHGA3 to the global optimum can decrease the search time, which is proved in the result appeared in terms of 'Time to Optimum.'

Figures 8 and 9 show the various behaviors of the crossover and mutation operators during the genetic search processes in each algorithm. The graph of the AHGA2 does not show any change at all, and that of AHGA1 shows a little change, during their genetic search processes. However, the graph of the AHGA3 shows various behaviors. This proves that the adaptive scheme used in the AHGA3 is more efficient than those in the AHGA1 and AHGA2.

Figure 10 shows various behaviors of the average fitness values among the three adaptive algorithms. The graph of AHGA3 converges more rapidly than those of the AHGA1 and AHGA2, though all the graphs of the three algorithms show various behaviors. By these analyses using Figures 8, 9 and 10, we can confirm that the AHGA3 get a better chance in locating the global optimum than the AHGA1 and AHGA2.

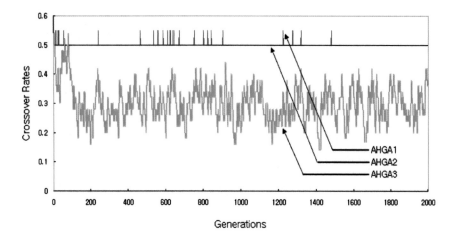

Figure 8. Evolutionary behaviors of crossover rate

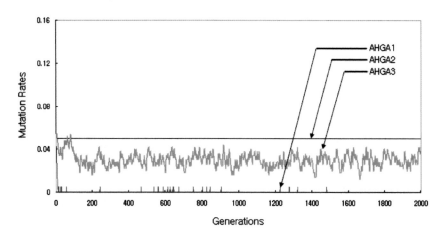

Figure 9. Evolutionary behaviors of mutation rate

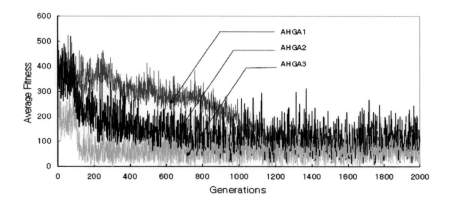

Figure 10. Evolutionary behaviors of average fitness

5.4   Test Problem 3

Test problem 3 considers a function with five continuous variables, called the Restrigin function (Hoffmeister and Bäck, 1991)

$$f(x_1, x_2, x_3, x_4, x_5) = 15 + \sum_{i=1}^{5} (x_i^2 - 3\cos(2\pi x_i))$$

This function has a global minimum of zero at $x_1 = x_2 = x_3 = x_4 = x_5 = 0.0$, and all of the variables should be considered as continuous values within the range from −5.12 to +5.12. The results obtained by the proposed four algorithms are shown in Table 5.

Table 5. The computational results for test problem 3

	No. of Getting Stuck	Gen. No. to Optimum	Time to Optimum
HGA	16	1359.2667	4.70
AHGA1	19	1388.8000	4.73
AHGA2	12	1083.3333	5.17
AHGA3	6	852.0667	3.37

In terms of 'No. of Getting Stuck' of Table 5, the AHGA3 has the lowest value when compared with the other algorithms, though all the algorithms cannot locate the global optimum in all the trials (in our case, 30 runs). By this result, we can confirm that the adaptive scheme used in the AHGA3 is more efficient than those in the AHGA1 and AHGA2. This have also influence on the search time and generation number to locate the global optimum, that is, the result of the AHGA3 has considerately a lower value than those of the AHGA1 and AHGA2 in terms of 'Gen. No. to Optimum,' and the search time of the AHGA3 is quicker than those of the AHGA1 and AHGA2 in terms of 'Time to Optimum.'

In the comparisons among the proposed adaptive algorithms (AHGA1, AHGA2, and AHGA3), the main difference depends highly on the crossover and mutation operators with adaptive schemes. The behaviors of these operators are shown in Figures 11 and 12. In the two Figures, the graph of the AHGA3 shows various behaviors as the generations are proceed, while those of the AHGA1 and AHGA2 show a little change behaviors. Especially, in Figure 11, the graph of the AHGA2 has considerably high values rather than that of the AHGA3, which makes the AHGA2 increase the search time, which also be proved in terms of 'Time to Optimum' of Table 5.

Figure 13 shows the various behaviors of the average fitness values. The AHGA3 has more various and quicker behaviors than the AHGA1 and AHGA2 in terms of the change rates of average fitness values and the convergence speed, which implies that the AHGA3 can search the solution space more variously than the AHGA1 and AHGA2. This definitely relies on the abilities of the adaptive crossover and mutation operators used in each algorithm.

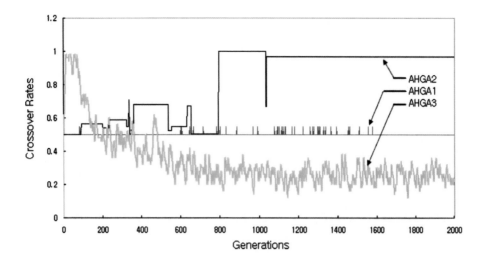

Figure 11. Evolutionary behaviors of crossover rate

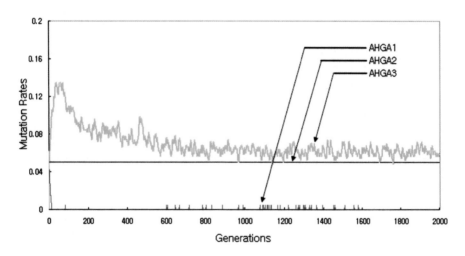

Figure 12. Evolutionary behaviors of mutation rate

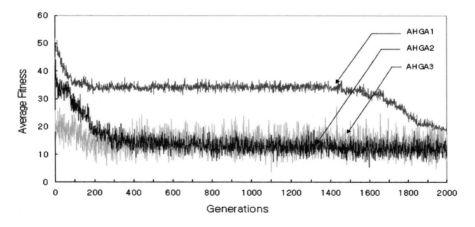

Figure 13.Evolutionary behaviors of average fitness

Based on the various results analyzed using the three test problems, we can conclude the following to be true in our experimental investigations:

(1) Overall comparison results among the four proposed algorithms have shown that the algorithms with the adaptive schemes are more efficient and more robust than the algorithm without it, though the HGA has better results than the AHGA1, AHGA2, and AHGA3 in some cases of Tables 4 and 5.

(2) Applying the adaptive schemes to the proposed algorithms (AHGA1, AHGA2, and AHGA3) enables the GA parameters (i.e., the rates of crossover and mutation operators) to be adaptively regulated during their genetic search processes, and also the average fitness values to be variously changed. Especially, in the comparison among each algorithm, the AHGA3 has showed the best results in terms of "No. of Getting Stuck," "Gen. No. to Optimum," and "Time to Optimum" presented in Tables 3, 4, and 5, except for the terms of "Gen. No. to Optimum" of Table 3, which implies that the FLC used in the AHGA3 is more efficient than the heuristics in the AHGA1 and AHGA2.

6   Conclusion

In this paper we have proposed one hybrid GA without any adaptive scheme and three hybrid GAs with various adaptive schemes. For constructing the proposed hybrid algorithms, we have combined a rough search technique, GA and improved iterative hill climbing technique. These hybrid algorithms have also employed one FLC and two heuristics for their adaptive schemes. All the algorithms have been tested and analyzed using three test problems.

In various comparisons among these algorithms, the algorithms with adaptive schemes are more efficient than that without it. Especially, the AHGA3 with the FLC adaptive scheme have shown considerately better performances than the AHGA1 and AHGA2 with the heuristic adaptive schemes. By this fact, we can conclude that applying the FLC, as adaptive ability, is more efficient than applying the heuristics for constructing an efficient hybrid GA.

References

Cheong, F. & Lai, R. (2000). Constraining the optimization of a fuzzy logic controller using an enhanced genetic algorithm. *IEEE Transactions on Systems, Man, and Cybernetics-Part B: Cybernetics*, **30**(1), 31-46.

Davis, L. (1991). Handbook of Genetic Algorithms, Van Nostrand Reinhold.

De Jong, K. A. (1975). Analysis of the behavior of a class of genetic adaptive systems. PhD Thesis, University of Michigan (University Microfilms No. 76-9381).

Gen, M. & Cheng, R. (1997). *Genetic Algorithms and Engineering Design.* John-Wiley & Sons.

Gen, M. & Cheng, R. (2000). *Genetic Algorithms and Engineering Optimization.* John-Wiley & Sons.

Herrera, F. & Lozano, M. (2001). Adaptive genetic operators based on coevolution with fuzzy behaviors. *IEEE Transactions on Evolutionary Computation*, **5**(2): 149-165.

Hoffmeister, F. & Bäck, T. (1991). Genetic algorithms and evolution strategies: Similarities and differences. *Proceedings of the 1st Workshop on Parallel Problem Solving from Nature* (PPSN1), 455-471.

Mak, K. L., Wong, Y. S. & Wang, X. X. (2000). An adaptive genetic algorithm for manufacturing cell formation. *International Journal of Manufacturing Technology*, **16**: 491-497.

Michalewicz, Z. (1994). Genetic algorithm + data structures = evolution programs. Springer-Verlag, Second Edition.

Lee, C. Y., Yun, S. Y. & Gen, M. (2002). Reliability Optimization Design for Complex Systems by Hybrid GA with Fuzzy Logic Control and Local Search. *IEICE Transaction on Fundamentals of Electronics Communications and Computer Sciences*, **E85-A**(4): 880-891.

Lee, M. & Takagi, H. (1993). Dynamic control of genetic algorithm using fuzzy logic techniques. *Proceedings of the 5th International Conference on Genetic Algorithms*, San Francisco, 76-83.

Li, B. & Jiang, W. (2000). A novel stochastic optimization algorithm. *IEEE Transactions on Systems, Man, and Cybernetics-Part B: Cybernetics*, **30**(1), 193-198.

Song, Y. H., Wang, G. S., Wang, P. T. & Johns, A. T. (1997). Environmental/economic dispatch using fuzzy logic controlled genetic algorithms. *IEEE Proceedings on Generation, Transmission and Distribution*, **144** (4): 377-382.

Srinvas, M. & Patnaik, L. M. (1994). Adaptive probabilities of crossover and mutation in genetic algorithms. *IEEE Transaction on Systems, Man and Cybernetics*, **24**(4): 656-667.

Subbu, R., Sanderson, A. C. & Bonissone, P. P. (1998). Fuzzy logic controlled genetic algorithms versus tuned genetic algorithms: An agile manufacturing application. *Proceedings of the 1999 IEEE International Symposium on Intelligent Control (ISIC)*, 434-440.

Wu, Q. H., Cao, Y. J. & Wen, J. Y. (1998). Optimal reactive power dispatch using an adaptive genetic algorithm. *Electrical Power & Energy Systems*, **20**(8): 563-569.

Yen, J., Liao, J. C., Lee, B. J. & Randolph, D. (1998). A hybrid approach to modeling metabolic systems using a genetic algorithm and simplex method. *IEEE Transactions on Systems, Man, and Cybernetics-Part B: Cybernetics*, **28**(2), 173-191.

Yun, Y. S. & Gen, M. (2003). Performance analysis of adaptive genetic algorithms with fuzzy logic and heuristics. *Fuzzy Optimization and Decision Making*. **2**(2): 161-175.

Yun, Y. S., Gen, M. & Seo, S. L. (2003). Various hybrid genetic algorithm based on a genetic algorithm with a fuzzy logic controller. *Journal of Intelligent Manufacturing*, **14**(3-4); 401-419.

In: Control and Learning in Robotic Systems
Editor: John X. Liu, pp. 281-319

ISBN 1-59454-356-9
©2005 Nova Science Publishers, Inc.

Chapter 11

RELIABLE FEATURE EVALUATION
IN CLASSIFICATION AND REGRESSION

Marko Robnik-Šikonja[*] *and Igor Kononenko*[†]
University of Ljubljana, Faculty of Computer and Information Science,
Tržaška 25, 1001 Ljubljana, Slovenia
tel.: + 386 1 4768459, fax: + 386 1 4768498

Abstract

Evaluation of the quality of attributes (features) is an important issue in the machine learning, data mining and in a research of intelligent systems. There are several important tasks in the process of data analysis e.g., feature subset selection, constructive induction, decision and regression tree building, and visualization, which contain an attribute evaluation procedure as their (crucial) ingredient.

Relief algorithms are general and successful attribute evaluators.They are able to detect conditional dependencies between attributes and provide a unified view on the attribute evaluation in regression and classification. In addition, their quality estimates have a natural interpretation. While they have commonly been viewed as feature subset selection methods that are applied in prepossessing step before a model is learned, they have actually been used successfully in a variety of settings, e.g., to select splits or to guide constructive induction in the building phase of decision or regression tree learning, as the attribute weighting method and also in the inductive logic programming.

This chapter describes how, why and when Relief algorithms work, their properties and parameters. It provides answers to questions: what kind of dependencies they detect, how do they scale up to large number of examples and features, how to sample data for them, how robust are they regarding the noise, how irrelevant and redundant attributes influence their output, and how different metrics influences them. It also lists some of their successful applications.

[*]E-mail address: Marko.Robnik@fri.uni-lj.si
[†]E-mail address: Igor.Kononenko@fri.uni-lj.si

1 Introduction

There are several important tasks in the process of data analysis e.g., feature subset selection, constructive induction, and decision and regression tree building, which contain the attribute evaluation procedure as their (crucial) ingredient. These task frequently occur in machine learning, data mining, robotics, and in the research of intelligent systems in general.

In many learning problems there are hundreds or thousands of (potential) features describing each input object. Many learning methods do not behave well in this circumstances because, from a statistical point of view, examples with many irrelevant, but noisy, features provide very little information. A feature subset selection is a task of choosing a small subset of features that ideally is necessary and sufficient to describe the target concept. To make a decision which features to retain and which to discard we need a reliable and practically efficient method for evaluation of their relevance to the target concept.

In the constructive induction we face a similar problem. In order to enhance the power of the representation language and construct a new knowledge we introduce new features. Typically many candidate features are generated and again we need to decide which features to retain and which to discard. To estimate the relevance of the features to the target concept is certainly one of the major components of such a decision procedure.

Decision and regression trees are popular description languages for representing knowledge in the machine learning. While constructing a tree the learning algorithm at each interior node selects the splitting rule (feature) which divides the problem space into two separate subspaces. To select an appropriate splitting rule the learning algorithm has to evaluate several possibilities and decide which would partition the given (sub)problem most appropriately. The evaluation of the quality of the splitting rules is here an important subproblem.

The problem of attribute evaluation has received much attention in the literature. There are several attribute evaluation measures proposed. If the target concept is a discrete variable (the classification problem) these are e.g., information gain [11], Gini index [2], distance measure [20], j-measure [34], Relief [14], ReliefF [15], MDL [16], and also χ^2 and G statistics are used. If the target concept is presented as a real valued function (numeric label and the regression problem) then the evaluation heuristics are e.g., the mean squared and the mean absolute error [2], and RReliefF [29].

The majority of the heuristic measures for evaluation of the quality of the attributes assume the conditional independence (upon the target variable) of the attributes and are therefore less appropriate in problems which possibly involve much feature interaction. Relief algorithms (Relief, ReliefF and RReliefF) do not make this assumption. They are efficient, aware of the contextual information, and can correctly evaluate the quality of attributes in problems with strong dependencies between attributes.

While Relief algorithms have commonly been viewed as feature subset selection methods that are applied in a prepossessing step before the model is learned [14] and are one of the most successful preprocessing algorithms to date [6], they are actually general feature evaluators and have been used successfully in a variety of settings: to select splits in the building phase of decision tree learning [18], to select splits and guide the constructive induction in learning of the regression trees [29], as attribute weighting method [36] and also in inductive logic programming [23].

In this chapter we describe how and why Relief algorithms work, their theoretical and practical properties, their parameters, what kind of dependencies they detect, how do they scale up to large number of examples and features, how to sample data for them, how robust are they regarding the noise, how irrelevant and redundant attributes influence their output, how different metrics influences them, and some of their successful applications. In Section 2 we present the Relief algorithms, discuss interpretation of their output, and their computational complexity. Section 3 presents the framework of our analytical approach: performance measures, compared heuristics, and the data sets used. In Section 4 we focus on practical issues on the use of ReliefF and try to answer the above questions. In Section 5 we discusses applicability of Relief algorithms for various tasks and conclude in Section 6.

We assume that examples $I_1, I_2, ..., I_n$ in the instance space are described by a vector of attributes A_i, $i = 1, ..., a$, where a is the number of explanatory attributes, and are labelled with the target value τ_j. The examples are therefore points in the a dimensional space. If the target value is categorical, we call the modelling task classification, and if it is numerical, we call the modelling task regression.

2   The Relief Algorithms

In this Section we describe the Relief algorithms and discuss their similarities and differences. First we present the original Relief algorithm [14] which was limited to classification problems with two classes. We give account on how and why it works. We discuss its extension ReliefF [15] which can deal with multiclass problems. The improved algorithm is more robust and also able to deal with incomplete and noisy data. Then we show how ReliefF was adapted for continuous class (regression) problems and describe the resulting RReliefF algorithm [29]. After the presentation of the algorithms we describe a comprehensible interpretation of their output and give account of their computational complexity.

2.1   Basic Ideas

A key idea of the original Relief algorithm [14], given in Figure 1, is to evaluate the quality of attributes according to how well their values distinguish between instances that are near to each other. For that purpose, given a randomly selected instance R_i (line 3), Relief searches for its two nearest neighbors: one from the same class, called *nearest hit H*, and the other from the different class, called *nearest miss M* (line 4). It updates the quality estimation $W[A]$ for all attributes A depending on their values for R_i, M, and H (lines 5 and 6). If instances R_i and H have different values of the attribute A then the attribute A separates two instances with the same class which is not desirable so we decrease the quality estimation $W[A]$. On the other hand if instances R_i and M have different values of the attribute A then the attribute A separates two instances with different class values which

Algorithm Relief

Input: for each training instance a vector of attribute values and the class value
Output: the vector W of evaluations of the qualities of attributes

1.   set all weights $W[A] := 0.0$;
2.   **for** i := 1 **to** m **do begin**
3.   randomly select an instance R_i;
4.   find nearest hit H and nearest miss M;
5.   **for** $A := 1$ **to** a **do**
6. $W[A] := W[A] - \text{diff}(A, R_i, H)/m + \text{diff}(A, R_i, M)/m$;
7.   **end**;

Figure 1: Pseudo code of the basic Relief algorithm

is desirable so we increase the quality estimation $W[A]$. The whole process is repeated for m times, where m is a user-defined parameter.

Function $\text{diff}(A, I_1, I_2)$ calculates the difference between the values of the attribute A for two instances I_1 and I_2. For nominal attributes it was originally defined as:

$$\text{diff}(A, I_1, I_2) = \begin{cases} 0 & ; value(A, I_1) = value(A, I_2) \\ 1 & ; otherwise \end{cases} \tag{1}$$

and for numerical attributes as:

$$\text{diff}(A, I_1, I_2) = \frac{|value(A, I_1) - value(A, I_2)|}{max(A) - min(A)} \tag{2}$$

The function diff is used also for calculating the distance between instances to find the nearest neighbors. The total distance is simply the sum of distances over all attributes (Manhattan distance).

The original Relief can deal with nominal and numerical attributes. However, it cannot deal with incomplete data and is limited to two-class problems. Its extension, which solves these and other problems, is called ReliefF.

2.2   ReliefF

The ReliefF (Relief-F) algorithm [15] (see Figure 2) is not limited to two class problems, is more robust and can deal with incomplete and noisy data. Similarly to Relief, ReliefF randomly selects an instance R_i (line 3), but then searches for k of its nearest neighbors from the same class, called nearest hits H_j (line 4), and also k nearest neighbors from each of the different classes, called nearest misses $M_j(C)$ (lines 5 and 6). It updates the quality estimation $W[A]$ for all attributes A depending on their values for R_i, hits H_j and misses $M_j(C)$ (lines 7, 8 and 9). The update formula is similar to that of Relief (lines 5 and 6 on Figure 1), except that we average the contribution of all the hits and all the misses. The contribution for each class of the misses is weighted with the prior probability of that class

Algorithm ReliefF
Input: for each training instance a vector of attribute values and the class value
Output: the vector W of evaluations of the qualities of attributes

1.   set all weights $W[A] := 0.0$;
2.   **for** i := 1 **to** m **do begin**
3.   randomly select an instance R_i;
4.   find k nearest hits H_j;
5.   **for** each class $C \neq class(R_i)$ **do**
6.   from class C find k nearest misses $M_j(C)$;
7.   **for** $A := 1$ **to** a **do**
8. $W[A] := W[A] - \sum_{j=1}^{k} \text{diff}(A, R_i, H_j)/(m \cdot k)$

9. $+ \sum_{C \neq class(R_i)} [\frac{P(C)}{1 - P(class(R_i))} \sum_{j=1}^{k} \text{diff}(A, R_i, M_j(C))]/(m \cdot k)$;

10.   **end**;

Figure 2: Pseudo code of ReliefF algorithm

$P(C)$ (estimated from the training set). Since we want the contributions of hits and misses in each step to be in $[0, 1]$ and also symmetric (we explain reasons for that below) we have to ensure that misses' probability weights sum to 1. As the class of hits is missing in the sum we have to divide each probability weight with factor $1 - P(class(R_i))$ (which represents the sum of probabilities for the misses' classes). The process is repeated for m times.

Selection of k hits and misses is the basic difference to Relief and ensures greater robustness of the algorithm concerning noise. User-defined parameter k controls the locality of the estimates. For most purposes it can be safely set to 10 (see [15] and discussion below).

To deal with incomplete data we change the diff function. Missing values of attributes are treated probabilistically. We calculate the probability that two given instances have different values for given attribute conditioned over class value:

- if one instance (e.g., I_1) has unknown value:

$$\text{diff}(A, I_1, I_2) = 1 - P(value(A, I_2)|class(I_1)) \tag{3}$$

- if both instances have unknown value:

$$\text{diff}(A, I_1, I_2) = 1 - \sum_{V}^{\#values(A)} \left(P(V|class(I_1)) \times P(V|class(I_2)) \right) \tag{4}$$

Conditional probabilities are approximated with relative frequencies from the training set.

2.3   RReliefF - in Regression

We finish the description of the algorithmic family with RReliefF (Regressional ReliefF) [29]. First we theoretically explain what Relief algorithm actually computes.

Relief's estimate $W[A]$ of the quality of attribute A is an approximation of the following difference of probabilities [15]:

$$
\begin{aligned}
W[A] \quad = \quad & P(\text{diff. value of A}|\text{nearest inst. from diff. class}) \\
- \quad & P(\text{diff. value of A}|\text{nearest inst. from same class}) \quad\quad (5)
\end{aligned}
$$

The positive updates of the weights (line 6 in Figure 1 and line 9 in Figure 2) are actually forming the estimate of probability that the attribute discriminates between the instances with different class values, while the negative updates (line 6 in Figure 1 and line 8 in Figure 2) are forming the probability that the attribute separates the instances with the same class value.

In regression problems the predicted value $\tau(\cdot)$ is continuous, therefore (nearest) hits and misses cannot be used. To solve this difficulty, instead of requiring the exact knowledge of whether two instances belong to the same class or not, a kind of probability that the predicted values of two instances are different is introduced. This probability can be modelled with the relative distance between the predicted (class) values of two instances.

Still, to estimate W[A] in (5), information about the sign of each contributed term is missing (where do hits end and misses start). In the following derivation Eq. (5) is reformulated, so that it can be directly evaluated using the probability that predicted values of two instances are different. If we rewrite

$$
P_{diffA} = P(\text{different value of A}|\text{nearest instances}) \quad\quad (6)
$$

$$
P_{diffC} = P(\text{different prediction}|\text{nearest instances}) \quad\quad (7)
$$

and

$$
P_{diffC|diffA} = P(\text{diff. prediction}|\text{diff. value of A and nearest instances}) \quad\quad (8)
$$

we obtain from (5) using Bayes' rule:

$$
W[A] = \frac{P_{diffC|diffA}P_{diffA}}{P_{diffC}} - \frac{(1 - P_{diffC|diffA})P_{diffA}}{1 - P_{diffC}} \quad\quad (9)
$$

Therefore, we can estimate $W[A]$ by approximating terms defined by Eq.s 6, 7 and 8. This can be done by the algorithm on Figure 3.

Similarly to ReliefF we select random instance R_i (line 3) and its k nearest instances I_j (line 4). The weights for different prediction value $\tau(\cdot)$ (line 6), different attribute (line 8), and different prediction & different attribute (line 9 and 10) are collected in N_{dC}, $N_{dA}[A]$, and $N_{dC\&dA}[A]$, respectively. The final estimation of each attribute $W[A]$ (Eq. (9)) is computed in lines 14 and 15.

Algorithm RReliefF
Input: for each training instance a vector of attribute values $\mathbf{x}$ and predicted value $\tau(\mathbf{x})$
Output: vector W of evaluations of the qualities of attributes

1.   set all N_{dC}, $N_{dA}[A]$, $N_{dC\&dA}[A]$, $W[A]$ to 0;
2.   **for** i := 1 **to** m **do begin**
3.   randomly select instance R_i;
4.   select $\mathbf{k}$ instances I_j nearest to R_i;
5.   **for** j := 1 **to** k **do begin**
6. $N_{dC} := N_{dC} + \text{diff}(\tau(\cdot), R_i, I_j) \cdot d(i,j)$;
7.   **for** A := 1 **to** a **do begin**
8. $N_{dA}[A] := N_{dA}[A] + \text{diff}(A, R_i, I_j) \cdot d(i,j)$;
9. $N_{dC\&dA}[A] := N_{dC\&dA}[A] + \text{diff}(\tau(\cdot), R_i, I_j) \cdot$
10. $\text{diff}(A, R_i, I_j) \cdot d(i,j)$;
11.   **end**;
12.   **end**;
13. **end**;
14. **for** A := 1 **to** a **do**
15. $W[A] := N_{dC\&dA}[A]/N_{dC} - (N_{dA}[A] - N_{dC\&dA}[A])/(m - N_{dC})$;

Figure 3: Pseudo code of RReliefF algorithm

The term $d(i,j)$ in Figure 3 (lines 6, 8 and 10) takes into account the distance between the two instances R_i and I_j. Rationale is that closer instances should have greater influence, so we exponentially decrease the influence of the instance I_j with the distance from the given instance R_i:

$$d(i,j) = \frac{d_1(i,j)}{\sum_{l=1}^{k} d_1(i,l)} \quad \text{and} \tag{10}$$

$$d_1(i,j) = e^{-\left(\frac{\text{rank}(R_i, I_j)}{\sigma}\right)^2} \tag{11}$$

where $\text{rank}(R_i, I_j)$ is the rank of the instance I_j in a sequence of instances ordered by the distance from R_i and σ is a user defined parameter controlling the influence of the distance. Since we want to stick to the probabilistic interpretation of the results we normalize the contribution of each of k nearest instances by dividing it with the sum of all k contributions. The reason for using ranks instead of actual distances is that actual distances are problem dependent while by using ranks we assure that the nearest (and subsequent as well) instance always has the same impact on the weights.

ReliefF was using a constant influence of all k nearest instances I_j from the instance R_i. For this we should define $d_1(i,j) = 1/k$.

Discussion about different distance functions can be found in following sections.

2.4   Interpretation of Relief's Evaluations

Another view on attributes' quality evaluations of ReliefF and RReliefF is possible with the portion of prediction values which the attribute helps to determine [32]. We say that attribute A is responsible for the change of the predicted value of the instance I to the predicted value $b(I)$ if the change of its values is one of the minimal number of changes required for changing the predicted value from I to $b(I)$. We denote this responsibility with $r_A(I,b(I))$. As the number of examples n goes to infinity, the quality evaluation $W(A)$ computed from m cases I from the sample s for each attribute converges to the number of changes in the predicted values the attribute is responsible for (R_A):

$$\lim_{n\to\infty} W(A) = \frac{1}{m}\sum_{I\in s} r_A(I,b(I)) = R_A \tag{12}$$

We interpret Relief's weights $W(A)$ as the contribution (responsibility) of each attribute to the explanation of the predictions. The actual quality evaluations for the attributes in the given problem are approximations of these ideal weights which occur only with abundance of data. On the basis of this interpretation we can present Relief's evaluations in a graphical form, for example as a pie chart or nomogram.

2.5   Computational Complexity

For n training instances and a attributes Relief (Figure 1) makes $O(m \cdot n \cdot a)$ operations. The most complex operation is selection of the nearest hit and miss as we have to compute the distances between R and all the other instances which takes $O(n \cdot a)$ comparisons.

Although ReliefF (Figure 2) and RReliefF (Figure 3) look more complicated their asymptotical complexity is the same as that of original Relief, i.e., $O(m \cdot n \cdot a)$. The most complex operation within the main **for** loop is selection of **k** nearest instances. For it we have to compute distances from all the instances to R, which can be done in $O(n \cdot a)$ steps for n instances. This is the most complex operation, since $O(n)$ is needed to build a heap, from which k nearest instances are extracted in $O(k \log n)$ steps, but this is less than $O(n \cdot a)$.

Data structure k-d (k-dimensional) tree [1, 33] is a generalization of the binary search tree, which instead of one key uses k keys (dimensions). The root of the tree contains all the instances. Each interior node has two successors and splits instances recursively into two groups according to one of k dimensions. The recursive splitting stops when there are less than a predefined number of instances in a node. For n instances we can build the tree where split on each dimension maximizes the variance in that dimension and instances are divided into groups of approximately the same size in time proportional to $O(k \cdot n \cdot \log n)$. With such tree called optimized k-d tree we can find t nearest instances to the given instance in $O(\log n)$ steps [8].

If we use k-d tree to implement the search for nearest instances we can reduce the complexity of all three algorithms to $O(a \cdot n \cdot \log n)$ [27]. For Relief we first build the optimized k-d tree (outside the main loop) in $O(a \cdot n \cdot \log n)$ steps so we need only $O(m \cdot a)$ steps in the loop and the total complexity of the algorithm is now the complexity of the preprocessing which is $O(a \cdot n \cdot \log n)$. The required sample size m is related to the problem complexity (and not to the number of instances) and is typically much more than $\log n$ so asymptotically we have reduced the complexity of the algorithm. Also it does not make sense to use sample size m larger than the number of instances n.

The computational complexity of ReliefF and RReliefF using k-d trees is the same as that of Relief. They need $O(a \cdot n \cdot \log n)$ steps to build k-d tree, and in the main loop they select t nearest neighbors in $\log n$ steps, update weights in $O(t \cdot a)$ but $O(m(t \cdot a + \log n))$ is asymptotically less than the preprocessing which means that the complexity has reduced to $O(a \cdot n \cdot \log n)$. This analysis shows that ReliefF family of algorithms is actually in the same order of complexity as multikey sort algorithms.

Several authors have observed that the use of k-d trees becomes inefficient with increasing number of attributes [8, 5, 21] and this was confirmed for Relief family of algorithms as well [27].

3   Environment for Performance Analysis

The goal of this chapter is to analyze different issues we run into when using ReliefF and RReliefF in practice.

Before we analyze these issues we define some useful measures and concepts for performance analysis, comparison and explanation. Also we present some data sets which will be used as indicators of the abilities of Relief algorithms and other measures compared with them.

3.1   Performance Measures

In our analysis below we run ReliefF and RReliefF on a number of different problems and observe

- if their estimates distinguish between important attributes (conveying some information about the concept) and unimportant attributes and

- if their estimates rank important attributes correctly (attributes which have stronger influence on prediction values should be ranked higher).

In the evaluation of the success of Relief algorithms we use the following measures:

Separability s is the difference between the lowest estimate of the important attributes and the highest estimate of the unimportant attributes.

$$s = W_{I_{worst}} - W_{R_{best}} \tag{13}$$

We say that a heuristics is successful in separating between the important and unimportant attributes if $s > 0$.

Usability u is the difference between the highest estimates of the important and unimportant attributes.

$$u = W_{I_{best}} - W_{R_{best}} \tag{14}$$

We say that estimates are useful if u is greater than 0 (we are getting at least some information from the estimates e.g., the best important attribute could be used as the split in tree based model). It holds that $u \geq s$.

3.2   Attribute Evaluation Measures for Comparison

For some problems we want to compare the performance of ReliefF and RReliefF with other attribute evaluation heuristics. We have chosen the most widely used. For classification this is the gain ratio (used in e.g., C4.5 [24]) and for regression it is the mean squared error (MSE) of average prediction value (used in e.g., CART [2]).

Note that MSE, unlike Relief algorithms and gain ratio, assigns lower weights to better attributes. To make s and u curves comparable to that of RReliefF we are actually reporting separability and usability with the sign reversed.

3.3   The Data Sets

We use mostly artificial data sets in the empirical analysis because we want to control the environment: in real-world data sets we do not fully understand the problem and the relation of the attributes to the target variable. Therefore we do not know what a correct output of the feature evaluation should be and we cannot evaluate the quality estimates of the algorithms. We mostly use variants of parity-like problems because these are the most difficult problems within the nearest neighbor paradigm. We try to control difficulty of the concepts and therefore we introduce many variants. We use also some non-parity like problems and demonstrate performances of Relief algorithms on them. We did not find another conceptually different class of problems on which the Relief algorithms would exhibit significantly different behavior.

Some characteristics of the data sets used are contained in Table 1. For each domain we present the name, reference to where it is defined, number of class values or label that it is regressional (r), number of important attributes (I), number of random attributes (R), and type of the attributes (nominal or numerical).

4   Using ReliefF and RReliefF

In this section we address different issues we run into when using ReliefF and RReliefF in practice: the impact of different distance measures, the use of numerical attributes, how distance can be taken into account, what kind of dependencies they detect, how do they scale up to large number of examples and features, how many iterations we need for reliable evaluation, how robust are they regarding the noise, and how irrelevant and duplicate attributes influence their output.

Table 1: Short description of data sets.

Name	Definition	#class	I	R	Attr. type
Bool-Simple	Eq. (22)	2	4	10	nominal
Modulo-2-2-c	Eq. (24)	2	2	10	nominal
Modulo-2-3-c	Eq. (24)	2	3	10	nominal
Modulo-2-4-c	Eq. (24)	2	4	10	nominal
Modulo-5-2-c	Eq. (24)	5	2	10	nominal
Modulo-5-3-c	Eq. (24)	5	3	10	nominal
Modulo-5-4-c	Eq. (24)	5	4	10	nominal
Modulo-10-2-c	Eq. (24)	10	2	10	numerical
Modulo-10-3-c	Eq. (24)	10	3	10	numerical
Modulo-10-4-c	Eq. (24)	10	4	10	numerical
MONK-1	Sect. 4.6.2	2	3	3	nominal
MONK-3	Sect. 4.6.2	2	3	3	nominal
Modulo-5-2-r	Eq. (24)	r	2	10	nominal
Modulo-5-3-r	Eq. (24)	r	3	10	nominal
Modulo-5-4-r	Eq. (24)	r	4	10	nominal
Modulo-10-2-r	Eq. (24)	r	2	10	numerical
Modulo-10-3-r	Eq. (24)	r	3	10	numerical
Modulo-10-4-r	Eq. (24)	r	4	10	numerical
Fraction-2	Eq. (25)	r	2	10	numerical
Fraction-3	Eq. (25)	r	3	10	numerical
Fraction-4	Eq. (25)	r	4	10	numerical
LinInc-2	Eq. (26)	r	2	10	numerical
LinInc-3	Eq. (26)	r	3	10	numerical
LinInc-4	Eq. (26)	r	4	10	numerical
LinEq-2	Eq. (27)	r	2	10	numerical
LinEq-3	Eq. (27)	r	3	10	numerical
LinEq-4	Eq. (27)	r	4	10	numerical
Cosinus-Lin	Eq. (23)	r	3	10	numerical
Cosinus-Hills	Eq. (28)	r	2	10	numerical

4.1 Metrics

The $\mathrm{diff}(A_i, I_1, I_2)$ function calculates the difference between the values of the attribute A_i for two instances I_1 and I_2. Sum of differences over all attributes is used to determine the distance between two instances in the nearest neighbors calculation.

$$\delta(I_1, I_2) = \sum_{i=1}^{a} \mathrm{diff}(A_i, I_2, I_2) \tag{15}$$

This looks quite simple and parameterless, however, in instance based learning there are a number of feature weighting schemes which assign different weights to the attributes in the total sum:

$$\delta(I_1, I_2) = \sum_{i=1}^{a} w(A_i)\mathrm{diff}(A_i, I_1, I_2) \qquad (16)$$

ReliefF's estimates of attributes' quality can be successfully used as such weights [36].

Another possibility is to form a metric in a different way:

$$\delta(I_1, I_2) = (\sum_{i=1}^{a} \mathrm{diff}(A_i, I_2, I_2)^p)^{\frac{1}{p}} \qquad (17)$$

which for $p = 1$ gives Manhattan distance and for $p = 2$ Euclidean distance. In our use of Relief algorithms we never noticed any significant difference using these two metrics. For example, on the regression problems from the UCI repository [22] (8 tasks: Abalone, Auto-mpg, Autoprice, CPU, Housing, PWlinear, Servo, and Wisconsin breast cancer) the average (linear) correlation coefficient is 0.998 and (Spearman's) rank correlation coefficient is 0.990.

4.2   Numerical Attributes

If we use diff function as defined by (1) and (2) we run into the problem of underestimating numerical attributes. Let us illustrate this by taking two instances with 2 and 5 being their values of attribute A_i, respectively. If A_i is the nominal attribute, the value of $\mathrm{diff}(A_i, 2, 5) = 1$, since the two categorical values are different. If A_i is the numerical attribute, $\mathrm{diff}(A_i, 2, 5) = \frac{|2-5|}{7} \approx 0.43$. Relief algorithms use results of diff function to update their weights therefore with this form of diff numerical attributes are underestimated.

Evaluations of the attributes in Modulo-8-2 data set (see definition by Eq. 24) by RReliefF in left hand side of Table 2 illustrate this effect. Values of each of 10 attributes are integers in the range 0-7. Half of the attributes are treated as nominal and half as numerical; each numerical attribute is exact match of one of the nominal attributes. The predicted value is the sum of 2 important attributes by modulo 8: $\tau = (I_1 + I_2)$ mod 8. We can see that nominal attributes get approximately double score of their numerical counterparts. This causes that not only important numerical attributes are underestimated but also numerical random attributes are overestimated which reduces the separability of the two groups of attributes.

We can overcome this problem with the ramp function as proposed by [9, 10]. It can be defined as a generalization of diff function for the numerical attributes (see Figure 4):

$$\mathrm{diff}(A, I_1, I_2) = \begin{cases} 0 & ; d \leq t_{eq} \\ 1 & ; d > t_{diff} \\ \frac{d - t_{eq}}{t_{diff} - t_{eq}} & ; t_{eq} < d \leq t_{diff} \end{cases} \qquad (18)$$

where $d = |value(A, I_1) - value(A, I_2)|$ presents the distance between attribute values of two instances, and t_{eq} and t_{diff} are two user definable threshold values; t_{eq} is the maximum

Table 2: Evaluations of attributes in Modulo-8-2 data set assigned by RReliefF. Left hand evaluations are for diff function defined by Eqs. (1) and (2), while the right hand evaluations are for diff function using thresholds (Eq. (18)).

Attribute	no ramp	ramp
Important-1, nominal	0.193	0.436
Important-2, nominal	0.196	0.430
Random-1, nominal	-0.100	-0.200
Random-2, nominal	-0.105	-0.207
Random-3, nominal	-0.106	-0.198
Important-1, numerical	0.096	0.436
Important-2, numerical	0.094	0.430
Random-1, numerical	-0.042	-0.200
Random-2, numerical	-0.044	-0.207
Random-3, numerical	-0.043	-0.198

distance between two attribute values to still consider them equal, and t_{diff} is the minimum distance between attribute values to still consider them different. If we set $t_{eq} = 0$ and $t_{diff} = \max(A) - \min(A)$ we obtain (2) again.

Evaluations of attributes in Modulo-8-2 data set by RReliefF using the ramp function are in the right hand side of Table 2. The thresholds are set to their default values: 5% and 10% of the length of the attribute's value interval for t_{eq} and t_{diff}, respectively. We can see that estimates for nominal attributes and their numerical counterparts are identical.

The threshold values can be set by the user for each attribute individually, which is especially appropriate when we are dealing with measured attributes. Thresholds can be learned in advance considering the context [25] or automatically set to sensible defaults [7]. The sigmoidal function could also be used, but its parameters do not have such straightforward interpretation. In general if the user has some additional information about the character of a certain attribute she/he can supply the appropriate diff function to (R)ReliefF.

We use the ramp function in results reported throughout this work.

4.3   Taking Distance into Account

In instance based learning it is often considered useful to give more impact to the near instances than to the far ones i.e., to weight their impact inversely proportional to their distance from the query point.

RReliefF is already taking the distance into account through Eqs. (10) and (11). By default we are using 70 nearest neighbors and exponentially decrease their influence with increasing distance from the query point. ReliefF originally used constant influence of k nearest neighbors with k set to some small number (usually 10). We believe that the former approach is less risky (as it turned out in a real world application [4]) because as we are

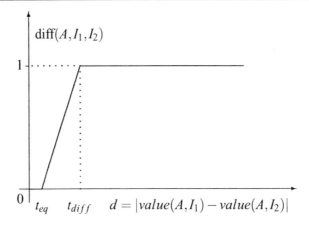

Figure 4: Ramp function

taking more near neighbors we reduce the risk of the following pathological case: we have a large number of instances and a mix of nominal and numerical attributes where numerical attributes prevail; it is possible that all the nearest neighbors are closer than 1 so that there are no nearest neighbors with differences in values of a certain nominal attribute. If this happens in a large part of the problem space this attribute gets zero weight (or at least small and unreliable one). By taking more nearest neighbors with appropriately weighted influence we eliminate this problem.

ReliefF can be adjusted to take distance into account by changing the way it updates it weights (lines 8 and 9 in Figure 2):

$$W[A] := W[A] - \frac{1}{m} \sum_{j=1}^{k} \text{diff}(A, R, H_j) d(R, H_j)$$

$$+ \frac{1}{m} \sum_{C \neq class(R)} \frac{P(C)}{1 - P(class(R))} \sum_{j=1}^{k} \text{diff}(A, R, M_j(C)) d(R, M_j(C)) \qquad (19)$$

The distance factor of two instances $d(I_1, I_2)$ is defined with Eqs. (10) and (11).

The actual influence of the near instances is normalized: as we want probabilistic interpretation of results each random query point should give equal contribution. Therefore we normalize contributions of each of its k nearest instances by dividing it with the sum of all k contributions in Eq. (10).

However, by using ranks instead of actual distances we might lose the intrinsic self normalization contained in the distances between instances of the given problem. If we wish to use the actual distances we only change Eq. (11):

$$d_1(i, j) = \frac{1}{\sum_{l=1}^{a} \text{diff}(A_l, R_i, I_j)} \qquad (20)$$

We might use also some other decreasing function of the distance, e.g., square of the sum in the above expression, if we wish to emphasize the influence of the distance:

$$d_1(i,j) = \frac{1}{(\sum_{l=1}^{a} \text{diff}(A_l, R_i, I_j))^2} \tag{21}$$

The differences in estimations can be substantial although the average correlation coefficients between estimations and ranks over regression data sets from UCI obtained with RReliefF are high as shown in Table 3.

Table 3: Linear correlation coefficients between estimations and ranks over 8 UCI regression data sets. We compare RReliefF using Eqs. (11), (20) and (21).

Problem	Eqs. (11) and (20)		Eqs. (11) and (21)		Eqs. (20) and (21)	
	ρ	r	ρ	r	ρ	r
Abalone	0.969	0.974	0.991	0.881	0.929	0.952
Auto-mpg	-0.486	0.174	0.389	-0.321	0.143	0.357
Autoprice	0.844	0.775	0.933	0.819	0.749	0.945
CPU	0.999	0.990	0.990	0.943	1.000	0.943
Housing	0.959	0.830	0.937	0.341	0.181	0.769
Servo	0.988	0.999	0.985	0.800	1.000	0.800
Wisconsin	0.778	0.842	0.987	0.645	0.743	0.961
Average	0.721	0.798	0.888	0.587	0.678	0.818

The reason for substantial deviation in Auto-mpg problem is sensibility of the algorithm concerning the number of nearest neighbors when using actual distances. While with expression (11) we exponentially decreases influence according to the number of nearest neighbors, Eqs. (20) and (21) use inverse of the distance and also instances at a greater distance may have a substantial influence. With actual distances and 70 nearest instances in this problem we get myopic estimate which is uncorrelated to non-myopic estimate. So, if we are using actual distances we have to use a moderate number of the nearest neighbors or test several settings for it.

4.4   Number of Nearest Neighbors

While the number of nearest neighbors used is related to the distance as described above there are still some other issues to be discussed, namely how sensitive Relief algorithms are to the number of nearest neighbors used (lines 4,5, and 6 in Figure 2 and line 4 in Figure 3). The optimal number of nearest neighbors used is problem dependent as we illustrate in Figure 5 which shows ReliefF's evaluation of four important and one of the random attributes in Boolean domain defined as:

$$\text{Bool} - \text{Simple}: \qquad C = (A_1 \oplus A_2) \vee (A_3 \wedge A_4). \tag{22}$$

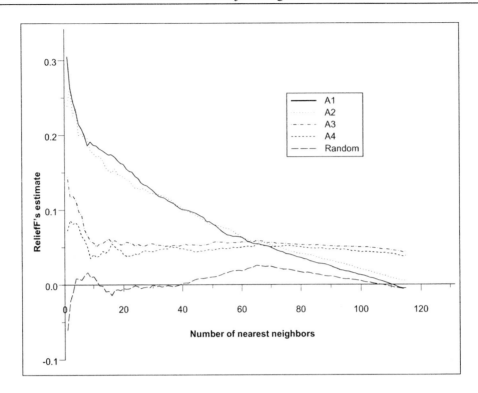

Figure 5: ReliefF's estimates and the number of nearest neighbors.

We know that A_1 and A_2 are more important for determination of the class value than A_3 and A_4 (the attributes' values are equiprobable). ReliefF recognizes this with up to 60 nearest neighbors (there are 200 instances). With more that 80 nearest neighbors used the global view prevails and the strong conditional dependency between A_1 and A_2 is no longer detected. If we increase the number of instances from 200 to 900 we obtain similar picture as in Figure 5, except that the crossing point moves from 70 to 250 nearest neighbors.

The above example is quite illustrative: it shows that ReliefF is robust in the number of nearest neighbors as long as it remains relatively small. If it is too small it may not be robust enough, especially with more complex or noisy concepts. If the influence of all neighbors is equal disregarding their distance to the query point the proposed default value is 10 [15]. If we do take distance into account we use 70 nearest neighbors with exponentially decreasing influence ($\sigma = 20$ in Eq. (11). In a similar problem with CM algorithm [10] it is suggested that using $\log n$ nearest neighbors gives satisfactory results in practice. However we have to emphasize that this is problem dependent and especially related to the problem complexity, the amount of noise and the number of available instances.

Another solution to this problem is to evaluate attributes for all possible numbers of nearest neighbors and take the highest score of each attribute as its final result. In this way we avoid the danger of accidentally missing an important attribute. Because all attributes receive somewhat higher score we risk that some differences would be blurred and we increase the computational complexity. The former risk can be resolved later on in the process of investigating the domain by producing a graph similar to Figure 5 showing dependencies of ReliefF's estimates on the number of nearest neighbors. The computational complexity increases from $O(m \cdot n \cdot a)$ to $O(m \cdot n \cdot (a + \log n))$ due to sorting of the instances with decreasing distance. In the algorithm we have to do also some additional bookkeeping, e.g., keep the score for each attribute and each number of nearest instances.

4.5 Sample Size and Number of Iterations

Output of Relief algorithms are actually statistical estimates i.e., the algorithms collect the evidence for (non)separation of similar instances by the attribute across the problem space. To provide reliable evaluation the coverage of the problem space must be appropriate. The sample has to cover enough representative boundaries between the prediction values. There is an obvious trade off between using more instances and the efficiency of computation. Wherever we have large data sets, sampling is one of possible solutions to make problem tractable. If the data set is reasonably large and we want to speed-up computations we suggest selection of all available instances (n in complexity calculations), and control the number of iterations with parameter m (line 2 in Figures 2 and 3). As it is non trivial to select a representative sample of the unknown problem space our decision is in favor of a (possibly) sparse coverage of the more representative space rather than a dense coverage of the (possibly) non-representative sample.

Figure 6 illustrates the behavior of RReliefF's estimates changing with the number of iterations on Cosinus-Lin data set. This problem consisting of 1000 examples is is a nonlinear dependency with cosine stretched over two periods, multiplied by the linear combination of two attributes:

$$\text{Cosinus} - \text{Lin}: \qquad f = \cos(4\pi A_1) \cdot (-2A_2 + 3A_3) \tag{23}$$

The attributes are numerical with values from 0 to 1, there are also 10 random attributes.

We see that after the initial variation at around 20 iterations the estimates settle to stable values, except for difficulties at detecting differences e.g., the quality difference between A_1 and A_3 is not resolved until around 300 iterations (A_1 within the cosine function controls the sign of the expression, while A_3 with the coefficient 3 controls the amplitude of the function). We should note that this is quite typical behavior and usually we get stable estimates after 20-50 iterations. However if we want to refine the estimates we have to iterate further on. The question of how much more iterations we need is problem dependent. We try to answer this question for some chosen problems.

We investigate how the number of available examples and the number of iterations influences the quality evaluations. For this purpose we generated a number of artificial domains with different number of examples, estimate the quality of the attributes and observe their s and u values. For each data size we repeat the experiment 10 times and test hypotheses that s and u are larger than zero. By setting the significance threshold to 0.01 we get the

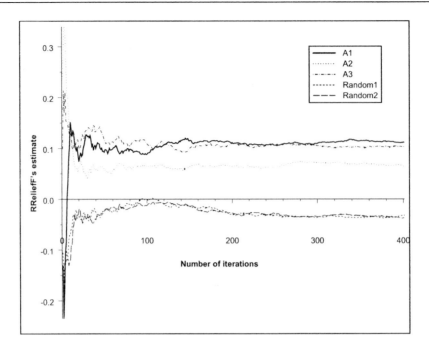

Figure 6: RReliefF's estimates and the number of iterations (m) on Cosinus-Lin data set.

limiting number of examples needed to separate all important from unimportant attributes and at least one important from unimportant attributes. We are reporting these numbers of examples in Table 4. Reported numbers are an indication of the robustness of the estimators.

We can observe that the number of required examples is increasing with complexity of problems within each group (e.g., Modulo-5, Modulo-10, LinInc, ...).

It is also interesting to compare Relief algorithms with other estimation measures (Gain ratio and MSE). Apart from the fact that some dependencies cannot be detected by these measures (sign '-' in Table 4), in other (easier) problems we see that Relief algorithms need approximately the same number of examples to detect the best important attribute and slightly more examples to detect the worst important attribute. This can be explained with the intrinsic properties of the estimators: while myopic attribute estimators evaluate each attribute independently, Relief algorithms evaluate them in the context of other attributes, i.e., all important attributes share the positive estimate and better attributes get more.

We check how many iterations are necessary for Relief algorithms to return good quality estimates by using the same setting as above. Table 5 gives a minimal number of iterations needed in each problem so that s and u are positive with probability > 0.99. The columns labelled '#Ex' are results for the number of examples fixed to the number of examples from the first s column in Table 4, and two columns labelled '$10\times$ #Ex' contain results for 10 times that many examples.

There are two interesting observations. The first is that even with the minimal number

Table 4: Results of varying the number of examples.

Name	ReliefF		Gain r.	
	s	u	s	u
Bool-Simple	100	50	-	40
Modulo-2-2-c	70	60	-	-
Modulo-2-3-c	170	110	-	-
Modulo-2-4-c	550	400	-	-
Modulo-5-2-c	150	130	-	-
Modulo-5-3-c	950	800	-	-
Modulo-5-4-c	5000	4000	-	-
Modulo-10-2-c	400	400	-	-
Modulo-10-3-c	4000	3000	-	-
Modulo-10-4-c	25000	22000	-	-
MONK-1	100	30	-	20
MONK-3	250	10	250	20
	RReliefF		MSE	
Modulo-5-2-r	60	50	-	-
Modulo-5-3-r	400	350	-	-
Modulo-5-4-r	2000	1600	-	-
Modulo-10-2-r	90	80	-	-
Modulo-10-3-r	800	600	-	-
Modulo-10-4-r	7000	4000	-	-
Fraction-2	80	70	-	-
Fraction-3	650	400	-	-
Fraction-4	4000	3000	-	-
LinInc-2	60	10	70	20
LinInc-3	400	20	150	10
LinInc-4	2000	40	450	10
LinEq-2	50	40	20	20
LinEq-3	180	50	50	20
LinEq-4	350	70	70	30
Cosinus-Lin	300	40	-	200
Cosinus-Hills	550	300	4000	2000

of examples needed for distinguishing two sets of attributes we do not need to do that many iterations and the actual number of iterations is sometimes quite low. And secondly, if we have an abundance of examples available than we can afford to do only very little iterations as seen from the '10× #Ex' columns in Table 5.

Table 5: Results of varying the number of iterations.

Name	#Ex		10× #Ex	
	s	u	s	u
Bool-Simple	83	1	21	1
Modulo-2-2-c	11	9	1	1
Modulo-2-3-c	144	15	2	1
Modulo-2-4-c	151	18	3	1
Modulo-5-2-c	51	35	1	1
Modulo-5-3-c	390	37	3	2
Modulo-5-4-c	3674	579	10	2
Modulo-10-2-c	31	26	1	1
Modulo-10-3-c	1210	263	3	1
Modulo-10-4-c	23760	3227	25	4
MONK-1	36	1	2	1
MONK-3	163	1	56	1
Modulo-5-2-r	15	12	2	1
Modulo-5-3-r	57	23	4	1
Modulo-5-4-r	975	182	17	3
Modulo-10-2-r	55	37	3	3
Modulo-10-3-r	316	164	16	9
Modulo-10-4-r	5513	929	126	17
Fraction-2	47	32	3	3
Fraction-3	379	51	29	11
Fraction-4	1198	78	138	13
LinInc-2	48	4	6	3
LinInc-3	109	4	52	3
LinInc-4	940	3	456	2
LinEq-2	17	9	3	2
LinEq-3	96	10	16	7
LinEq-4	215	13	45	9
Cosinus-Lin	188	10	16	7
Cosinus-Hills	262	51	110	30

4.6 Detecting Dependencies

Let we answer the question what sort of dependencies the Relief algorithms can detect. For that matter we define different high order dependencies and test how ReliefF and RReliefF work on them.

4.6.1   Sum by Modulo Concepts

We start our presentation of abilities of ReliefF and RReliefF with the concepts based on summation by modulo. Sum by modulo p problems are integer generalizations of parity concept, which is a special case where attributes are Boolean and the class is defined by modulo 2. In general, each Modulo-p-I problem is described by a set of attributes with integer values in the range $[0, p)$. The predicted value $\tau(X)$ is the sum of I important attributes by modulo p.

$$\text{Modulo}-\text{p}-\text{I}: \quad \tau(X) = (\sum_{i=1}^{I} X_i) \bmod p \tag{24}$$

We use also Fraction-I problem which is the same as the Modulo-∞-I with prediction and attributes normalized to $[0, 1]$. It can be described as floating point generalization of the parity concepts of order I with predicted values being the fractional part of the sum of I important attributes:

$$\text{Fraction}-\text{I}: \quad f = \sum_{j=1}^{I} A_j - \lfloor \sum_{j=1}^{I} A_j \rfloor \tag{25}$$

Let us start with the base case i.e., Boolean problems ($p = 2$). As an illustrative example we will show problems with parity of 2-8 attributes ($I \in [2, 8]$) on the data set described with 9 attributes and 512 examples (a complete description of the domain). Figure 7 shows s curve for this problem (u curve is identical as we have a complete description of a domain). In this and all figures below each point on the graph is an average of 10 runs.

We can see that separability of the attributes is decreasing with increasing difficulty of the problem for parity orders of 2, 3, and 4. At order 5 when more than half of the attributes are important the separability becomes negative i.e., we are no longer capable of separating the important from unimportant attributes. The reason is that we are using more than one nearest neighbor (one nearest neighbor would always produce positive s curve on this noiseless problem) and as the number of peaks in the problem increases with 2^I, and the number of examples remains constant (512) we are having less and less examples per peak. At $I = 5$ when we get negative s the number of nearest examples from the neighboring peaks with distance 1 (different parity) surpasses the number of nearest examples from the target peak. An interesting point to note is when $I = 8$ (there is only 1 random attribute left) and s becomes positive again. The reason for this is that the number of nearest examples from the target peak and neighboring peaks with distance 2 (with the same parity!) surpasses the number of nearest examples from neighboring peaks with distance 1 (different parity).

A sufficient number of examples per peak is crucial for reliable attribute evaluation with ReliefF as we show in Figure 8. The bottom s and u curves show exactly the same problem as above (in Figure 7) but in this case the problem is not described with all 512 examples but rather with 512 randomly generated examples. The s scores are slightly lower than in Figure 7 as we have in effect decreased the number of different examples (to 63.2% of the total). The top s and u curves show the same problem but with 8 times more examples (4096). We can observe that with that many examples the separability for all problems is positive.

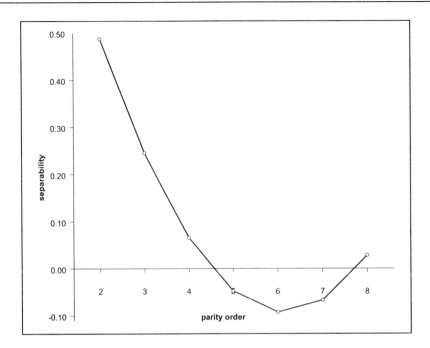

Figure 7: Separability on parity concepts of orders 2 to 8 and all 512 examples.

In the next problem p increases while the number of important attributes and the number of examples are fixed (to 2 and 512, respectively). Two curves at the bottom of Figure 9 show separability (usability is very similar and is omitted due to clarity) for the classification problem (there are p classes) and thus we can see the performance of ReliefF. The attributes can be treated as nominal or numerical, however, the two curves show similar behavior i.e., separability is decreasing with increasing modulo. This is expected as the complexity of problems is increasing with the number of classes, attribute values, and peaks. The number of attributes values and classes is increasing with p, while the number of peaks is increasing with p^l (polynomial increase). Again, more examples would shift positive values of s further to the right. A slight but important difference between separability for nominal and numerical attributes shows that numerical attributes convey more information in this problem. Function diff is 1 for any two different nominal attributes while for numerical attributes diff returns the relative numerical difference which is more informative.

The same modulo problems can be viewed as regression problems and the attributes can be again interpreted as nominal or numerical. Two curves at the top of Figure 9 shows separability for the modulo problem formulated as regression problem (RReliefF is used). We get positive s values for larger modulo compared to the classification problem and if the attributes are treated as numerical the separability is not decreasing with modulo at all. The reason is that classification problems were actually more difficult. We tried to predict p separate classes (e.g., results 2 and 3 are completely different in classification) while in

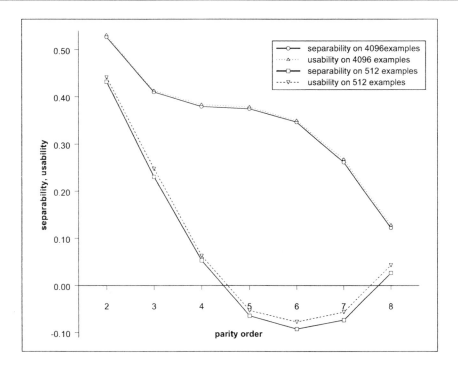

Figure 8: Separability and usability on parity concepts of orders 2 to 8 and randomly sampled 512 or 4096 examples.

regression we model numerical values (2 and 3 are different relatively to the scale).

Another interesting problem arises if we fix modulo to a small number (e.g., $p = 5$) and vary the number of important attributes. Figure 10 shows s curves for 4096 examples and 10 random attributes. At modulo 5 there are no visible differences in the performance for nominal and numerical attributes therefore we give curves for nominal attributes only. The s curves are decreasing rapidly with increasing I. Note that the problem complexity (number of peaks) is increasing with p^I (exponentially).

Modulo problems are examples of difficult problems. Note that impurity-based measures such as Gain ratio, are not capable of separating important from random attributes for any of the above described problems.

4.6.2   MONK's Problems

We present results of attribute evaluation on well known and popular MONK's problems [35] which consist of three binary classification problems based on common description by six attributes. A_1, A_2, and A_4 can take the values of 1, 2, or 3, A_3 and A_6 can take the values 1 or 2, and A_5 can take one of the values 1, 2, 3, or 4. Altogether there are 432 examples but we randomly generated training subsets of the original size to evaluate the attributes in each of the tasks, respectively.

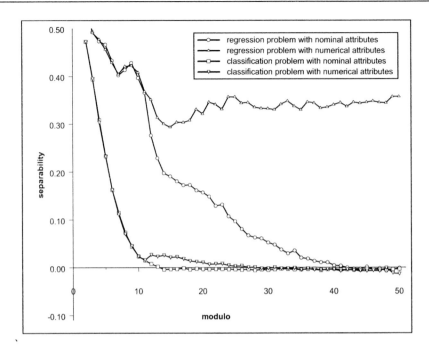

Figure 9: Separability for ReliefF and RReliefF on modulo classification and regression problems with changing modulo.

- **Problem M_1:** 124 examples for the problem: $(A_1 = A_2) \vee (A_5 = 1)$

- **Problem M_2:** 169 examples for the problem: exactly two attributes have value 1

- **Problem M_3:** 122 examples with 5 % noise (misclassifications) for the problem: $(A_5 = 3 \wedge A_4 = 1) \vee (A_5 \neq 4 \wedge A_2 \neq 3)$.

We generated 10 random samples of specified size for each of the problems and compared estimates of ReliefF and Gain ratio. Table 6 reports results.

For the first problem we see that ReliefF separates the important attributes (A_1, A_2, and A_5) from unimportant ones while Gain ratio does not recognize A_1 and A_2 as important attributes in this task.

In the second problem where all the attributes are important ReliefF assigns them all positive weights. It favors attributes with less values as they convey more information. Gain ratio does the opposite: it favors attributes with more values.

In the third problem ReliefF and gain ratio behave similarly: they separate important attributes from unimportant ones and rank them equally (A_2, A_5, and A_4).

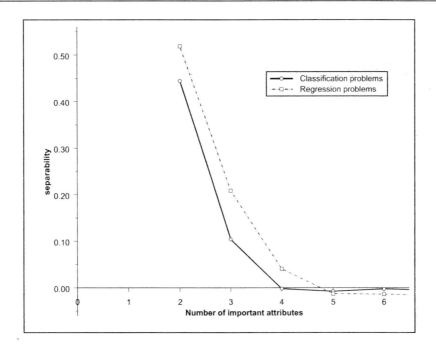

Figure 10: Separability for ReliefF and RReliefF on modulo 5 problems with changing the number of important attributes.

4.6.3 Linear and Nonlinear Problems

In typical regression problems linear dependencies are mixed with some nonlinear dependencies. We investigate problems of such type. The problems are described with numerical attributes with values from the $[0,1]$ interval and 1000 instances. Besides I important attributes there are also 10 random attributes in each problem.

We start with linear dependencies and create problems of the form:

$$\text{LinInc}-\text{I}: \qquad \tau = \sum_{j=1}^{I} j \cdot A_j \qquad (26)$$

The attributes with larger coefficient have stronger influence on the prediction value and should be estimated as more important.

Table 7 reports quality evaluations of attributes for RReliefF and MSE. We see that for small differences between importance of attributes both RReliefF and MSE are successful in recognizing this and ranking them correctly. When the differences between the importance of attributes become larger (in LinInc-4 and LinInc-5) it is possible due to random fluctuations in the data one of the random attributes is estimated as better than the least important informative attribute (A_1). This happens in LinInc-4 for RReliefF and in LinInc-5 for MSE. The behavior of s and u are illustrated in Figure 11.

Table 6: Evaluations of attributes in three MONK's databases for ReliefF and Gain ratio. The results are averages over 10 runs.

Attribute	M_1		M_2		M_3	
	ReliefF	Gain r.	ReliefF	Gain r.	ReliefF	Gain r.
A_1	0.054	0.003	0.042	0.006	-0.013	0.004
A_2	0.056	0.004	0.034	0.006	0.324	0.201
A_3	-0.023	0.003	0.053	0.001	-0.016	0.003
A_4	-0.016	0.007	0.039	0.004	-0.005	0.008
A_5	0.208	0.160	0.029	0.007	0.266	0.183
A_6	-0.020	0.002	0.043	0.001	-0.016	0.003

Table 7: Evaluations of the best random attribute (R_{best}) and all informative attributes in LinInc-I problems for RReliefF (RRF) and MSE. RReliefF assigns higher scores and MSE assigns lower scores to better attributes.

Attr.	LinInc-2		LinInc-3		LinInc-4		LinInc-5	
	RRF	MSE	RRF	MSE	RRF	MSE	RRF	MSE
R_{best}	-0.040	0.424	-0.023	1.098	-0.009	2.421	-0.010	4.361
A_1	0.230	0.345	0.028	1.021	-0.018	2.342	-0.007	4.373
A_2	0.461	0.163	0.154	0.894	0.029	2.093	0.014	4.181
A_3			0.286	0.572	0.110	1.833	0.039	3.777
A_4					0.180	1.508	0.054	3.380
A_5							0.139	2.837

Another form of linear dependencies we are going to investigate is

$$\text{LinEq}-\text{I}: \qquad \tau = \sum_{j=1}^{I} A_j \qquad (27)$$

Here all the attributes are given equal importance and we want to see how many important attributes can we afford. Figure 12 shows separability and usability curves for RReliefF and MSE.

We see that separability is decreasing for RReliefF and becomes negative with 10 important attributes. This is not surprising considering properties of Relief: each attribute gets its weight according to the portion of explained function values (see Section 2.4) so by increasing the number of important attributes their weights decrease and approaches zero. The same is true for RReliefF's usability which, however, becomes negative much later. MSE estimates each attribute separately and is therefore not susceptible to this kind of defects, however, by increasing the number of important attributes the probability to assign

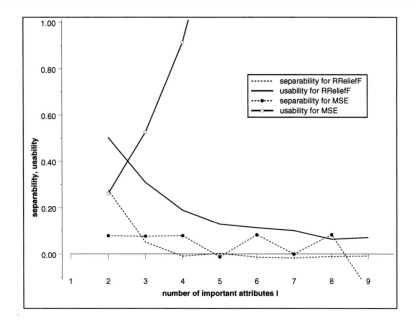

Figure 11: Separability and usability on LinInc concepts for 1000 examples.

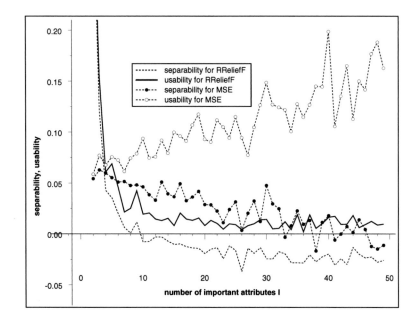

Figure 12: Separability and usability on LinEq concepts for 1000 examples.

one of them a low score increases and so s curve becomes negative. If we increase the number of examples to 4000, RReliefF's s curve becomes negative at 16 important attributes while the behavior of MSE does not change.

We end our analysis with non-linear dependencies where the prediction is defined as

$$\text{Cosinus} - \text{Hills}: \qquad \tau = \cos 2\pi(A_1^2 + 3A_2^2). \tag{28}$$

In this problem there are 1000 examples and the attributes have values in the range $[-1, 1]$. Beside two important attributes there are 10 random attributes. Figure 13 visualizes this domain. We see that prediction values are symmetric along both important attributes with 3 times more hills along A_2.

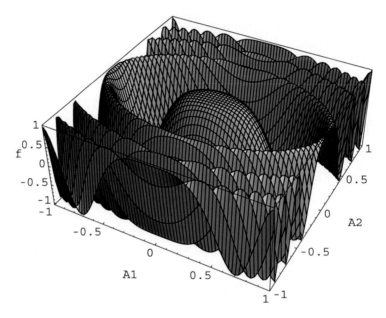

Figure 13: Visualization of Cosinus-Hills problem where: $f = \cos 2\pi(A_1^2 + 3A_2^2)$

The quality evaluations of RReliefF and MSE are contained in Table 8. We see that RReliefF recognizes A_1 and A_2 as the important attributes and separates them from the random attributes, while MSE is not successful in this task.

4.7 Adding Noise by Changing Predicted Value

We check robustness of Relief algorithms concerning by using the same setting as before with the number of examples from the first s column in Table 4. We added noise to the data sets by changing certain percent of predicted values to a random value. We varied the noise from 0 to 100% in 5% steps. Table 9 gives the maximal percentage of corrupted prediction values where s and u were still positive with high probability (> 0.99).

Table 8: Evaluations of two important attributes and the best of random attributes (R_{best}) in Cosinus-Hills problem for RReliefF (RRF) and MSE. RReliefF assigns higher scores and MSE assigns lower scores to better attributes.

Attr.	RRF	MSE
A_1	0.0578	0.5001
A_2	0.0398	0.5003
R_{best}	0.0194	0.4992

We can observe that Relief algorithms are quite robust concerning the noise in all the problems. When we increase the number of examples 10 times we are able to randomly set more than 80% of values in every problem. Comparison with myopic measures (Gain ratio and MSE) is included in two right-hand columns. We see that in problems where these measures are successful their noise tolerance is comparable with Relief algorithms.

4.8   Adding More Random Attributes

ReliefF and RReliefF are context sensitive and take all attributes into account when evaluating their quality. They are therefore more sensitive to abundance of random attributes than myopic measures, which evaluate each attribute separately. We test this sensitivity with similar settings as before (default parameters, the number of examples from the first s column in Table 4) and add various numbers of randomly generated attributes.

Table 10 summarizes results (we consider MONK's data sets fixed and did not include them in this test). The two columns give the maximal number of random attributes that can be added before s and u are no more positive with probability > 0.99. We see that while we can afford only a moderate number of random attributes to separate all important attributes from unimportant ones, this number is much higher for the best estimated important attribute. Again this confirms the intrinsic behavior of Relief algorithms: they evaluate each attribute in the context of other attributes and better attributes get higher score. This can be clearly seen in Lin-Inc and Cosinus-Lin domains where the important attributes are not all equally important.

Since MSE and Gain ratio are less sensitive to this kind of noise (only to the extent that the probability of virtually 'good' random attribute increases with more random attributes) we did not include them into this experiment.

4.9   Number of Important Attributes

Although the answer to the question, how many important attributes in the domain Relief algorithms can handle, is already contained in previous sections we will summarize the results here and give some explanations for this important practical issue.

Table 9: Results of adding noisy prediction value.

Name	ReliefF		Gain r.	
	s	u	s	u
Bool-Simple	10	45	-	20
Modulo-2-2-c	30	50	-	-
Modulo-2-3-c	5	50	-	-
Modulo-2-4-c	20	50	-	-
Modulo-5-2-c	5	5	-	-
Modulo-5-3-c	15	35	-	-
Modulo-5-4-c	10	50	-	-
Modulo-10-2-c	20	20	-	-
Modulo-10-3-c	25	55	-	-
Modulo-10-4-c	2 10	40	-	-
MONK-1	30	70	-	55
MONK-3	25	75	5	75
	RReliefF		MSE	
Modulo-5-2-r	25	50	-	-
Modulo-5-3-r	20	45	-	-
Modulo-5-4-r	15	40	-	-
Modulo-10-2-r	25	40	-	-
Modulo-10-3-r	10	40	-	-
Modulo-10-4-r	5	40	-	-
Fraction-2	20	40	-	-
Fraction-3	15	35	-	-
Fraction-4	10	35	-	-
LinInc-2	15	50	10	50
LinInc-3	10	65	20	70
LinInc-4	5	80	35	85
LinEq-2	30	60	25	45
LinEq-3	25	40	30	50
LinEq-4	10	40	30	45
Cosinus-Lin	15	50	-	40
Cosinus-Hills	5	20	-	-

If important attributes are not equally important then Relief algorithms ideally would share the credit among them in proportion of the explained concept (see Eq. (12)). In practice (with limited number of examples) less important attributes get less than this. The reason for this is that the differences in predicted value caused by less important attribute are overshadowed by larger changes caused by more important attributes. The example of such

Table 10: Results of adding random attributes.

Name	s	u
Bool-Simple	12	150
Modulo-2-2-c	20	45
Modulo-2-3-c	14	25
Modulo-2-4-c	12	20
Modulo-5-2-c	17	30
Modulo-5-3-c	20	45
Modulo-5-4-c	13	25
Modulo-10-2-c	110	160
Modulo-10-3-c	50	85
Modulo-10-4-c	11	30
Modulo-5-2-r	19	20
Modulo-5-3-r	16	20
Modulo-5-4-r	11	19
Modulo-10-2-r	14	14
Modulo-10-3-r	13	19
Modulo-10-4-r	13	25
Fraction-2	18	25
Fraction-3	14	20
Fraction-4	13	18
LinInc-2	20	> 10000
LinInc-3	16	> 10000
LinInc-4	14	> 10000
LinEq-2	45	140
LinEq-3	25	170
LinEq-4	16	300
Cosinus-Lin	20	3200
Cosinus-Hills	12	20

behavior can be seen in Figure 11 and Table 7 where attributes are not equally important. The separability was no longer positive for RReliefF with four attributes in these conditions. Usability on the other hand is robust and remains positive even with several hundred of important attributes.

If all attributes are equally important then we can afford somewhat larger number of attributes still to get positive separability as illustrated in Figure 12. Note, however, that the exact number is strongly dependent on sufficient number of examples to cover the problem space adequately. Again the usability stays positive even for several hundreds of important attributes.

In short, Relief algorithms tend to underestimate less important attributes while recognition of more important attributes in not questionable even in difficult conditions.

4.10 Duplicate Attributes

We demonstrate behaviors of Relief algorithms in the presence of duplicate attributes. As a baseline we take a simple Boolean XOR problem with 2 important and 10 random attributes with 1000 examples for which ReliefF returns the following estimates:

I_1	I_2	R_{best}
0.445	0.453	-0.032

Now we duplicate the first important attribute I_1 and get:

I_{1,C_1}	I_{1,C_2}	I_2	R_{best}
0.221	0.221	0.768	-0.015

Both copies of I_1 now share the estimate and I_2 gets some additional credit. Before we explain these estimates we try with 3 copies of the first important attribute:

I_{1,C_1}	I_{1,C_2}	I_{1,C_3}	I_2	R_{best}
0.055	0.055	0.055	0.944	-0.006

and also with 10 copies:

I_{1,C_1}	I_{1,C_2}	...	$I_{1,C_{10}}$	I_2	R_{best}
0.000	0.000	...	0.000	1.000	0.006

The reason for this behavior is that the additional copies of the attribute change the problem space in which the nearest neighbors are searched. Recall that for each instance the algorithm searches nearest neighbors with the same prediction and nearest neighbors with different prediction and then updates the estimates according to the values of diff function (see Eq. 1). The updates occur only when values of diff is non-zero i.e., when the attribute's values are different. The difference in c times multiplied attribute causes that the instances are now at least on the distance c which causes that this instance will slip out of the near neighborhood of the observed instance. As a consequence the odds of the multiplied attribute to get any update are diminishing and its quality evaluations converges towards zero. The reverse is true for non-multiplied attributes: they get more frequently into the near neighborhood and are more frequently updated so the important attributes get higher score as deserved.

If all important attributes are multiplied (not very likely in practical problems) then this effect disappears because the distances are symmetrically stretched. Here is an example with 10 copies of both important attributes:

I_{1,C_1}	...	$I_{1,C_{10}}$	I_{2,C_1}	...	$I_{2,C_{10}}$	R_{best}
0.501	...	0.501	0.499	...	0.4999	-0.037

The conclusion is that Relief algorithms are sensitive to duplicated attributes because they change the problem space. We can also say that Relief algorithms give credit to attributes according to the amount of explained concept and if several attributes have to share the credit it is smaller for each of them. Myopic evaluators such as Gain ratio and MSE, are not sensitive to duplicate attributes.

5   Applications

In previous sections we browsed through the practical issues on the use of Relief algorithms. In this section we try to give a short overview of their applications. We begin with the feature subset selection and feature weighting which were the initial purposes of Relief algorithm and then continue with the use of ReliefF and RReliefF in tree based models, cost-sensitive problems, discretization of attributes, and in inductive logic programming. Since the list of applications is by no means exhaustive, we end the sections with a few ideas for further work.

5.1   Feature Subset Selection and Feature Weighting

Originally, the Relief algorithm was used for feature subset selection [14, 13] and it is considered one of the best algorithms for this purpose [6]. Feature subset selection is a problem of choosing a small set of attributes that ideally is necessary and sufficient to describe the target concept.

To select the set of the most important attributes [13] introduced significance threshold θ. If weight of a given attribute is below θ it is considered unimportant and is excluded from the resulting set. Bounds for θ were proposed i.e., $0 < \theta \leq \frac{1}{\sqrt{\alpha m}}$, where α is the probability of accepting an irrelevant feature as relevant and m is the number of iterations used (see Figure 1). The upper bound for θ is very loose and in practice much smaller values can be used.

Feature weighting is an assignment of a weight (importance) to each feature and can be viewed as a generalization of feature subset selection in the sense that it does not assign just binary weights (0-1, include-exclude) to each feature but rather an arbitrary real number can be assigned to it. If Relief algorithms are used in this fashion then we do not need a significance threshold but rather use their weights directly. ReliefF was tested as the feature weighting method in the lazy learning [36] and was found to be very useful.

5.2   Building Tree Based Models

In learning tree based models (decision or regression trees) in a top down manner we need an attribute evaluation procedure in each node of the tree to determine the appropriate split. Commonly used evaluators for this task are impurity based e.g., Gini index [2] or Gain ratio [24] in classification and mean squared error [2] in regression. While these evaluators are myopic and cannot detect conditional dependencies between attributes they also have inappropriate bias concerning multi-valued attributes [16]. ReliefF was successfully employed in classification [17, 18] and RReliefF in regression problems [28, 29]. Relief algorithms perform as good as myopic measures if there are no conditional dependencies among the attributes and clearly surpass them if there are strong dependencies. We claim that when faced with an unknown data set it is unreasonable to assume that it contains no strong conditional dependencies and rely only on myopic attribute evaluators. Furthermore, using impurity based evaluator near the fringe of the decision tree leads to suboptimal splits concerning accuracy and switch to accuracy has been suggested as a remedy [3, 19]. It was shown [30] that ReliefF in decision trees as well as RReliefF in regression trees do not need such switch as they contain it implicitly.

We employed Relief algorithms to guide the constructive induction process during growing of the trees. Only the most promising attributes were selected for construction and various operators were applied on them (conjunction, disjunction, summation, product). The results are good and in some domains the obtained constructs provided additional insight into the domain [4].

We also observed in our experiences with machine learning applications (medicine, ecology) that trees produced with Relief algorithms are more comprehensible for human experts. Splits selected by them seem to mimic human's partition of the problem which we explain with the interpretation of Relief's weights as the portion of the explained concept (see Section 2.4).

5.3   Cost-Sensitiveness

Typically, machine learning algorithms are not designed to take the non-uniform cost of misclassification into account. A general account of attribute evaluation with non-uniform cost od misclassification has been given in [31]. For Relief a number of solutions have been proposed based on different diff functions and their weighting.

5.4   Discretization of Attributes

Discretization divides the values of a numerical attribute into a number of intervals. Each interval can then be treated as one value of the new nominal attribute. The purpose of discretization, which is viewed as an important preprocessing step in machine learning is multiple: some algorithms cannot handle numerical attributes and need discretization, it reduces computational complexity and the splits may convey important information. While Relief algorithms are used for discretization in our tree learning system, the discretization capability was analyzed separately for ReliefF in [26]. The discretization on artificial data sets showed that conditionally dependent attributes might have important boundaries, which cannot be detected by myopic measures.

5.5   Use in ILP and with Association Rules

In inductive logic programming a variant of ReliefF algorithm was used to evaluate the quality of literals, which are candidates for augmenting the current clause under construction when inducing the first order theories with a learning system [23]. To adjust to specifics of estimating literals diff function has to be changed appropriately. The modified diff function is asymmetric. The learning system, which uses this version of ReliefF, obtains good results on many learning problems.

The use of ReliefF together with an association rules based classifier [12] is connected also with feature subset selection. The adaptation of the algorithm to association rules changes diff function in a similar way as ReliefF in ILP.

5.6   Further Work

Although there has been a lot of work done with Relief algorithms we think that there are plenty left to do. We discuss some ideas concerning parallelization, use as anytime algorithm, applications in time series, and in meta learning.

5.6.1   Parallelization of Relief Algorithms

While Relief algorithms are computationally more complex than some other (myopic) attribute evaluation measures they also have a possible advantage hat they can be naturally split into several independent tasks which is a prerequisite for successful parallelization of an algorithm. Each iteration of the algorithm is a natural candidate for a separate process, which would turn Relief into the fine-grained parallel algorithm.

5.6.2   Anytime Algorithm and Time Series

Anytime algorithm is an algorithm, which has a result available at any time and with more time or more data it just improves the result. We think that Relief algorithms can be viewed as anytime algorithms, namely in each iteration they refine the result. As our experiments show (see Table 5) even after only few little iterations Relief algorithms already produce sensible results. The problems we may encounter if we add more data during the processing, are in proper normalizations of updated weights.

Another idea worth considering is to observe changes in quality weights when we add more and more data. In this way it might be possible to detect some important phases in the time series e.g., a change of context.

5.6.3   Meta Learning

One of the approaches to meta learning is to obtain a number of features concerning the data set and available learning algorithms and then try to figure out which algorithm is the most appropriate for the problem. Some of these features are obvious e.g., the number and type of attributes, the number of examples and classes, while others are more complex stemming from different data characteristics. We think that quality evaluations of the attributes produced by Relief algorithms could be a very useful source of meta data. The estimate of the best attribute, difference between the best and the worst estimate, correlation between the Relief's estimate and estimates of myopic functions are examples of features, which could help to appropriately describe the data set at hand.

6   Conclusion

We have investigated features, parameters and uses of the Relief family of algorithms. Relief algorithms are general and successful attribute evaluators and are especially good in detecting conditional dependencies. They provide a unified view on attribute evaluation in regression and classification and their quality estimates also have a natural interpretation.

We explained how and why they work, presented some of their properties and analyzed their parameters and issues which may occur during their practical use. To help at their successful application we showed what kind of dependencies they detect, how do they scale up to large number of examples and features, how to sample data for them, how robust are they regarding the noise and how irrelevant and duplicate attributes influence their output. In short, Relief algorithms are robust and noise tolerant. Their somewhat larger computational complexity can, in huge tasks, be alleviated by their nature: as anytime algorithms and intrinsic parallelism.

And finally, a learning system which incorporates ReliefF and RReliefF is freely available from author's web site http://lkm.fri.uni-lj.si/rmarko/

References

[1] J. L. Bentley. Multidimensional binary search trees used for associative searching. *Communications of the ACM*, **15**(9):509–517, 1975.

[2] L. Breiman, J. H. Friedman, R. A. Olshen, and C. J. Stone. Classification and regression trees. Wadsworth Inc., Belmont, California, 1984.

[3] C. E. Brodley. Automatic selection of split criterion during tree growing based on node location. In A. Prieditis and S. Russell, editors, *Machine Learning: Proceedings of the Twelfth International Conference (ICML'95)*, pages 73–80. Morgan Kaufmann, San Francisco, 1995.

[4] A. Dalaka, B. Kompare, M. Robnik-Šikonja, and S. Sgardelis. Modeling the effects of environmental conditions on apparent photosynthesis of Stipa bromoides by machine learning tools. *Ecological Modelling*, **129**:245–257, 2000.

[5] K. Deng and A. W. Moore. Multiresolution instance-based learning. In *Proceedings of the International Joint Conference on Artificial Intelligence (IJCAI'95)*, pages 1233–1239. Morgan Kaufmann, 1995.

[6] T. G. Dietterich. Machine learning research: Four current directions. *AI Magazine*, **18** (4):97–136, 1997.

[7] P. Domingos. Context-sensitive feature selection for lazy learners. *Artificial Intelligence Review*, **11**:227–253, 1997.

[8] J. H. Friedman, J. L. Bentley, and R. A. Finkel. An algorithm for finding best matches in logarithmic expected time. Technical Report STAN-CS-75-482, Stanford University, 1975.

[9] S. J. Hong. Use of contextual information for feature ranking and discretization. Technical Report RC19664, IBM, July 1994.

[10] S. J. Hong. Use of contextual information for feature ranking and discretization. *IEEE transactions on knowledge and data engineering*, **9**(5):718–730, 1997.

[11] E. B. Hunt, J. Martin, and P. J. Stone. Experiments in Induction. Academic Press, New York, 1966.

[12] V. Jovanoski and N. Lavrač. Feature subset selection in association rules learning systems. In M. Grobelnik and D. Mladenič, editors, *Prooceedings of the Conference Analysis, Warehousing and Mining the Data (AWAMIDA'99)*, pages 74–77, 1999.

[13] K. Kira and L. A. Rendell. The feature selection problem: traditional methods and new algorithm. In *Proceedings of AAAI'92*, 1992.

[14] K. Kira and L. A. Rendell. A practical approach to feature selection. In D. Sleeman and P. Edwards, editors, *Machine Learning: Proceedings of International Conference (ICML'92)*, pages 249–256. Morgan Kaufmann, San Francisco, 1992.

[15] I. Kononenko. Estimating attributes: analysis and extensions of Relief. In L. De Raedt and F. Bergadano, editors, *Machine Learning: ECML-94*, pages 171–182. Springer Verlag, Berlin, 1994.

[16] I. Kononenko. On biases in estimating multi-valued attributes. In *Proceedings of the International Joint Conference on Artificial Intelligence (IJCAI'95)*, pages 1034–1040. Morgan Kaufmann, 1995.

[17] I. Kononenko and E. Šimec. Induction of decision trees using ReliefF. In G. Della Riccia, R. Kruse, and R. Viertl, editors, *Mathematical and Statistical Methods in Artificial Intelligence, CISM Courses and Lectures* No. **363**. Springer Verlag, 1995.

[18] I. Kononenko, E. Šimec, and M. Robnik-Šikonja. Overcoming the myopia of inductive learning algorithms with RELIEFF. *Applied Intelligence*, 7:39–55, 1997.

[19] D. J. Lubinsky. Increasing the performance and consistency of classification trees by using the accuracy criterion at the leaves. In A. Prieditis and S. Russell, editors, *Machine Learning: Proceedings of the Twelfth International Conference (ICML'95)*, pages 371–377. Morgan Kaufmann, San Francisco, 1995.

[20] R. L. Mantaras. ID3 revisited: A distance based criterion for attribute selection. In *Proceedings of Int. Symp. Methodologies for Intelligent Systems*, October 1989.

[21] A. W. Moore, J. Schneider, and K. Deng. Efficient locally weighted polynomial regression predictions. In D. H. Fisher, editor, *Machine Learning: Proceedings of the Fourteenth International Conference (ICML'97)*, pages 236–244. Morgan Kaufmann, 1997.

[22] P. M. Murphy and D. W. Aha. UCI repository of machine learning databases, 1995. http://www.ics.uci.edu/ mlearn/MLRepository.html.

[23] U. Pompe and I. Kononenko. Linear space induction in first order logic with ReliefF. In G. Della Riccia, R. Kruse, and R. Viertl, editors, *Mathematical and Statistical Methods in Artificial Intelligence, CISM Courses and Lectures* No. **363**. Springer Verlag, 1995.

[24] J. R. Quinlan. *C4.5: Programs for Machine Learning*. Morgan Kaufmann, San Francisco, 1993.

[25] F. Ricci and P. Avesani. Learning a local similarity metric for case-based reasoning. In *Proceedings of the international conference on case-based reasoning (ICCBR-95)*, 1995.

[26] M. Robnik. Constructive induction in machine learning. *Electrotehnical Review*, **62**(1):43–49, 1995. (in Slovene).

[27] M. Robnik Šikonja. Speeding up Relief algorithm with k-d trees. In F. Solina and B. Zajc, editors, *Proceedings of Electrotehnical and Computer Science Conference (ERK'98)*, pages B:137–140. Slovene section of IEEE, Ljubljana, 1998.

[28] M. Robnik Šikonja and I. Kononenko. Context sensitive attribute estimation in regression. In M. Kubat and G. Widmer, editors, *Proceedings of ICML'96 workshop on Learning in context sensitive domains*, pages 43–52. Morgan Kaufmann, 1996.

[29] M. Robnik Šikonja and I. Kononenko. An adaptation of Relief for attribute estimation in regression. In D. H. Fisher, editor, *Machine Learning: Proceedings of the Fourteenth International Conference (ICML'97)*, pages 296–304, San Francisco, 1997. Morgan Kaufmann.

[30] M. Robnik Šikonja and I. Kononenko. Attribute dependencies, understandability and split selection in tree based models. In I. Bratko and S. Džeroski, editors, *Machine*

Learning: Proceedings of the Sixteenth International Conference (ICML'99), pages 344–353. Morgan Kaufmann, 1999.

[31] M. Robnik-Šikonja. Experiments with cost-sensitive feature evaluation. In N. Lavrač, D. Gamberger, H. Blockeel, and L. Todorovski, editors, *Machine Learning: Proceedings of ECML 2003*, pages 325–336. Springer, Berlin, 2003.

[32] M. Robnik-Šikonja and I. Kononenko. Theoretical and empirical analysis of ReliefF and RReliefF. *Machine Learning Journal*, **53**:23–69, 2003.

[33] R. Sedgewick. Algorithms in C. Addison-Wesley, 1990.

[34] P. Smyth and R. M. Goodman. Rule induction using information theory. In G. Piatetsky-Shapiro and W. J. Frawley, editors, Knowledge Discovery in Databases. MIT Press, 1990.

[35] S. B. Thrun, J. W. Bala, E. Bloedorn, I. Bratko, B. Cestnik, J. Cheng, K. De Jong, S. Džeroski, S. E. Fahlman, D. H. Fisher, R. Hamann, K. A. Kaufman, S. F. Keller, I. Kononenko, J. Kreuziger, R. S. Michalski, T. Mitchell, P. W. Pachowicz, Y. Reich, H. Vafaie, W. Van de Welde, W. Wenzel, J. Wnek, and J. Zhang. The MONK's problems - a performance comparison of different learning algorithms. *Technical Report CS-CMU-91-197*, Carnegie Mellon University, December 1991.

[36] D. Wettschereck, D. W. Aha, and T. Mohri. A review and empirical evaluation of feature weighting methods for a class of lazy learning algorithms. *Artificial Intelligence Review*, **11**:273–314, 1997.

INDEX